Palgrave Studies in Literary Journalism

Literary journalism is a vibrant and fast growing field, increasingly attracting international academic attention due to the unique way narrative journalism communicates matters of national and international importance to readers. Palgrave Studies in Literary Journalism provides a pioneering home to scholarship about the field in the first series of its kind. The book series is organised under the multi-disciplinary umbrella of literary journalism studies into four recognisable types: theoretical monographs, historical monographs, edited collections, monographs about practice and pedagogy.

Pablo Calvi
Editor

The Literary Journalist as a Naturalist

palgrave
macmillan

Editor
Pablo Calvi
School of Communication and Journalism
Stony Brook University
New York, NY, USA

ISSN 2731-9539 ISSN 2731-9547 (electronic)
Palgrave Studies in Literary Journalism
ISBN 978-3-031-56633-2 ISBN 978-3-031-56634-9 (eBook)
https://doi.org/10.1007/978-3-031-56634-9

Cover illustration: Pablo Calvi

This Palgrave Macmillan imprint is published by the registered company Springer Nature Switzerland AG.
The registered company address is: Gewerbestrasse 11, 6330 Cham, Switzerland

To the Sápara people of Ecuador.

Contents

Notes on Contributors

Tessa Ditner Amorosi is a British-French writer and philosopher of language. A Cambridge philosophy graduate, she completed a Master's in Creative and Professional Writing at Roehampton University. Her fiction/nonfiction writing and teaching spans conservation, aesthetics, and comedy. Editorial work has championed writing communities and creative collaboration. She is Associate Blog Editor for the *Journal of Ecology* and formerly the city of Portsmouth's Literature Worker. Her current doctoral research at Central Saint Martins, University of the Arts London, is on conceptual ethics in ecological innovation. **Instagram**: @TessaAmorosi

Deb Anderson is an academic based in Melbourne/Naarm. After a decade spent writing and editing for *The Age* and *The Sydney Morning Herald*, she joined Monash University where she teaches local and international units in journalism and social change. Born in the wet tropics, her research draws on oral history to explore the experience of extreme weather for ordinary people in an era of politicized knowledge on climate. She is the author of *Endurance: Australian Stories of Drought* (CSIRO Publishing, 2014).

Pablo Calvi is an Argentine-American writer and journalist. His long form, which appears in *The Believer*, *Guernica Magazine*, *The Nation*, and *El Mercurio* (Chile), has been listed as notable in *Best American Essays*, *Best American Travel Writing*, and *Best American Nonrequired Reading*. He teaches global journalism at Stony Brook University, where he is the associate director for Latin America at the Marie Colvin Center for

International Reporting. In 2019 he published *Latin American Adventures in Literary Journalism*, a cultural history of literary journalism in the Americas. **Twitter**: @plcalvi

Manuel Carvalho Coutinho is a researcher and an invited assistant professor at the Research Centre for Communication and Culture (CECC) at the Faculty of Human Sciences (FCH), Universidade Católica Portuguesa (UCP). He holds a Bachelor's in Philosophy, a Master's in Philosophy of Education, and a Master's and a PhD in Communication Sciences—Media and Journalism Studies. His main research subject is contemporary Portuguese literary journalism and the relationship between literature and journalism throughout the twentieth and twenty-first centuries. His investigative interests also cover a number of other topics such as sports journalism, comics journalism, and cultural journalism.

John M. Coward is Professor Emeritus of Media Studies at the University of Tulsa, where he served as department chair for more than a decade. His research on Native American representations in the nineteenth- and twentieth-century press has been published in *American Journalism*, *Journalism History*, *Visual Communication Quarterly*, *Journalism & Mass Communication Quarterly*, and *Literary Journalism Studies*. He is the author of two books, *The Newspaper Indian: Native American Identity in the Press, 1820–90* and *Indians Illustrated: The Image of Native Americans in the Pictorial Press*, both from the University of Illinois Press.

David O. Dowling is a professor in the School of Journalism & Mass Communication at the University of Iowa. He is co-editor (with Jacqueline Marino) of *The Art of Fact in the Digital Age: An Anthology of New Literary Journalism* (Bloomsbury) and author of ten books and numerous articles on cultural production and creative industries from print to digital media. Among his most recent books are *Podcast Journalism: The Promise and Perils of Audio Reporting* (Columbia), *A Delicate Aggression: Savagery and Survival in the Iowa Writers' Workshop* (Yale), and *Immersive Longform Storytelling: Media, Technology, and Audience* (Routledge).

L. Sasha Gora is a writer and cultural historian working at the intersection between food studies, contemporary art, and the environmental humanities. Her doctoral thesis at the Ludwig Maximilian University of Munich chronicled the history of Indigenous restaurants in the lands known as Canada and received the Bavarian American Academy

Dissertation Award. Since 2023, she has been based at the University of Augsburg where she is the project director of the "Off the Menu: Appetites, Culture, and Environment" research group. Her scholarship focuses on restaurant politics and cultural representation, the connections between eating and ecology, and all things fishy and salty.

Sue Joseph taught journalism and creative writing, particularly creative nonfiction, at the University of Technology Sydney for more than 20 years. Now as an associate professor, she is a senior research fellow at the University of South Australia and a doctoral supervisor at the University of Sydney, Central Queensland University, and the University of Technology Sydney. She is currently Joint Editor of *Ethical Space: The International Journal of Communication Ethics* and co-editor of the Literary Journalism Palgrave book series.

Callie Long is an organizational communicator, with a background in journalism and media development practice in the public health sector. Her primary research focuses on literary journalism and fictional narratives and how they can be employed as unsettling agents within social contexts, specifically related to HIV and the environment. Her scholarly work includes "Bearing Witness to Epistemic Injustice" and "Apprehending HIV Stigma," and she is the co-editor of *Dignity, Freedom and Grace*—a search for common ground in combatting HIV.

Steffen Moestrup (PhD) is a senior associate professor at the Danish School of Media and Journalism in Aarhus, Denmark. His research areas include persona-driven journalism, documentary film, and digital storytelling. Moestrup has a background in print and television journalism and has produced a number of documentary films. He is a film and media critic for the Scandinavian edition of *Le Monde diplomatique*. Furthermore, he is the host of the film show *film:syn* aired on national Danish television, an international voter for Golden Globe Awards, and a jury member at film festivals such as Cannes, Toronto, Mumbai, and Palm Springs. **Twitter**: @ steffmoes

Dolors Palau-Sampio (PhD, Autonomous University of Barcelona) is Lecturer in Journalism at the University of Valencia (Spain). Prior to her academic career, she worked as a journalist for nine years. Her research focuses on journalistic genres and styles, accountability, quality, and narrative journalism.

Mileta Roe is an American professor, writer, and translator. A former staff editor at the *Atlantic*, her work has been published in the *Phoenix Literary Supplement*, the *Boston Herald*, the *Boston Book Review*, *Scope: An Online Film Journal*, *Literary Journalism Studies*, and *Canadian Review of Comparative Literature*. She teaches nonfiction writing and literary journalism at Bard College at Simon's Rock, where she is Professor of Comparative Literature and division head of Languages and Literature. Her current work analyzes historical and contemporary connections between medicine and nonfiction writing.

Pepi Ronalds is a researcher, writer, and tutor based in Melbourne/Naarm. Her work appears in *Meanjin*, *The Lifted Brow*, *Kill Your Darlings*, *Arena Magazine*, the Wheeler Centre's *Notes*, and more. She holds a PhD in Creative Writing (specializing in journalism practice) from Monash University. Her research considers the precedents and application of creative nonfiction writing about disaster recovery with a current focus on recovery from the 2011 Great East Japan Earthquake. She tutors at Monash University in journalism, communications, and Japanese studies. **Twitter**: @pepironalds

Bret Schulte is Professor of Journalism and chair of the School of Journalism and Strategic Media at the University of Arkansas. He has reported for *The New York Times* and *The Washington Post* as well as magazines such as *Politico Magazine*, *U.S. News & World Report*, and *Wilson Quarterly*. He covered the 2004 and 2008 presidential campaigns as a member of the White House travel pool and earned several Society of Professional Journalists (SPJ) Green Eyeshade awards for the best reporting in the South. **Twitter**: @bretschulte

Isabel Soares is an associate professor (with Habilitation) at Instituto Superior de Ciências Sociais e Políticas at Universidade de Lisboa (Portugal) and senior research fellow at Conselho de Associações Académicas Portuguesas (CAPP). She was President of the International Association for Literary Journalism Studies from 2016 to 2018. Her teaching and research cover the fields of literary journalism and modern languages. Internationally published in such outlets as Palgrave Macmillan or Routledge, her research has been the recipient of multiple awards, including the Merit Award Caixa-Geral de Depósitos and ISCSP. Outside academe, Soares is a novelist under the pen name Isabel Tallysha-Soares. **Instagram**: @tallysha_soares

Susan E. Swanberg is a former geneticist who now teaches science and environmental journalism at the University of Arizona where she is an associate professor. Her current research examines the history of science and environmental journalism; the public understanding of science; and the intersection of law, science, and journalism. She has published academic articles on genetics as well as academic articles on the history of science journalism. She has also published science articles for the public in newspapers, magazines, and online. **Twitter**: @seswanberg

Aleksandra Katarzyna Wiktorowska is a Polish-Spanish researcher, writer, and literary translator with 14 novels to her credit. Her articles, interviews, and reportage have been featured in the Spanish newspaper *La Vanguardia*, Poland's *Gazeta Wyborcza*, and various magazines. She holds a PhD from the Universitat de Barcelona and an MA in Cultural Studies from Uniwersytet Warszawski. She co-edited *Literary Journalism and Latin American Wars* (2020) with Margarita Navarro Pérez and Mateus Yuri Passos. Immersed in teaching and research focused on Polish reportage and Spanish literature and culture, she received a research grant from the Mercè Rodoreda Foundation in 2023.

List of Figures

Revisiting Places and People: How Immersion and Cohesion Are Created in Danish Digital Long-Form Journalism That Deals with Climate Change

List of Tables

Urgency and Its Downfalls: How Have We Covered Nature and Have We Done It Right?

Pablo Calvi

When this book was conceived, in May of 2018, there was a clear sense of urgency propelling it. That month, the United Nations issued a report warning that one million animal species had just been listed as endangered, in risk of extinction due to human activity and the increase in global temperatures. Europe and Asia were amidst unprecedented heat waves. Rains in the Mid-Atlantic USA, Japan, and China had flooded crops and cities, and droughts in the American South West had drained farmlands on the other half of the world. Devastating forest fires in Queensland, Australia, were only a foreshadowing of the mass destruction that would take place in the Amazon basin less than a year later. In August of 2018, the *New York Times* published an entire issue of its Sunday magazine dedicated to climate change. 2018 was a catastrophic year. And then it got worse.

P. Calvi (✉)
School of Communication and Journalism, Stony Brook University, Stony Brook, NY, USA
e-mail: Pablo.calvi@stonybrook.edu

P. Calvi (ed.), *The Literary Journalist as a Naturalist*, Palgrave Studies in Literary Journalism,
https://doi.org/10.1007/978-3-031-56634-9_1

The Literary Journalist as a Naturalist is the urgent byproduct of an urgent time. It was first imagined at a conference of the same name that took place at Stony Brook University, New York, in May of 2018. But five years between the idea and the book, with a global pandemic that stopped everything in its tracks for more than 20 months, haven't made this anthology any less urgent. In July 2023, two months before I sat down to write this introduction, amidst the hottest July on record in the history of climate data, with wildfires raging across Southern Europe and North Africa, the Secretary General of the United Nations, António Guterres, announced the dawn of a new time: "the era of global boiling."

The essays that make up this collection span across more than one hundred years and cover five continents. They discuss the history of environmental journalism and its connection to environmental resistance; they dive into social ecology and deep ecology; ecofeminism and ecoterrorism; they explore food journalism and travel journalism; they study the different mediums and platforms where journalism about nature has made its appearance, from books and magazines to television and online news productions; they study in detail the work of the founding fathers of eco journalism in the Global North and the Global South. In a sense, this is the first academic collection dedicated to explore literary journalism and its links to environmentalism spanning almost two centuries and five continents.

As I read these pieces once again, an attribute that I barely noticed at first now jumps off their pages. These essays are all written with kindness, care, and a deep sense of responsibility for finding out what went wrong. If the message of ecological tragedy was so clearly, potently, and beautifully laid out, if the stories here analyzed were so compelling and heartbreaking, why didn't these masterpieces of nature writing lead to more meaningful collective action? What else could have been done?

Frames of Reference

In August of 2018, the New York Times Magazine dedicated a full issue to Nathaniel Rich's long-form piece "Losing Earth: The Decade We Almost Stopped Climate Change." This marked a significant shift from 2013, when the newspaper of record dismantled the nine-person Environment Desk established in 2009 to cover the climate catastrophe.[1]

[1] Katherine Bagley. "New York Times Dismantles Its Environment Desk." *Inside Climate News*, January 11, 2013. https://insideclimatenews.org/news/11012013/new-york-times-dismantles-environmental-desk-journalism-fracking-climate-change-science-global-warming-economy/

Rich's 30,000-word piece explores the ten-year period from 1979 to 1989, a time when "humankind first came to a broad understanding of the causes and dangers of climate change" but ultimately failed to prevent the catastrophe. The piece was one of the first serious attempts to narrativize the manmade climate downfall.[2] Rich's story met all the criteria proposed by Norman Sims and others for a narrative piece to be considered literary journalism: facts presented with literary quality and resources, episodic narratives built on scenes, full sections of reconstructed dialogue, deep characterization of both sources and relevant actors in the drama, and a clear narrative arc.[3]

In many respects, Rich's piece followed the trajectory of narrative successes such as Elizabeth Kolbert's *The Sixth Extinction* (2014) and anticipated David Wallace Wells' *The Uninhabitable Earth* (originally a *New York Magazine* article in July of 2017, later published as a book in 2019).[4] Decades before these works, Rachel Carson's *Silent Spring* (1962), Sandra Steingraber's *Living Downstream* (1997), and more recently Naomi Klein's *This Changes Everything: Capitalism vs. the Climate* (2014) laid the foundations for these alarming and illuminating masterpieces of literary environmental journalism to reach mass readerships.[5]

Over the past fifty years, climate writing has flourished as a dark reminder of the impending catastrophe looming over humanity, reflecting the audience's desire to understand the pathways for climate action. Yet in hindsight, these extraordinary stories, rich in accuracy and detail about ecosystems in decay, human carbon release, environmental pollutants, ocean acidification, deforestation, and the depletion of animal populations, have achieved little beyond documenting this man-made catastrophe.

Also in hindsight, it is possible to understand the cold detachment that many of these journalistic pieces have ultimately elicited in their readers of

[2] Nathaniel Rich. "Losing Earth: The Decade We Almost Stopped Climate Change." New York Times Magazine, August 1, 2018. https://www.nytimes.com/interactive/2018/08/01/magazine/climate-change-losing-earth.html

[3] Sims. *The Literary Journalists,* 1-2.

[4] Elizabeth Kolbert. *The Sixth Extinction.* (Henry Holt & Co.: New York, New York, 2014) and David Wallace Wells. *The Uninhabitable Earth: Life After Warming.* New York: Tim Duggan Books, 2019.

[5] Rachel Carson. *Silent Spring,* Boston: Houghton Mifflin, 1962 [1994]; Sandra Steingraber. *Living Downstream: An Ecologist's Personal Investigation of Cancer and the Environment.* Cambridge, Massachusetts: Da Capo, 1997; Klein, Naomi. *This Changes Everything: Capitalism vs. the Climate,* New York: Simon & Schuster, 2014.

the past decades in the framing mechanism that's explicit in Rich's "Losing Earth."

"Human beings have altered Earth's atmosphere through the indiscriminate burning of fossil fuels," Rich proposes in his article. "[But] in the decade that ran from 1979 to 1989, we had an excellent opportunity to solve the climate crisis." Yet *we* didn't. Almost *nothing stood in our way — nothing except ourselves.*"[6]

Rich's account, as Klein and others point out, is only partially true. His perspective underscores the notion that humanity had the knowledge and capacity to mitigate climate change but collectively failed to act due to innate selfishness, greed, and lack of empathy inbuilt in our DNA. And while the larger portions of Rich's narrative may be on point, and most of his facts check, this framing device is insidiously deceptive.

A few days after Rich's piece appeared in the *New York Times*, Naomi Klein offered a pointed rebuttal on *The Intercept*,[7] critiquing his failure to consider deeper systemic issues, particularly the influence of capitalism and corporate greed in the climate catastrophe. Klein argued that blaming *human nature*, as Rich had done, was not just disingenuous but inaccurate, and it overlooked the historical context and the power dynamics that had shaped environmental policies for at least five decades.

Blaming the climate downfall squarely on "human nature," emphasizing the collective responsibility of individuals above organizations, governments, and economic systems, Rich's and other Global North narratives, have led to passivity, stasis, and a defeatist mentality.

The clash between the neoliberal economic paradigm, climate science imperatives, and the lack of corporate regulations, are key to understanding the crisis, Klein proposed. "Because what becomes clear when you look back at this juncture is that just as governments were getting together to get serious about reining in the fossil fuel sector, the global neoliberal revolution went supernova."

By reframing this narrative, Klein directs our attention to the disproportionate impact of certain industries and governments on greenhouse gas emissions, environmental pollution, and ecosystem destruction,

[6] Nathaniel Rich. "Losing Earth: The Decade We Almost Stopped Climate Change." New York Times Magazine, August 1, 2018. https://www.nytimes.com/interactive/2018/08/01/magazine/climate-change-losing-earth.html, emphasis is mine.

[7] Naomi Klein, "Capitalism Killed Our Climate Momentum, Not 'Human Nature.'" *The Intercept*, August 3, 2018. https://theintercept.com/2018/08/03/climate-change-new-york-times-magazine/

highlighting a small number of corporations and the US military-industrial complex—responsible for most of global carbon emissions—while undermining the idea of a general, collective, human responsibility. It is precisely in the Global South where these forces of extraction and capitalism are boldly seen, running rampant and unhinged.

Just as Klein's critique challenges the notion of general human culpability, the essays in this book aim to show to what extent literary journalism in the Global South incessantly points to the Global North and its extractive practices to assign responsibilities for the damage caused by unfettered corporate capitalism.

Capitalist development has had the most impact on both pollution and climate change, with the United States claiming the largest cumulative impact on greenhouse gas emissions since the Industrial Revolution. Only one hundred companies are responsible for 70% of the climate crisis, yet personal responsibility could affect less than 10% of all emissions, as established by Klein and others.

This collection follows Klein's footsteps, analyzing from a Global South perspective, some efforts in literary journalism that sometimes question Western humanism, the neutrality of science, and the perceived power of central governments as the only levers for climate action.

Contents

Anna Krien's *Into the Woods* (2012) focuses on her eighteen months in Tasmania and her study of the Australian island's environmental conundrum posed by logging companies, the eco activism against them, and a local community torn between the two. This essay by Manuel Coutinho considers the implications of a "journalist going into the woods," figuratively and literally. Krien becomes involved with the activists by hearing their stories, but the book still shows different but essential sides of this environmental issue. Coutinho's essay is an attempt to explore the impact of Anna Krien's work and how she portrays nature, the human interactions with it, and the real-life outcome of these exchanges.

In Chap. 3, "John Joseph Mathews: The 'Blackjack Discourse' of an Osage Naturalist," John Coward examines the distinctive literary voice of John Joseph Mathews, the part-Osage historian, novelist, and amateur naturalist who built a one-man sandstone cabin on the prairies of Oklahoma and lived there for ten years. Mathews recounts his prairie experiences in *Talking to the Moon*, published in 1945. Although the book was largely overlooked at the time, it was reissued in 1981, helping establish Mathews as a Native American Thoreau.

"Edible Armageddon. Insect Superheroes and Other Villains," Chap. 4 in this collection, looks at the literary devices used in "insectaggedon" narratives. It references two stories: Brooke Jarvis' 2018 "The insect apocalypse is here, what does it mean for the rest of life on Earth?" appeared in the *New York Times* and Pulitzer Prize winner Kathryn Schulz's stinkbug piece for *The New Yorker*. This essay, by Tessa Ditner, breaks up these narratives piece by piece and analyzes the literary devices their authors resort to, in order to present the issue, including writing within a genre, polarity, conflation, comparison, echoing cultural stereotypes, exaggeration, as well as the creation of a viable villain within a hero-villain dynamic.

Chapter 5, "Toward an Activist Aesthetic of Environmental Literary Journalism: Deep Ecology in Thoreau, Carson, and Jenkins," goes back to Rachel Carson's *Silent Spring*, and McKay Jenkins' 2016 *ContamiNation: My Quest to Survive in a Toxic World*, to draw on tensions between these works and Henry David Thoreau's deep ecology. The connections between these works of literary journalism oscillate between exultation and activism, ecstasy in nature's beauty, and the horror in its industrial abuse. David Dowling shows in this chapter how these three key authors leverage their own signature journalistic storytelling through subjective narration and felt detail, dramatizing the human experience missing from abstract scientific discourse on the environment. Critics of deep ecology, Dowling argues, failed to forecast the *anthropocene*, a systematic destruction of nature that has accelerated since the Industrial Revolution.

Chapter 6, "The Bitter Taste of Extinction: Writing the Environmental Crisis Through Food," is an exploration on food writing as a means with which to document environmental change. Beginning with Jonathan Gold's 2011 op-ed in the *Los Angeles Times*—"Shark fin soup: The taste of extinction"—Sasha Gora's essay charts the environmental turn in American food writing. Like with Gold's piece, Gora finds examples of articles that rally particular causes, and others document and mourn the loss that has already happened. This chapter considers how American food writing represents, activates, or seeks to preserve the environment by surveying the genre's growing environmental consciousness.

"Silent Spring: The Rise of the Environmental Movement," Chap. 7 in the collection, puts American writer and scientist Rachel Carson at the center of the canon of literary journalism as naturalism. In this essay, Sue Joseph recognizes Carson as the "scientist-poet of the sea," and her final text, *Silent Spring* (1962), as the masterpiece that cements her name in the global psyche. Joseph argues that Carson's long-form text helped launch the modern environmental movement. And that ten years before the term "ecofeminism" was coined, Carson had already displayed a vivid intersectionality in her writing.

Chapter 8, "Alive to the Slow Terror of the Megadams: The Literary Journalism of Jacques Leslie's Deep Water," is Callie Long's attempt to recognize Jacques Leslie's work on megadams *Deep Water*, Leslie's 2005 masterpiece, exemplifies this immersive approach to storytelling, through an account of the every-day, mostly non-spectacular, yet deadly consequences that come with environmental wrongs brought on by the building of megadams. In a nod to John McPhee, Leslie selects three members of the independent World Commission on Dams to tell the story of megadams. Yet, what also surfaces is Leslie's voice. In this chapter, Long shows how Leslie's writing, tangentially informed by McPhee's, brings these excised bodies back into view.

"Revisiting Places and People. A Case Study on Two Danish Digital Longform Journalism Projects that Deal with Climate Change" delivers a comparative analysis of the Danish newspaper *Jyllands-Posten* series of long-form investigative reports on the environment between 2004 and 2019. The essay, by Steffen Moestrup, focuses on the use of digital elements to create a cohesive and immersive experience across time and space. In 2004, the *Jyllands-Posten* had made a series of long-form investigative reports on environmental changes in places such as Tuvalu, Delhi, and Copenhagen. In 2019, the newspaper repeated the reporting by revisiting the same places spanning five different continents. The new series of long-form digital productions named "The Climate Tour 2019" was produced by the same journalists as the 2004 reporting series, but the 2019 series, naturally, made further use of new digital storytelling tools.

Chapter 10, "Environmental activism and resistance in Latin America: Portraying the fight and murder of environmental leaders from literary journalism," uses qualitative methodology to analyze the narrative resources and journalistic treatment of a dozen of long-form reportages and profiles dealing with activism and resistance in Honduras, Perú, Colombia, Mexico, Brazil, Ecuador, and Argentina. The essay, by Dolors Palau, starts with the murder of the Brazilian environmental activist Chico Mendes in 1988. Palau argues that Mendes' assassination awakened environmental consciousness after years of his defense against Amazonian deforestation. His assassination, however, did not lead to further protection of environmentalists but rather to an increasing impunity in the attacks against them, particularly in Latin America, the deadliest region for local communities fighting extractive industries, deforestation, or large infrastructure projects.

In Chap. 11, "An Ecologist's Personal Investigations: Sandra Steingraber on Danger, Ecology, and Writing," Mileta Roe discusses the work of ecologist and science writer Sandra Steingraber and her activist stance regarding the environment. In her essay, Roe argues that it is difficult to write about an imperiled or endangering environment from a neutral stance. Dispassionate description seems disingenuous. Objective observation cannot convincingly convey devastation or real threats to humanity. Maybe it just has to get personal, Roe proposes. And that is what Steingraber has done to great effect in her numerous articles and books on environmental topics. Armed with a doctorate in biology, a mind attuned to both science and poetry, and a far-reaching ecological awareness, Steingraber stands out amidst a field of women writers who have followed in Rachel Carson's footsteps. Steingraber embodies a form of scientific literary journalism that is racing to get our attention.

"How Marjory Stoneman Douglas Saved Environmental Reporting (and maybe the Everglades)," Bret Schulte's essay, discusses how literary reportage bears much of the credit for the success of the environmental movement in the nineteenth and twentieth centuries, helping spur the American public to demand protections of sensitive land and natural wonders, endangered species, and their own health—threatened by polluted air and water. The chapter examines the historical role of investigative, long-form literary journalism in the conservation movement to assess why some environmental reportage carried such public resonance. Examples include Henry David Thoreau's influence in the creation of the Cape Cod National Seashore and Marjory Stoneman Douglas's crusade to save the Everglades that culminated in her book, *Everglades: River of Grass.*

Chapter 13, "The Naturalist as Literary Journalist: David Attenborough's sixty years of documentary film-making," focuses the camera on Attenborough's narrative work as part of the literary journalism universe. Addressing the UN Climate Summit in 2018 as representative of the People of the World, ninety-two-year-old David Attenborough urged global leaders to take effective action against climate change. The moment, Isabel Soares recalls, was solemn. Attenborough's stature as the "inventor" of the natural world documentary in the 1950s and the genre's consolidation over the past six decades have made him a privileged witness of the ways nature has changed at the hands of man and also a respected authority on matters concerning the natural world. In her chapter, Soares analyzes the genesis of the natural history documentary and the part played by Attenborough in creating its narrative style.

"Interviewing Nature: The Dangers and Delights of the Pathetic Fallacy and Anthropomorphism in Nature Writing and Environmental Journalism" discusses the ways in which literary journalists translate Nature's narratives when so much of the natural world is under threat. Susan Swanberg explores the trademark techniques of the genre—such as immersive reporting, digressive narration, effective use of dialogue, and the twin virtues of truth and accuracy—and whether they can or cannot engage a reluctant audience. Should the rules of literary journalism be augmented when a writer's main source, Nature, cannot be interviewed? Swanberg wonders. Her chapter identifies and examines selected works of nature writers and environmental journalists who pushed the boundaries of literary journalism's canons with their, often unrecognized, use of the pathetic fallacy (the figurative attribution of human emotions and behaviors to non-human agents) and anthropomorphism (the literal attribution of human characteristics and behaviors to animals and other non-human agents). Examples include Rachel Carson and John McPhee—as well as lesser-known or forgotten authors who might yet be inducted into the literary journalism pantheon, including nineteenth-century French author and entomologist, Jean-Henri Fabré.

In Chap. 15, "Thanatos syndrome: Literary forms in Domosławski's Death in the Amazon," Aleksandra Wiktorowska aims to show how Artur Domosławski—a Polish reporter known to English- and Spanish-speaking readers only as a biographer of Ryszard Kapuściński—deals in his work with the question of responsibility for the environmental crisis. Domosławski is the author of *Death in the Amazon* (2013), an over 300-page long book on severe environmental damage to the Amazon, namely deforestation and the assassination of ecologists in Brazil, water pollution caused by gold mining in Peru, and environmental poisoning caused by oil extraction in Ecuador.

In her analysis, Wiktorowska compares Domosławski's book-length reportage to two other accounts of investigative journalism, showing how his literary journalistic account of events differs considerably while explaining what makes it so special in the context of an environmental crisis.

"It's Personal. Women's Writing on Weather Disaster in the Context of Climate Crisis in Australia" delineates the state of environmental and nature writing in Australia through a thorough discussion of some of the most significant Australian environmental writers. Women lead the way there by providing "a direct response to climate crisis *and* the deficits of journalism that is both personal and distinctively Australian." Deb Anderson and Pepi Ronals essay asks an important question about the deficits inherent in environmental journalism. "How well has objectivity served journalism in this era of climate catastrophe and how have Australian women met the challenge of writing convincingly about this crisis without violating journalistic objectivity?"

In Chap. 17, "The Big Picture and the Small Scene. Anna Tsing's Assemblages and Capitalist Survivalism vs. Paul Engle Iowa's Writer's Workshop, and the Nature In Between," I discuss the work of anthropologist Anna Tsing as she follows several groups of mushroom foragers across the Pacific Northwest in her 2015 ethnography *The Mushroom at the End of the World*. The world of precarity that she discovers couples itself with the capitalist economy as an appendage, bringing into it the fruits of a labor process that, despite being pre-industrial, succeeds in creating value in a post-industrial society. In the final essay of this collection, I discuss the success of this anthropological, big-picture approach to writing about nature and compare it with the individualistic, monadic take on long-form writing proposed by the Iowa Writer's Workshop and its director Paul Engle. Through the tools of a season storyteller, anthropologist Anna Tsing tells this tale of end-and regrowth of nature, presenting before us the most viable literary alternative to a deserted, post-apocalyptic dystopia. Is this the new way to nature-centric literary journalism?

* * *

This collection aspires to explore and understand the ways in which literary journalists have written and talked about nature, the environment, the agents that modify and disrupt it, and the ways in which humans have survived and may continue to do so as we enter a phase of extreme climate events. Its ambition is to find a thread, the connective tissue between different authors, with different styles and ideas, writing about the natural world from different places, perspectives, and times, but with a common goal: to understand the natural world humans are a part of. And to assess whether these efforts—our collective, intellectual attempts to explain nature through science, through journalism—have made a difference or not, and why.

References

Alexander, Rob. "Literary Journalism and Ecocriticism." *The Routledge Companion to American Literary Journalism*. Ed. Dow, W. and Maguire, R. London: Routledge, 2021.

Bagley, Katherine. "New York Times Dismantles Its Environment Desk." *Inside Climate News*, January 11, 2013. https://insideclimatenews.org/news/11012013/new-york-times-dismantles-environmental-desk-journalism-fracking-climate-change-science-global-warming-economy/

Carson, Rachel *Silent Spring*, Boston: Houghton Mifflin, 1962 [1994].

Klein, Naomi. *This Changes Everything: Capitalism vs. the Climate*, New York: Simon & Schuster, 2014.

Klein, Naomi. "Capitalism Killed Our Climate Momentum, Not 'Human Nature'" *The Intercept*, August 3, 2018.

Kolbert, Elizabeth. *The Sixth Extinction*. New York: Henry Holt & Co., 2014.

Rich, Nathaniel. "Losing Earth: The Decade We Almost Stopped Climate Change." *New York Times Magazine*, August 1, 2018. https://www.nytimes.com/interactive/2018/08/01/magazine/climate-change-losing-earth.html

Sims, Norman. The Literary Journalists. New York: Ballantine Books, 1984.

Steingraber, Sandra. *Living Downstream: An Ecologist's Personal Investigation of Cancer and the Environment*. Cambridge, Massachusetts: Da Capo, 1997.

Wallace Wells, David. *The Uninhabitable Earth: Life After Warming*. New York: Tim Duggan Books, 2019.

Anna Krien's *Into the Woods* and How to Report on Environmental Activism

Manuel Carvalho Coutinho

Into the Woods and the Tasmanian Island

I open my laptop one day and press the ON button like I have done a million times before. As usual the computer takes its time to wake up. While I hear the turbines working hard to cool the aging device, I am confronted with the lock screen asking for my password. Then, suddenly and unannounced, I see it. My laptop's lock screen displays a different photograph every time, a random picture provided by the algorithm of the Windows spotlight image of the day. While these are often beautiful photographs of natural landscapes, I tend to disregard them quite quickly since they usually are the last hurdle to overcome before logging in and going on with my day. But this photograph in particular caught my attention. It showed an enormous mountain sitting atop a large body of water that perfectly mirrored the same image below. I was mesmerized and instead of inputting the password and moving on to my scheduled work, I was compelled

M. C. Coutinho (✉)
Research Centre for Communication and Culture, Faculty of Human Sciences, Universidade Católica Portuguesa, Lisboa, Portugal
e-mail: mccoutinho@ucp.pt

P. Calvi (ed.), *The Literary Journalist as a Naturalist*, Palgrave Studies in Literary Journalism,
https://doi.org/10.1007/978-3-031-56634-9_2

to know more. I discovered that the picture showed a view of Cradle Mountain and Lake St Clair, and it was one of the many national parks of Tasmania. While writing this I was almost tempted not to include its location for fear that others would find it like I did, but I still share it with a different hope and intention. Because, as I found out, what might endanger Tasmania's wilderness, nature, and beauty is not necessarily the tourism sector and those that like you might one day seek to visit such a place.

For context purposes let us note that Tasmania is a large island with an area of 68,401 km^2 (26,410 square miles), and along with its 334 smaller islands, it constitutes one of Australia's states. It is located to the southeast of mainland Australia and the total population is slightly larger than half a million. Due to the island's vast forests and natural resources, Tasmania is promoted as a natural state thanks to its various national parks and world heritage sites. Still the island's relationship with its forests seems somewhat complex and a look at the last fifty years reveals a troubling sight that many have dubbed as the "forest wars." While logging has been an important source of income for the island, tourism has progressively become one of, if not the main, pillars of its economy. Yet an analysis of Tasmania's political debates of the past few decades would seem to indicate otherwise. With so many forests in Tasmania being worthy of world heritage listing—one of the main reasons for its booming tourism—the island's politics however have continuously made an effort to change the status of "protected forests" to that of "reserved forests" so that they can be logged. This type of policies has led to a continuous political debate on the island, directly impacting the growth of the logging corporations and the presence of environmental groups seeking to delay them. Perhaps one of the best ways to explore the complexity surrounding Tasmania is to refer to journalist Anna Krien's *Into the Woods* (2012) and her 18-month research journey into the topic.

Forest Wars and Protesters: Structure and Sympathizing with One's Subject of Analysis

Anna Krein's work on Tasmania is an in-depth analysis divided into five distinct sections, each one organized in such a way that the reader becomes aware of different sides of the argument concerning the island and its "forest wars." And yet, as the book further explores its subject matter, it also draws the reader in by sharing the author's personal perspective and establishing closeness to its characters at different moments throughout. As a work of investigative journalism, Krien's book was praised in *Literary Journalism Studies Journal* as being journalistic, self-aware, and literary, with Krien herself seen as a willing student, a reporter, and an important addition to the tradition of female literary journalists.[1] And indeed, Krien's personal voice is clear from the start of the book as she sets foot in Tasmania and begins to describe how the pervasive debate and conflict concerning the forests is present everywhere, even in small details such as bumper stickers:

> "The Natural State," I read on returning vehicles' number plates as I boarded the ferry, the Tasmanian state slogan accompanied by a picture of the now-extinct Tassie tiger peering out from beneath the exhaust pipes. On the rear windows of almost half the cars I see, there are stickers in eternal argument with one another. "Tasmania: The Corrupt State" and "Save the Styx" versus "Greens Tell Lies," "Greens Cost Jobs" or simply "Green Scum" – slightly tamer versions of older stickers that read "Keep Warm This Winter: Burn a Greenie."[2]

While it was in Tasmania that the world's first green party was established back in March of 1972, it is important to note that the word "greens" seen in the stickers also refers to the environmental groups that seek to protect the island's forests. One such group is the focus of the first section of the book, the "ratbags" as the journalist refers to them: a new group of protesters who live in or near a logging site to actively and peacefully disrupt the logging operations. The techniques and strategies used by them are ingenious in their own manner:

> "Lock-on" contraptions are built well in advance, as the longer cement is left to set, the stronger it is. ... Using padlocks, handcuffs, thumb cuffs, chains and climbing karabiners, the activists fasten themselves to difficult to reach "lock-on" points ... Bicycle D-locks can be placed around the throat of an activist, securing them to an obstacle and making it particularly hairy for them to be removed ... A suitable spot is chosen by mapping out where a proposed logging road will likely go; the idea is to make it as difficult as possible for trucks to get into the forest.[3]

[1] Lindsay Morton. "Discovering a New Voice," *Literary Journalism Studies* Vol. 4, No. 1, Spring 2012, 127 & 130.

[2] Anna Krien. *Into the Woods – The Battle for Tasmania's Forests* (Black Inc.: Australia, 2012), 138–145.

[3] Krien, 414–418 & 423–424.

Their efforts are generally met by angry loggers who often react violently since they see them as a hindrance to their job and projected quota. On several instances throughout the book the author describes how intense these confrontations can be and at times these clashes also occur away from the logging sites and the areas of activism. On one occasion the author describes a troubling moment at night when several loggers invaded the ratbags' improvised encampment to destroy and burn it to the ground.

> "Wakey, wakey," one man called out in a singsong voice as they stepped into the blockade, holding tools and lugging jerry cans full of petrol. A deep sleeper was still curled up in his sleeping bag in the back of a car. An axe blunting into the roof above him did the trick ... he fell out the door and half-ran, half-crawled into the bush. They didn't follow him into the forest. Some activists climbing trees, others crouching beneath ferns, they stayed quiet as the men taunted them to come into the open. ... Flicking a handful of lit matches into the dark, they got back into their cars and drove away. And as the flames licked up the petrol, the hiding protesters slowly emerged, calling out to one another and up to the people in the tree-sits, checking to see that everyone was okay. In a huddle they watched the burning vehicles and hut, jumping each time something – the fuel tanks, gas cooker, bottles of cheap cider – exploded in the heat.[4]

In these moments it is normal to be unnerved by such violence knowing that the activists strive to be peaceful in their protests, and yet the journalist does not shy away from criticizing the activists' demeanor and their way of presenting themselves and their movement:

> I shake my head. In a world of sound-bites and five-second grabs of footage, I can't help thinking that if they're going to go to the effort of living out here and getting the media to visit, they could smarten up. A woman with blue eyes and perfect teeth reads my mind. She holds out one of her sandy-colored dreadlocks to me; it's the length of her arm. "But we live in the forest at a blockade," she says, gesturing at her muddy black hooded jumper with patches safety-pinned to it. "This is what happens."[5]

While the journalist acquaints herself with the blockade and the activists that support it, we, the readers, are for the first section of the book naturally inclined to sympathize with their movement. Indeed, it is difficult not

[4] Krien, 258–262 & 265–268.
[5] Krien, 620–624.

to when their efforts seek to peacefully delay the logging process and bring attention to positive dialogue, as former blockader Jess Wright presents it: "We knew we won't save the forest by hanging banners from branches. What we're trying to do is press pause so this place has a chance to survive while the people with power can think about what is happening here."[6] At the same time, Krien makes an effort to understand the protesters by spending time with them and getting to know their routines and daily life in the encampment. While doing this, Krien not only offers a minute account of what she sees, but she does so while further informing us about the protesters themselves. The choice of approaching this research and narrative in this matter further enhances the overall story being told and rightfully seems to fit what writer Jon Franklin states: "Description alone isn't helpful. A reporter who doesn't understand a subject has no way of knowing what the details of the surroundings mean. To put it another way, information that explains motive goes into the piece."[7]

The Loggers and the Forestry Business: Giving Context to Convey Meaning

As *Into the Woods* begins its second section, the focus shifts to the loggers and the individuals behind the machines and so we, the readers, become further enthralled in the humanity behind each side. The choice to change focus is not in itself revolutionary or unexpected. Indeed, Krien's choice to present each side of the argument goes in line with any other work of journalism that seeks to inform in a complete and thoughtful manner. And while it would be easy to fault the loggers as the villains of this story, the blockaders and ratbags are the first to recognize their reasoning:

> Most of the activists seem to forgive the loggers for their violent lapses. "It's not their fault," says one blockader. "They get psyched up by people telling them bullshit." "It's a shit feeling watching the logging trucks turn the corner, only to find us in their way," says Bridget. "All those clichés about putting food on the table and having mortgages, they're true."[8]

[6] Krien, 1297–1299

[7] Jon Franklin. "A Story Structure," in *Telling True Stories* (Mark Kramer & Wendy Call eds.) (Penguin Books: New York, 2007), 127.

[8] Krien, *Into the Woods*, 138–145.

In an effort to portray both sides the journalist visits a pub whose customers are mainly loggers. And so we are introduced to John and the "other side" is given a face and voice:

> When my meal arrives … the man beside me … introduces himself as John, and tells me he is a tree-faller … Fallers are among the few who still touch the trees before they fall. If a tree is too difficult to cut with the jerky outstretched arm of a machine, the faller steps in with a handheld chainsaw. … I ask him about the activists. "It's the weirdest thing," he says slowly, two beers between us. "You come to work and there they are. One dressed as an eagle, another looking like she wants to tear your eyes out, and then they try to talk to us as if this is normal."[9]

The choice to leave room for the loggers to speak and show emotion, something that prevails throughout most of the book's second section, further adds to the complexity of the subject matter and how Krein's work and narrative choices fall in line with Mark Kramer's definition of literary journalism of combining cold fact with personal experience through the author's caring voice: "And that broadens readers' scans, allows them to behold others' lives … The process moves readers, and writers, toward realization, compassion, and in the best of cases, wisdom."[10] And so as the second section of the book develops, we are presented with a different side of the equation and we are not surprised to find that the loggers are merely a means to an end and part of the machinations of politics and money. Interestingly in one interview to promote the book, when asked about her decision to talk to loggers and portray that side, Krien was quick to explain the complexity of the subject and her reasoning:

> I knew all along that I'd have to get to know the other side. I just didn't realize that … I was going to have to talk to politicians from the last twenty years and go all the way back into company archives and all those kinds of things. The general idea is you talk to a logger you talk to a ratbag activist and then you go home and you write the story. But it turned out to be that they were in a sense a false battleground and they were interesting and colorful but they weren't the real story.[11]

[9] Krien, 1040–1042 & 1051–1054.

[10] Mark Kramer, "Breakable Rules for Literary Journalists" in *Literary Journalism* (Norman Sims and Mark Kramer eds.) (Ballantine Press: New York, 1995), 34.

[11] The Monthly, "Into the Woods: Anna Krien in conversation," Youtube Video, 23:37, May 3, 2013, https://youtu.be/tINXZdUR-EE?si=OkSStjTDf_ufMvCl.

The real story, or part of it as she notes in the same interview, came to her unexpectedly:

> The loggers had been done over by two movements and one was the environmental movement and the other was the woodchipping movement. … They were being beaten down by both sides and … [it's] something that is expressed not nearly enough. I think it's partially their fault because they are eager to blame the Greenies and the environmentalists for what is going on.[12]

To understand this matter more clearly and at the same time seeking to reinforce the arguments behind the environmental movement, one should first define woodchipping. By woodchips we refer to small- to medium-sized pieces of wood formed by chipping or cutting larger pieces of wood like trees, branches, wood waste, etc. The problem with woodchipping, as explained in Krien's book, is that you don't need high-quality wood to make it and therefore using native forests of Tasmania with their centuries-old trees is a clear waste and an environmental disaster. And yet the numbers are staggering:

> In 2007, according to the Wood Resource Quarterly, Australia exported a record quantity of woodchips, over six million tonnes, about 70 per cent of which was eucalyptus, mostly from native forests. Tasmania manages to export more native forest woodchips than all the other states and territories put together, despite making up less than 1 per cent of Australia's landmass.[13]

As the narrative unfolds, the reasoning for logging the unique forests of Tasmania becomes murkier, and as we follow Krien's investigation, the financial interest, lobbies, and politics become clearer and clearer. The journalist spends time researching each of these subjects as she pieces together the complexity of the matter and how entangled the problem really is. And this is where her book again proves its resourcefulness by providing different moments of her research and further emphasizing her work in getting each detail right. A smaller work of journalism—a single article in a newspaper, for example—would not have been able to thoroughly address each subject in this book. And while not all examples of in-depth reportage need to be as long as a book, Krien's *Into the Woods* justifies its size and structural choice to span into five different sections as

[12] The Monthly.
[13] Krien, *Into the Woods*, 1327–1330.

an effort to accurately report on different aspects of the environmental situation in Tasmania. The decision to approach her work in such a slow and thorough manner fits Robert Alexander's perspective on the three advantages of literary journalism to address environmental writing:

> First, the slow, immersive reporting techniques of literary journalism are well suited to the incremental pace of environmental change. Second, the genre's concern with everyday subjects of modest scale attracts it to the seemingly small, supposedly non-newsworthy environmental events that are collectively transforming the planet. And third, literary journalism's capacity to link those instances to larger contexts and to do so in a discourse that combines factual rigor with emotional impact makes it a potent force in the efforts to address the current ecological crisis.[14]

And so, as we become aware of Krein's journey, we are given time to follow her footsteps as the journalist pieces together the data and the different—and supposedly independent—studies that support the argument for logging native forests. As the issue becomes clearer, she is led to the proverbial reasoning which proposes to "follow the money," admitting at one point: "All of these self-proclaimed independent bodies are linked by one thing, directly or indirectly: Gunns' money."[15] Gunns Limited is one of the biggest logging companies operating in Tasmania at the time of the book's publishing, and because of their operations on the island, they are mentioned continuously in the third section of *Into the Woods*. As more and more facts come to light it is difficult not to negatively judge this corporation and their questionable agenda (among others, such as Forestry Tasmania, the state's official forestry body). When in 2004 the Gunns Corporation announced a large investment on a new mill project coincidently, Krien notes that Tasmania had been ranked as one of the world's hotspots to visit with more than one million visitors spending "more than a billion dollars on the island each year. And in comparison to the elusive 6000-odd jobs in the island's timber industry, more than 30,000 Tasmanians are employed in tourism."[16]

It is difficult not to notice that if the logging companies continue to prioritize old forests instead of implementing a sustainable and reasonable

[14] Robert Alexander, "Literary Journalism and Ecocriticism," in *The Routledge Companion to American Literary Journalism* (William Dow & Roberta Maguire eds.) (Taylor & Francis: New York, 2020), 482.

[15] Krien, *Into the Woods*, 2236–2237.

[16] Krien, 3253–3257.

reforestation process, then they are in fact destroying a significant part of the tourism sector and the attractiveness of visiting Tasmania in the first place. And yet when visiting Maydena, one of the many small villages close to the logging sites, Krien notes that: "Despite the millions of dollars now passing through Maydena at all hours on the back of logging trucks, these royalties barely touch the town. And yet locals seem as loyal as ever to the midnight haulers."[17] In spite of this, logging continues to be used in Tasmania as a bargaining chip with the promise of jobs and prospects for the locals, even though economically it pales in comparison with the real opportunities of the booming tourism sector. With the advancement of machinery, logging continues to change as there is less demand for large numbers of workers, consequently reinforcing the need to rethink reforestation and the sector overall including its often void political and socio-economic promises.

Reforestation, Pesticides, and Logging: Showing a Journalist's Emotions

However, as explored in the fourth section of Krien's book, the myth of reforestation and regrowth assumes that planted trees will grow as fast as the demand behind them, and new trees will deliver as much as previously untouched forests. But reforestation is expensive and new trees are natural targets for wild animals who lost their natural habitat in the first place due to logging. And so, to meet quotas and achieve the unreasonable expectations of regrowth in a short period of time, the forestry industry in Tasmania often employs dangerous techniques. One of them was using 1080, a synthetic pesticide banned in most countries that spreads quickly and is toxic to all living species. In one passage of her book, Krien reveals the damages the pesticide has caused:

> ... over 200,000 wild animals including wallabies, quolls, potoroos and wombats were killed between early 2002 and April 2004 by 1080. But ... that is only the number of dead animals found. When the Co-operative Research Centre for Forestry undertook a 'poison lethality' study ... they found that ... 75 per cent of the animals managed to hide themselves after consuming the poison, their carcasses ending up in dens, burrows, bales of hay, hollow tree trunks, and even buried beneath fallen leaves.[18]

[17] Krien, 1178–1179.

[18] Krien, 3058–3064.

When a local sent a letter to the political representative of the region inquiring on the solutions for the situation, "he wrote back saying there was no need to worry, that 1080 doesn't harm humans."[19] The disregard for the situation was so evident that even when locals saw their dogs dying, the answer was to require muzzles and move on. By 2010 the pesticide stopped being used in most areas of Tasmania but its impact had already dealt a tragic blow to the island's wildlife.

In the fifth and last section of the book and its following epilogue, we are confronted not only with more facts and developments on the fight for Tasmania's forests, but we are also made acutely aware of the journalist's voice and the impact of her research on a personal level. Indeed, as we approach the final chapters, we become closer and closer to Krien, her sadness and disbelief. By showing us her emotions, the journalist seeks to describe what she witnessed in a way akin to what Professor Cindy Royal describes as a desire to reach a deeper truth by working within the boundaries of dialogue and the description of scenarios.[20] Indeed, Krein's literary style conveys the complexity of witnessed events, particularly in one moment where we are again following alongside the ratbag activists, as their blockade is lifted once again and loggers approach:

> A few local journalists arrive and the police inspector comes … to escort them through. One reporter still has the plastic price tags on his Blundstone boots and a couple of the young female journalists are wearing high heels. They totter up the hill with the inspector. It is quiet for a moment. Then the buzz of the chainsaw starts up again until, finally, a crack. It is a big tree, some 70 meters tall and about 300 years old. When it starts to fall, I can hear its canopy swish, falling into the arms of the surrounding forest … taking other smaller trees down with it to the ground. … Workers have dismantled the activists' blockade and tossed it into a nearby quarry. Afterward, Forestry Tasmania took photos of the quarry and released them to the media with the caption "ferals rubbishing our forests."[21]

When Anna Krien revisits the scenery later, she is distraught:

> It's gone. It's all gone. It feels as though I'm at the edge of the world. The loggers were only in here a little over a month. Logistically it is incredible: a

[19] Krien, 3044–3046.

[20] Cindy Royal, "The Future of Literary on the Internet," University of Texas, April 1, 2000, http://cindyroyal.com/litjour_croyal.doc.

[21] Krien, *Into the Woods*, 3728–3738.

> vast forest, thousands of years old, gone in six weeks. The gravel widens into the size of a two-lane highway … The trees left behind are precarious. They creak in the wind. One blockader told me she could hear the lone leftover trees cracking and falling over during the night, unable to withstand the wind without the surrounding forest … it's all gone, a stadium-sized pile of mud, rubble and nothing. On my way back, I slip. I sink into the grey mud, the wet oozing inside my boots, socks and jeans … I hold onto a branch to pull myself out. The branch snaps and one boot comes off. I plunge my foot back down into it. The mud makes slurping sounds. I feel like I am in the Swamp of Sadness from The Never-ending Story, the wet sucking me down, its sadness swallowing me.[22]

The Years After *Into the Woods*' Publication and the Future of Tasmania's Forests

By the end of the book and in the aftermath of its publication one would hope that some progress has been made towards establishing a reasonable dialogue for the future of the logging industry business. In the afterword to the second edition of *Into the Woods*, however, Anna Krien is the first to admit that peace for Tasmania's forests and wildlife remains elusive. A closer look at the last few years and the developments that took place after 2012, the year Krien's book was published, truly reveal the depth of the ongoing environmental crisis. In September 2014, the liberal party, six months after winning the elections that year, was able to repeal the forestry peace deal, an achievement that took the Labor-green government years to achieve. To add insult to injury the repeal bill reclassified 400,000 hectares of native forest that are now on the list for potential logging (this almost tripled the hectares of reserved forest for logging as a result).[23] The irony and danger of this decision to the island's stability was not lost even on those working in the sawmills. In 2017, after more announcements of reserved forests now being permitted to be logged, the head of Tasmania's largest native sawmill warned of the dangers of returning to the forest wars. In his statement he even added that the proposals to opening more native forests were unnecessary and they might even go unclaimed by the logging companies. This obviously enraged the Labor party leader and

[22] Krien, 3756–3761 & 3763–3768.

[23] Annah Fromberg, "Forest peace deal repeal: Early vote in Tasmania's Legislative Council flags end of accord," *ABC News*, Augst 28, 2014. https://www.abc net.au/news/2014-08-27/early-vote-flags-end-for-tasmania27s-forest-peace-deal/5699812.

caused bad publicity by implying that this impractical decision was merely used for political advantage.[24] Even the state-owned Forestry Tasmania, one of the major entities in this process, was surprised since they had not requested to log previously protected forests and the local sawmillers and woods processors had no demand or need for such a substantial increase.

As the years pass, the political rhetoric, the logging, and the protests continue to be part of the island's daily routine. And yet, while the environmental groups seek non-violent protests and dialogue in an age of climate change awareness, the Australian government's response to them and to Tasmania's nature and wildlife has become harsher and harsher. In November 2019, the House of Assembly was able to pass a set of problematic anti-protest laws[25] based on an original legislation first overturned by the High Court back in 2017 due to the efforts of former Greens leader Bob Brown.[26] The government of Tasmania was quick to add that the polemical new regulations were not instituted "against peaceful protest" but in reality they were meant "to protect the rights of workers and businesses from disruptions from activists."[27] When months later the Bob Brown Foundation sought to protest peacefully in Tasmania, the environmental group was swiftly banned and threatened with massive fines that could reach half a million dollars.[28] While it was reasonable to consider that the pandemic crisis that started in 2020 could in some way stifle the logging business, like it did for so many other businesses, the news proved the opposite. In May 2020, Tasmania's resources minister Guy Barnett reinforced the importance of the forestry sector to combat the crisis. The persistence of the Tasmanian Government in ramping up the

[24] Georgie Burgess, "Sawmill boss warns against return to Tasmanian 'forest wars', says wood may go unclaimed." *ABC News*. February 27, 2017. https://www.abc.net.au/news/2017-02-27/neville-smith-forest-products-warning-tasmanian-forest-wars/8307064.

[25] Matt Maloney, "Tasmania's House of Assembly passes country's toughest anti-protest laws." *The Examiner*. November 28, 2019. https://www.examiner.com.au/story/6517184/anti-protest-laws-pass-the-house-of-assembly/.

[26] Michael Slezak, "Bob Brown wins high court challenge to Tasmanian anti-protest laws." *The Guardian*. October 18, 2017. https://www.theguardian.com/australia-news/2017/oct/18/bob-brown-wins-high-court-challenge-to-tasmanian-anti-protest-laws.

[27] Maloney, "Tasmania's House of Assembly."

[28] James Dunlevie, "Bob Brown Foundation banned from Tasmanian forest protest due to 'safety concerns' over actions at Tarkine." *ABC News*. February 21, 2020. https://www.abc.net.au/news/2020-02-21/bob-brown-protesters-banned-from-tasmanian-forests/11986996.

logging industries—despite a drop in demand—and their obsession in keeping business as usual, was not lost on the Greens leader Cassy O'Connor when she reminded minister Barnett that the climate crisis "had not simply disappeared."[29]

Still, there are some glimpses of hope for a change with the recent news that the tourism sector has been pressuring the Tasmanian government to end native forest logging,[30] and with some new data showing that the closure of lumbering mills has had a clear effect on Tasmania's carbon emissions.[31] While this news seems to show signs of change, Tasmania's political track record is enough evidence that the fight to preserve the island's natural resources is still ongoing. And while it is easy to get lost on Tasmanian politics and the worrying developments taking place on such a distant island, hopefully accounts such as Anna Krien's *Into the Woods* might serve as a possible guideline on how to effectively report on activism. With any luck, maybe her work will encourage others to give their time and attention to Tasmania's and many other man-made natural disasters.

References

Alexander, Robert. "Literary Journalism and Ecocriticism." In *The Routledge Companion to American Literary Journalism*, 482–497, edited by William Dow and Roberta Maguire. New York: Taylor & Francis, 2020.

Beavis, Laura. "Environmentalists say logging in Tasmania has been illegal since 1987. Here's why law changes are being rushed through." *ABC News*. May 3, 2022. https://www.abc.net.au/news/2022-05-04/tasmania-1987-forestry-law-row-protesters/101035222.

Burgess, Georgie. "Sawmill boss warns against return to Tasmanian 'forest wars', says wood may go unclaimed." *ABC News*. February 27, 2017. https://www.abc.net.au/news/2017-02-27/neville-smith-forest-products-warning-tasmanian-forest-wars/8307064.

[29] Emily Jarvie. "Greens call on Resources Minister Guy Barnett to rethink the future of the state's forestry industry." *The Examiner*. May 3, 2020. https://www.examiner.com.au/story/6743637/greens-call-for-forestry-rethink/.

[30] Adam Langenberg, "Tourism pressure mounts on Tasmanian government to end native forest logging." *ABC News*. November 9, 2021. https://www.abc.net.au/news/2021-11-10/tourism-businesses-call-for-end-tas-native-logging/100607752.

[31] Laura Beavis. "Environmentalists say logging in Tasmania has been illegal since 1987. Here's why law changes are being rushed through." *ABC News*. May 3, 2022. https://www.abc.net.au/news/2022-05-04/tasmania-1987-forestry-law-row-protesters/101035222.

Cindy Royal, "The Future of Literary Journalism on the Internet." University of Texas. April 1, 2000. http://cindyroyal.com/litjour_croyal.doc.

Dunlevie, James. "Bob Brown Foundation banned from Tasmanian forest protest due to 'safety concerns' over actions at Tarkine." *ABC News.* February 21, 2020. https://www.abc.net.au/news/2020-02-21/bob-brown-protesters-banned-from-tasmanian-forests/11986996.

Franklin, Jon. "A Story Structure" in *Telling True Stories*, edited by Mark Kramer and Wendy Call. New York: Penguin Books, 2007.

Fromberg, Annah. "Forest peace deal repeal: Early vote in Tasmania's Legislative Council flags end of accord." *ABC News.* August 28, 2014. https://www.abc.net.au/news/2014-08-27/early-vote-flags-end-for-tasmania27s-forest-peace-deal/5699812.

Jarvie, Emily. "Greens call on Resources Minister Guy Barnett to rethink the future of the state's forestry industry." *The Examiner.* May 3, 2020. https://www.examiner.com.au/story/6743637/greens-call-for-forestry-rethink/.

Kramer, Mark. "Breakable Rules for Literary Journalists" in *Literary Journalism*, 21–34, edited by Norman Sims and Mark Kramer. Ballantine Press: USA, 1995.

Krien, Anna. *Into the Woods – The Battle for Tasmania's Forests.* Australia: Black Inc. Kindle, 2012.

Langenberg, Adam. "Tourism pressure mounts on Tasmanian government to end native forest logging." *ABC News.* November 9, 2021. https://www.abc.net.au/news/2021-11-10/tourism-businesses-call-for-end-tas-native-logging/100607752.

Maloney, Matt. "Tasmania's House of Assembly passes country's toughest anti-protest laws." *The Examiner.* November 28, 2019. https://www.examiner.com.au/story/6517184/anti-protest-laws-pass-the-house-of-assembly/.

Morton, Lindsay. "Discovering a New Voice." *Literary Journalism Studies* Vol. 4, n. 1 (2012): 127–130.

Slezak, Michel. "Bob Brown wins high court challenge to Tasmanian anti-protest laws." *The Guardian.* October 18, 2017. https://www.theguardian.com/australia-news/2017/oct/18/bob-brown-wins-high-court-challenge-to-tasmanian-anti-protest-laws.

The Monthly. "Into the Woods: Anna Krien in conversation." Youtube Video, 23:37. May 3, 2013. https://youtu.be/tINXZdUR-EE?si=OkSStjTDf_ufMvCl.

John Joseph Mathews: The "Blackjack Discourse" of an Osage Naturalist

John M. Coward

John Joseph Mathews was an extraordinarily complicated man. Born in 1894, "Jo," as he came to be known, was the precocious child of a prosperous family in Pawhuska, Indian Territory (now Oklahoma). As a boy, he explored the hills and valleys of the vast Osage prairie, becoming a keen observer of birds and other wildlife. He went on to become a popular student, football player, and fraternity boy at the University of Oklahoma, a flight instructor during World War I, a hard-drinking Oxford student following the war, an avid hunter, traveler, newspaper correspondent, ladies' man, failed husband and absent father, and, eventually, an acclaimed

J. M. Coward (✉)
University of Tulsa, Tulsa, OK, USA

P. Calvi (ed.), *The Literary Journalist as a Naturalist*, Palgrave Studies in Literary Journalism,
https://doi.org/10.1007/978-3-031-56634-9_3

Osage man of letters, historian, and tribal representative.[1] Jo Mathews was also a self-taught naturalist who lived for several decades in a small sandstone cabin called The Blackjacks, named for the tough species of oak that covered the ridges of the Osage Hills not far from his childhood home. Among the blackjacks, Mathews observed the changing colors of the seasons, studied bird behavior, planted a garden, raised chickens and pheasants, hunted coyotes, hosted dinner parties that lasted well into the morning hours, and, not least, mused on the problems of modern life and the mysteries of existence. Drawing on his Osage heritage, Mathews turned his life at the cabin into a book-length chronicle of his prairie experiences, *Talking to the Moon*, a book revered by his fans and compared to Henry David Thoreau's *Walden*.[2] Mathews's book, still in print today, is organized around the twelve moons of the Osage year, beginning with Just-Doing-That Moon (March) and ending with Light-of-Day-Returns Moon (February). Early in his narrative, Mathews explains the reason why he returned to the natural world of his remembered childhood. The seasons and their days flowed by, he writes, no two alike, and

> I wanted to become a part of that flow in so far as I was able; to learn something of the moods of the little corner of the earth which had given me being; to learn something of the biological progression and the mysterious urge which inspired it, until the biological changes within myself dimmed the romance of it.[3]

As an Oxford-educated Osage man from small town Oklahoma, Mathews was often critical of modern civilization, another factor that explains his retreat to the Oklahoma plains. Mathews wanted to escape the speed and noise of twentieth-century urban life and, as he put it, "the artificialities and the crowding and elbowing of men in Europe and

[1] Michael Snyder, *John Joseph Mathews: Life of an Osage Writer* (Norman: University of Oklahoma Press, 2017), 1–2; and Robert Allen Warrior, *Tribal Secrets: Recovering American Indian Intellectual Traditions* (Minneapolis: University of Minnesota Press, 2005), 14–19.

[2] A. LaVonne Brown Ruoff, "John Joseph Mathews's *Talking to the Moon*: Literary and Osage Contexts," in *Multicultural Autobiography: American Lives*, ed. James Robert Payne (Knoxville: University of Tennessee Press, 1992), 5. Writing in the "Foreward" to the 1981 edition of *Talking to the Moon*, Mathews's widow Elizabeth wrote, "This is John Joseph Mathews' *Walden*. It is a book that a Thoreau or a Muir might write, but it is a *Walden* of the plains and prairies, of the 1930s and 1940s, by a Native American."

[3] John Joseph Mathews, *Talking to the Moon* (Norman: University of Oklahoma Press, 1981), 2.

America...."[4] The Blackjacks offered a measure of solitude and an intimate, renewed connection to nature. Mathews had more personal reasons as well. His marriage was failing and his real estate career had stalled. He was adrift, so lost and desperate that he abandoned his wife and two small children in California. Returning to the prairie, Mathews took refuge in a place far from his personal and professional troubles, a place where he could take stock of his life and, with perseverance and a little luck, set down roots, renew his spirit, and fulfill his ambitions as an Osage writer and intellectual.

This essay offers a brief exploration of Mathews's nature writing from The Blackjacks, a years-long project that turned into *Talking to the Moon*, published in 1945, and praised as his "masterpiece."[5] Although the book attracted little attention when it was published, it is a multilayered text, rich in symbol and meaning, and its reputation has grown over the years. Writing in 1992, critic A. LaVonne Brown Ruoff praised the book as "a highly sophisticated literary autobiography."[6] Yet *Talking to the Moon* is something more than a traditional autobiography. It might be better described as a "spiritual autobiography," observations about and meditations on human existence and the natural world filtered through Mathews's unique worldview—part naturalist, part Osage cultural historian, part hunter, part educated elitist and social critic.[7] All of this makes *Talking to the Moon* a unique American achievement in literary nonfiction and one that deserves greater critical and popular attention. The purpose of this essay is twofold: first, to describe and highlight Mathews's achievements as a singular observer and interpreter of the natural world; and, second, to analyze and elucidate Mathews's literary voice as an Osage writer who found worth and meaning in nature and who offered an Indigenous alternative to the hollow values of twentieth-century American society.

A Privileged Osage

Jo Mathews was born in Pawhuska, the capital of the Osage Nation, the son of William S. Mathews, a prosperous rancher, merchant, banker, and influential Osage official, and Pauline E. Girard, a devout woman whose

[4] Mathews, *Talking*, 2.

[5] Warrior, *Tribal Secrets*, xiv.

[6] Ruoff, "John Joseph Mathews's *Talking to the Moon*," 1.

[7] Ruoff, "Mathews," 1.

Missouri family was French Catholic. Jo's Osage line can be traced to his paternal great-grandmother, A-Ci'n'Ga, a member of the Big Hill Band and the Buffalo clan.[8] In 1813, A-Ci'n'Ga married William Sherley Williams, an early missionary to the tribe.[9] Williams hunted, trapped, and danced with the tribe, becoming known as the "Red-Headed Shooter." He became more widely known as "Old Bill," a legendary mountain man whose Western adventures likely fired Jo's boyhood imagination.[10]

Mathews led a charmed life as a child, more of "village" Osage than a traditional Osage. As biographer Michael Snyder notes, his family's relative wealth and status meant that Mathews "enjoyed security, privilege, and leisure in an exciting time and place to be a boy."[11] Mathews grew to be an attractive, athletic teenager. But he was something of a loner, awkward around girls, and happiest when he was riding his horse Bally over the ridges of the surrounding prairie. After completing high school in 1914, Mathews enrolled at the University of Oklahoma. Mathews reported for the student newspaper and contributed to the university magazine, an early indication of his interest in writing. His plans shifted with the US entry into World War I. He enlisted in 1917 and was assigned to the Air Corps. Snyder attributes Mathews's interest in flying to his tribal background, his "Osageness." In this view, Mathews was fascinated by birds and flight, drawing on the Osage belief in the "interrelation with and connection to animals and birds."[12] Whatever the case, Mathews became a flight instructor at Ellington Field near Houston. Although the army offered him a post after the war, Mathews returned to OU in 1919. He resumed writing for student publications and made contacts with faculty members who helped shape his career as a budding writer. Mathews graduated with a degree in geology in 1920 and briefly considered working as a geologist for an Oklahoma oil company. He chose instead to continue his studies, applying for and—with the help of his OU faculty mentor, writer, and historian Walter Campbell—getting accepted to Merton

[8] Snyder, *Mathews,* 8.
[9] Ruoff, "Mathews's," 6; Snyder, *Mathews* 8.
[10] Snyder, *Mathews,* 8–10.
[11] Snyder, *Mathews,* 17.
[12] Snyder, *Mathews,* 34.

College, Oxford.[13] Mathews was an inconsistent student at Merton—he spent a great deal of time drinking in Oxford pubs and chasing women—but he graduated with a second bachelor's degree in 1923.

With funds from the generous oil headrights from his allotted land, Mathews spent his vacations hunting in Scotland.[14] He traveled on the continent, partied in Paris, and hunted in Algeria where, in 1922, he had an epiphany. One day his hunting party was surprised—and frightened—by a group of "wild" Kabyle tribesmen who rushed upon them, rifles firing. The incident was entirely peaceful, yet it stirred something in Mathews.[15] As he explained to interviewer Guy Logsdon in 1972:

> That night I got to thinking about it, and I thought that's exactly what happened to me one day when I was a little boy, riding on the Osage prairies. Osage warriors with only their breechclouts and their guns had come up and surrounded me—firing. Of course, I knew some of them, about them; they knew me, who I was. That's what we called joy shooting, you see, just joy.[16]

The encounter reminded Mathews that his own people were interesting and might be a worthy subject for his creative endeavors. Mathews asked himself: "What am I doing over here? Why don't I go back and take some interest in my people? Why not go back to the Osage? They've got a culture."[17]

There's surely some truth to this story, but as Snyder points out, it is misleading. Mathews did go back to the Osage, but his return to Pawhuska took more than seven years, and it was another three years before he moved to The Blackjacks. Nevertheless, it seems likely that Mathews's encounter with the Algerian tribesmen did inspire him to see his own people in a new light, stimulating a new appreciation for the Osage, their

[13] Snyder, *Mathews*, 39. Mathews did not attend Oxford as a Rhodes Scholar, though that assertion has been widely published. In fact, Mathews was able to pay his Oxford tuition with the oil royalties he received as a member of the Osage tribe. Mathews's mentor, Walter Campbell, was a Rhodes Scholar.

[14] Snyder, *Mathews*, 50. Because the Osage retained rights to the mineral wealth beneath their reservation land, tribal members received substantial royalties in the first decades of the twentieth century. In *The Osages*, Mathews wrote that oil royalties peaked in 1925 at $13,000 per capita, diminishing considerably in the 1930s.

[15] Snyder, *Mathews*, 52.

[16] Quoted in Richard Higgs, "Searching for John Joseph Mathews," *This Land* 4, No. 16 (August 15, 2013), 12

[17] Quoted in Higgs, "Searching for John Joseph Mathews," 12.

culture, and their lands. From the sands of the Algerian desert, he could see—perhaps for the first time in his life—that his people and his place were worthy of his literary imagination.

Home Among the Blackjacks

Mathews opens *Talking to the Moon* with two introductory chapters that lay out his prairie project and describe the building of his cabin and the surrounding landscape. The cabin was small, he noted, built for one man and made from local sandstone. This modest house served Mathews well. He lived at The Blackjacks from 1932 until the 1970s; his first decade there was the basis for *Talking to the Moon*. He spent his time observing nature, reading, writing, hunting, and entertaining his family and literary friends, including journalist Ernie Pyle, OU botanist Paul Sears, and European authors Emil Ludwig and Andre Maurois.[18] A portrait of Mathews at The Blackjacks about 1945 shows him sitting in his red leather easy chair in front of the fireplace, pipe in hand. This was the place where he contemplated the natural life of the Osage prairie and, importantly, established a place for himself as an Indigenous writer pursuing his own ideas and living life on his own terms.

Most of the chapters in *Talking to the Moon* follow a pattern. Mathews opens with Osage stories about the weather and the natural world during each month. Writing in Buffalo-Breeding Moon (July), Mathews begins by noting that the month is almost always dry and hot. The breeding buffalo, once common on the prairies, have been replaced now by white-faced bulls, "their mating frenzy causing their sides to work in a bellows action and their tempers to become even shorter in the heat."[19] In this chapter, as in many others, Mathews then comments on the flora and fauna that draw his interest that month, especially birds:

> The bird chorus has almost ceased, and only a few of the more independent singers ignore the blazing sun and the hot shade. The dickcissels sit on the fence posts or on a swaying weed and sing with some cheer but weakly, the cuckoo croaks with petulance, and the flycatchers—the kingbird, crested, and scissor-tailed—build their nests on the ridge, but only the crested is loudly vocal in anything that resembles song.[20]

[18] Snyder, *Mathews*, 101–102.
[19] Mathews, *Talking*, 95.
[20] Mathews, *Talking*, 95.

After such initial observations, Mathews typically turns to other topics, including observations and critiques of modern Euro-American society, which is—not coincidently—the very society he rejected to take up life at The Blackjacks. Importantly, too, Mathews provides a detailed accounting of his daily activities on the prairie, including stories about coyote and bear hunting, reflections on Osage culture and religion, and conversations with Osage neighbors and local cowboys. These accounts are sometimes small, sometimes instructive, sometimes philosophical. Taken together, these details place Mathews at the center of a distinctive physical and intellectual life in the natural and human world of the Osage prairie. To put it another way, Mathews's move to The Blackjacks and his activities and observations about his life on the prairie constitute an intentional act of self-preservation and literary independence—or, more precisely, "intellectual sovereignty," as Robert Allen Warrior writes.[21] This was Mathews's way of creating a physical and spiritual place for himself as an Osage writer in what he saw as a hostile and dehumanized world. Early in *Talking to the Moon*, he explains: "So after years spent in many other parts of the earth, I had come back to that very spot where I had lain as a boy, watching the circling of the red-tailed hawks and actually shedding tears over the fate that had made me earth-bound...."[22] This was home for Mathews, the place of his boyhood dreams and the center of his Osage world.

Indeed, the Osage and their beliefs are a significant part of *Talking to the Moon*. Mathews devotes most of his Buffalo-Pawing-Earth Moon (June) chapter to the Osage dances held every year at that time. Mathews remembered visiting the Osage camps with his father, watching the activity "among the rounded lodges, rounded in harmony with the rounded hills of the reservation."[23] He also learned about the Osage religion, explaining the tribe's beliefs as a product of their existence on the prairie. "Their religion," he writes, "their concept of God, came out of the blackjacks, out of the fears inspired by the elements...."[24] Although Mathews did not practice the old religion, known as Wah-Kon-Tah, he appreciated the dances and the singing. "I am always impressed when the singers sing the 'Song of the Red Horse.' In this song there is simplicity and joy and a

[21] Warrior, *Tribal Secrets*, xxii.
[22] Mathews, *Talking*, 4.
[23] Mathews, *Talking*, 77.
[24] Mathews, *Talking*, 77.

poignant fervency which affects me deeply."[25] Clearly, Mathews's fondness for Osage ceremonies was connected to his childhood memories, but this passage suggests that the mature Mathews found something deeper in their religion. With his return to the Osage land, Mathews—a highly assimilated Osage—came to understand and appreciate Osage culture and practices intensely and personally. He found, in his words, "simplicity," "joy," and "poignant fervency."[26] As with his renewed connection to the natural world, Mathews's embrace of Osage culture can be seen as part of his spiritual renewal and the foundation for his life as a thinker and writer. To find his distinctive voice, Mathews immersed himself in nature at a place that was important to his identity; from that place, he rediscovered and engaged Osage beliefs and spiritual traditions.

Writing the Prairie

Mathews often wrote lyrically about the prairie, mining its changing patterns in intimate detail. In Buffalo-Pawing-Earth Moon (June), for example, Mathews describes summer on the plains:

> This is the season of shining green, whispering breezes, and distant, lilac-scented hills that turn mauve after sunset. During this season each day is very much like the one that preceded it; days filled with throbbing life and earth-scents; days when the sun is not yet too hot and the winds not yet too persistent.[27]

In this passage and many others, Mathews describes the "feel" of the prairie—its colors, its motion, and its odors. This is especially significant because some outsiders perceive the prairie landscape as unremarkable,

[25] Mathews, *Talking*, 83.

[26] Lee Schweninger, "Irony and the 'Balance of Nature on the Ridges' in Matthews's *Talking to the Moon*," *Studies in American Indian Literature* 9, no. 2 (Summer 1997), 41. Curiously, Mathews does not explore his Osage roots in *Talking to the Moon*. The book's narrator, different from Mathews himself, "assiduously avoids exploring or even identifying his own Indian ethnicity," Schweninger argues. Thus Mathews writes "not as an Osage but as an American of Osage descent." This was a useful position for Mathews because it allowed him to offer a more complicated critique of American society and to highlight the contradictions surrounding his life on the prairie.

[27] Mathews, *Talking*, 77.

even dull.[28] Yet Mathews deploys his poetic sensibilities to allow the reader to see the "shining green" of the prairie grasses, to hear the breeze, to smell the lilacs. His ability to evoke the natural life of the prairie—Mathews's focus on the small, sensual facts of the land—is one of the strengths of *Talking to the Moon*. It is also a means by which Mathews renders his life at The Blackjacks as memorable, lived experience. By naming and identifying the contours of his prairie days, Mathews creates impressions and images of the Osage prairie in language that links him (and his readers) to the land, its flora and fauna, and to the seasons and ever-changing weather. This literary immersion into the daily details and rhythms of the prairie has the effect of universalizing this place, making it intimate and vivid for readers who have never set foot on the plains.

Some of the book's most charming scenes involve Mathews's encounters with wildlife. He recalled, for example, a surprise meeting with a curious skunk that wandered into the cabin one day while Mathews was reading. For a few moments, Mathews and the skunk eyed each other warily. Fighting his fears, Mathews remained calm and allowed the creature to explore in peace, which it did.

> Assured that I was harmless, he went into the kitchen, and I could hear him tearing the paper from the garbage box, the bacon wrapping. He stayed so long that I continued reading for some time, then I heard his claws ticking against the cement, and he passed on out into the yard. I got up and closed the door.[29]

As this story suggests, Mathews sought to live in harmony with the natural world, a goal he sometimes achieved—or thought he achieved—at The Blackjacks. In his years at the cabin Mathews noted that the nearby wildlife became accustomed to him and his two Irish setters. "The fox squirrels were the first to ignore my presence. They lay along the limbs of the yard trees in catlike poses or even played over the roof," he wrote.

[28] Ian Frazier, *Great Plains* (New York: Farrar, Straus and Giroux, 1989), 3. In the opening paragraph of *Great Plains*, Frazier refers to the "immense Western short-grass prairie now mostly plowed under" and the "still-empty land beyond newsstands and malls." The prairies, he continues—playing the devil's advocate— are home to "fields of wheat and milo and sudan grass and flax and alfalfa and nothing!"

[29] Mathews, *Talking*, 59.

Even the coyotes became more accustomed to seeing him around the cabin, trotting "nonchalantly over the ridge at dusk…."[30]

Yet living in harmony on the prairie was complicated and nature itself could be harsh, as Mathews soon discovered. To begin with, the very act of building a house and drilling a well was a disruption to prairie life. He was, in critic Lee Schweninger's terms, a "settler," an invader who brought elements of white, American civilization to the prairie environment.[31] His house, his car, his garden, and especially his chickens, pheasants, and guineas upset the natural balance at The Blackjacks, bringing conflict, not harmony. Mathews recognized the irony of his situation: "Perhaps my position was unnatural, living as I did, not from the ridge, but feeding myself artificially from cans brought from town and food from the ranch," he wrote.[32]

Mathews concluded that his animals and his activities at The Blackjacks created a new kind of balance, less harmonious but more in tune with nature's predatory ways. He needed to protect his game animals, he realized, "against my predacious neighbors, which was probably a more natural state and in the end more satisfying than the 'friends and neighbors' idea."[33] As a result, Mathews became important to the wild animals on the ridge. His flocks "sharpened their cunning" and changed the relationship between himself and the predators. It was a hard truth about man and nature:

> We learned more about each other; we…pitted our wits against each other and saw no more of each other in repose. We had greater mutual respect, and I became a part of the balance through my strength to protect my flocks. Thus, I achieved a greater harmony with my environment and found that there is no place for dreams in natural progression, and it seems to me that I had realized for the first time that with responsibility come enemies.[34]

Mathews's realization reflected his general philosophy of life, which was overtly masculine, Darwinian, and Nietzschean.[35] Accordingly, he celebrated strength and vigor. Mathews's daughter Virginia told biographer

[30] Mathews, *Talking*, 59.
[31] Schweninger, "Irony," 41–42.
[32] Mathews, *Talking*, 60.
[33] Mathews, *Talking*, 60.
[34] Mathews, *Talking*, 60.
[35] Snyder, *Mathews*, 107.

Snyder that her father was "very scornful, very impatient of weakness, especially in himself."[36] This muscular view of nature gave Mathews a way to rationalize his disruption of prairie life, positioning himself not as a settler but as another natural force on the prairie that nature had to account for and accommodate.

This view of nature also explains Mathews's actions when nature seemed to scheme against him. In contrast to his endearing skunk story discussed earlier, Mathews unleashed his wrath on a skunk that killed some of his expensive English chickens. Finding the satiated skunk asleep among its victims, Mathews acted:

> I didn't hesitate. I was so annoyed that I held the muzzle of my Smith & Wesson to his head and emptied the cylinder, glorying in the nauseating musk odor that hung on the heavy air of night, transforming its glory with the sharp explosions that broke the silence of the ridge into a symbol of the mighty powers of *Homo sapiens* when aroused and announcing his entrance into the struggle.[37]

In this incident, Mathews's romantic reveries about nature fell before the Darwinian struggle for dominance. He recognized, too, that he had created this particular struggle by bringing his birds to the prairie. The skunk was acting naturally, Mathews decided, though he insisted that this skunk had let its "lust" for killing run amok, killing far more chickens than it needed to survive. He concluded that this skunk was "abnormal" and he linked this behavior to abnormal human behavior, again eschewing the romantic view of nature for the harsh reality of competition.

Despite this realization, many sections of *Talking to the Moon* reveal a deeper understanding of the natural world. Mathews lived close to the prairie and he observed it carefully, even when it intruded on his comfort. One night, for example, a centipede dropped from the walls of his porch onto his bed, stinging him while he was asleep. He recovered, but not before his arm swelled and turned bluish green.[38] Then there were ever-present spiders. "I am always poking my face into spiders' webs as I come out of my doors...," he wrote in Deer-Hiding Moon (September). The webs bothered him, but Mathews celebrated their beauty: "Some of these [spider webs] are beautiful, even iridescent, with the early-morning sun

[36] Snyder, *Mathews*, 107.
[37] Mathews, *Talking*, 65.
[38] Mathews, *Talking*, 146.

making jewels of the collected dew. Stretched between the house and the locust trees, they are as large and as perfect as pieces of Swiss lace."[39] Mathews's embrace of nature—its beauty but also its cruelty and pain—was part of his spiritual engagement with the landscape, a way of staking his claim on the Osage prairie and making his experiences meaningful for himself as well as his readers.

Mathews's experiences at The Blackjacks, combined with his Darwinian philosophy, led him to develop a set of ideas about human life and the biological forces of the natural world. Early in *Talking to the Moon*, Mathews identified two "primal" biological laws, reproduction and survival. "Primarily," he wrote, "man is so constituted by nature that she seems interested in him only as a reproducer of himself and as a protector of his offspring until they can care for themselves." Old men—like Mathews himself—were of little use, he concluded, "unless through their experience they might have a natural importance in the tribe and in the larger units of society."[40] *Talking to the Moon*, perhaps, was Mathews's way of sharing his experience with his tribe and society and, in so doing, justify his life as a writer.

In addition, Mathews believed that all species had a deep-seated urge toward "ornamentation," a term he used to describe the impulse to expression not tied to survival. Ornamentation, he wrote, was that "biological urge which inspires the wood thrush to sing and the coyote to talk to the moon."[41] Ornamentation, too, was present in modern Euro-American society and, according to Warrior, ornamentation was "the key to [Mathews's] understanding the situation modern humans find themselves in.[42] For Mathews, there was a conflict between the ornamental dreams of civilization—social and political ideologies and the creation of beauty—and the "primal laws of survival and reproduction."[43] This conflict represented a serious flaw in the political and social order of twentieth-century life, the very problems that led Mathews to move to The Blackjacks. Mathews's perceptions were, no doubt, sharpened by the economic and social problems of the Great Depression, and he was quick to note the destructive effects of business and commerce on society and the environ-

[39] Mathews, *Talking*, 146.
[40] Mathews, *Talking*, 12.
[41] Mathews, *Talking* 3.
[42] Warrior, *Tribal Secrets*, 61.
[43] Mathews, *Talking*, 16.

ment. On the prairie and in industrial America, Mathews had witnessed the exploitation of natural resources, which had reached a saturation point "under the berserk methods of economically, socially and politically liberated Europeans called Americans...."[44] The buffalo had been replaced by white-faced cattle, the prairie grasses had given way to industrial agriculture, and the oil fields had fouled the land and waterways of the plains. There were other losses too: "There are no baby bears on the old reservation now. I imagine there are no wild bears in all the former domain of the Osage," Mathews wrote.[45]

In contrast to the "Europeans called Americans," Mathews noted that the Osage lived in and with nature, not against it. Mathews believed the Osage were part of the "balance" of the prairie, having lived in harmony with native plants and animals for centuries. European Americans gave little thought to such matters. They pillaged the land for quick profits. They killed off the buffalo. They introduced alien species like the English sparrow which, Mathews wrote, "now overruns every city from coast to coast and has even seeped into Death Valley."[46] For Mathews, the lesson was clear: balance could protect the ridges of the Osage prairie. Unbridled growth and exploitation of mid-century capitalism, on the other hand, would bring ruin:

> [If] some foul dough-belly, chewing an unlit cigar and squinting his eyes, decided that there would be greater economic benefit in dividing my pastures into town lots, with a belching factory chimney as the Gothic tower of the religion of mechanism, in the center, the peace of the blackjack would be destroyed.[47]

Mathews's philosophy—his "biologically deterministic fatalism" in Warrior's phrase—gave him a way to draw a sharp contrast between Indigenous and Euro-American views of the land.[48] From his home on the prairie, Mathews recognized the devasting results of arrogant and short-sighted economic practices.

[44] Mathews, *Talking*, 15.
[45] Mathews, *Talking*, 193.
[46] Mathews, *Talking*, 151.
[47] Mathews, *Talking*, 232.
[48] Warrior, *Tribal Secrets*, 98.

Refuge and Renewal on the Prairie

Mathews's escape to The Blackjacks had a strong aesthetic and philosophical bent. He wanted—*needed* might be a more accurate term—a place to settle, a place to reflect and forge a literary life. As noted above, he also wanted to escape the larger social and economic forces of Depression-era America, especially the work-a-day world of American business and industry. The relentless flow of contemporary life had been sweeping him onward, preventing him from seeing and engaging the world in a fully human—or fully Osage—way. Among the blackjacks, Mathews could slow down and begin to think independently and deeply about nature, time, society, and his place in the world. In fact, Osage scholar Robert Allen Warrior writes that Mathews became "obsessed with self-critical reflection."[49] Mathews makes this clear throughout *Talking to the Moon*, as in this passage that expands his "flow" metaphor:

> I came to the blackjacks as a man who had pulled himself out of the roaring river of civilization to rest for a while; out of the flood where formerly only his head had been above the surface. Stopping for a time and looking back, he could better appreciate the sweep of the river and the spectrum in the mist above the falls which had battered him. From his island he could perhaps see the natural *raison d'etre* of the mad whirlpools with their deadwood and drift, endlessly gyrating, as well as the naturalness of the eddies and stagnant bayous.[50]

Out of the symbolic river and secure on the actual prairie, Mathews was able to live deeply in the natural world and reconnect to his people and their culture. Although he was doubtful that his words would prevail, Mathews expressed faith in nature and its enduring beauty. "Word symbols couldn't satisfy me as an expression; there was too much that was inexpressible…," he wrote, listing a few moments and places of transcendence: "flying in the moonlight, the Grand Canyon, desert sunsets, riding over the Osage prairie in June."[51] He found peace, too, in physical activity, especially "masculine" activities such as hunting, which he maintained was part of the grand scheme of nature. "Physical action and living every day

[49] Robert Allen Warrior, "Intellectual Sovereignty and the Struggle for an American Indian Future," *Wicazo Sa Review* 8, no. 1, (Spring 1992): 7.

[50] Mathews, *Talking*, 3–4.

[51] Mathews, *Talking*, 16.

in harmony with the life about me were exhausting and therefore completely satisfactory," he wrote.[52]

A major literary accomplishment of Mathews's self-reflection involves his efforts to create a space for himself as a man and as an Osage intellectual. His ambitions were modest but significant: "I didn't presume that I could come out of the blackjack with the banner of truth flying," Mathews wrote, "but that I might be able to devote a few years to pleasant, undisturbed living."[53] By rejecting mainstream American social and business values, Mathews created a home for himself on the prairie, a place from which he could live his life his own way and establish his own Indigenous voice. This was no small matter, and Mathews worked diligently to make sense of his place as an Osage writer living on the plains. From his time among the blackjacks, Mathews observed birth, decay, and sudden death on the prairie. He interacted with his neighbors, cooked for his literary friends, contemplated world affairs, and explored Osage beliefs. He pondered the anxieties of solitude and aging and took comfort in the patterns of the seasons and the movements of rabbits, coyotes, and pack rats. These activities—some trivial, some important—mattered. Musing on the songs of the prairie birds during Baby-Bear Moon (December) Mathews writes, "Every thing and every action on my ridges and among the blackjacks have meaning, and the better I understand the *raison d'etre* of these things and actions, the more intense my pleasure...."[54] In the end, Mathews's search for a deeply realized and independent existence in the natural and cultural world of the Osage prairie is the enduring achievement of *Talking to the Moon* and a major reason that his experience resonates with readers—both Indigenous and not—in the twentieth-first century.

Mathews denied that he was escaping anything when he moved to the prairie. He argued that he was seeking only "to get my feet on my own bit of earth, as one might arrange one's body comfortably in a bed or in a chair."[55] But his move to the prairie was deeper than this statement suggests. Indeed, Mathews's attachment to nature was a way of living well and standing against the void:

[52] Mathews, *Talking*, 16.
[53] Mathews, *Talking*, 3.
[54] Mathews, *Talking*, 207.
[55] Mathews, *Talking*, 16.

> I love every manifestation of nature, though now, like an aging player, I am not so keen to get into the game to become part of the struggle; and though during the last years, when I take things more as matters of fact rather than as miracles, there is still never a moment of boredom, and the tree shadows tell me that I must live intensely if I am to get the last drop of juice from the glowing orange.[56]

Conclusion

As many observers have noted, Mathews's motives for moving to his prairie cabin have similarities to Henry David Thoreau's motives for building his cabin at Walden Pond.[57] In a *Walden* passage that has become iconic, Thoreau proclaimed, "I came to the woods because I wished to live deliberately, to confront only the essential facts of life and see if I could not learn what it had to teach, and not, when I came to die, discover that I had not lived."[58] Similarly, Mathews writes, "I came to the blackjacks as a man who had pulled himself out of the roaring river of civilization to rest for a while; out of the flood where formerly only his head had been above the surface."[59]

Like Thoreau, Mathews sought to live a more basic existence, free from society's obligations and rules. Like Thoreau, Mathews was engrossed in nature. Naturalist and scholar Joseph Wood Krutch observed that Thoreau spent his days "walking in the woods observing the birds and small animals, noting the phenomena of the seasons, talking with a simple Canadian wood chopper who worked happily felling trees for Concord fireplaces, and cultivating his little garden, or as he put it, 'making the earth say beans instead of grass.'"[60] This description applies equally to Mathews at his sandstone cabin, where he walked the ridges and woodlands, observed all manner of birds and other wildlife, planted a garden, and chronicled the changes of the seasons. Like Thoreau, Mathews also talked with the local workers and tradesmen, who in his case were not Canadian wood choppers but cowboys and Osage friends. Beyond mere declarations of the

[56] Mathews, *Talking*, 124.

[57] Ruoff, *John Joseph Mathews's*, 8, writes that Mathews's literary models were Thoreau and John Muir.

[58] Quoted in Joseph Wood Krutch, ed., *Thoreau: Walden and Other Writings* (New York: Bantam Books, 1962), 172.

[59] Mathews, *Talking*, 3–4.

[60] Krutch, "Introduction," *Thoreau,* 8.

simple life, Mathews was the rare literary figure who actually practiced Thoreau's gospel of simplicity.

But Mathews was not Thoreau, as Warrior has pointed out, and his views of nature and Indigenous life in North America were informed by a history of oppression, dispossession, and loss, including the loss of flora and fauna to the forces of commerce and industrialization.[61] Mathews makes this point explicit when he introduces Buffalo-Pawing-Earth Moon (June), noting that the buffalo no longer graze freely across the Osage plains. "Here, in my country, they have been displaced by the white-faced bulls, and the roaring of those great curly-headed brutes as they walk arrogantly across the prairie is the dominant voice of June," he writes.[62]

Mathews was well aware, too, that the Osage themselves had been displaced by waves of European-American settlers and the relentless forces of commerce and development that accompanied their arrival. Given this circumstance, Mathews was not optimistic about the survival of Osage culture and traditions. Nevertheless, he did his part to preserve tribal culture, proposing and then working to establish the Osage Tribal Museum in the 1930s, and, during his time at The Blackjacks, fostering a series of twelve portraits of Osage leaders to hang in the new museum. Mathews began the project with a list of twenty-four subjects but soon found that some Osage were reluctant to sit for their portraits. In fact, Mathews found it necessary to accompany the artist to each Osage home, where he sometimes needed to explain the purpose of the project. As he told the daughter of one chief, "This picture is for the Osage people and their children for a hundred years."[63] These paintings were "for posterity," Mathews wrote, and thus constituted "real conservation, since these old men would soon pass away and with them the Osage as he was, the era and type passing with the individuals."[64]

In a very real sense, Jo Mathews was a writer who lived a complicated life in two worlds, a cultural and intellectual division that allowed him to tell a complex story about his relationship to his tribe and to nature. As literary scholar Lee Schweninger has argued, the narrator of *Talking to the Moon* is different from Mathews himself.[65] Schweninger notes, for exam-

[61] Warrior, *Tribal Secrets*, 114.
[62] Mathews, *Talking*, 75.
[63] Mathews, *Talking*, 127.
[64] Mathews, *Talking*, 126.
[65] Schweninger, "Irony," 41.

ple, that Mathews was more involved in Osage politics and tribal affairs than the narrator of *Talking to the Moon*, who mentions these activities only briefly. Similarly, Mathews the author was much more tied to the oil industry and its many threats to the Osage prairie than the narrator, who ignores these issues. Schweninger argues that this distance between the book's narrator and Mathews the author creates a space for irony in *Talking to the Moon*. This irony, he continues, allows the narrator to construct stories about Osage life and the harmony of his prairie life that downplay Mathews's own role in the disruption of nature on the prairie. It also allows Mathews to present a critique of American society while glossing over the fact that Mathews himself remains deeply connected to that society. Thus Schweninger writes that irony in *Talking to the Moon* "names the unspoken compact between the historical John Joseph Mathews and the reader, *against* the narrator (as it were) who tells the story."[66] To put it another way, irony in *Talking to the Moon* allows Mathews to tell a multifaceted story about his people and their land, a story that—between the lines—reveals the deep-seated contradictions and conflicts between man and nature in mid-twentieth-century America generally, and more specifically, between this Osage man and the modern world. For Mathews, this was a literary space where he could be at home in nature while also serving his people as an historian, politician, and advocate.

For all these reasons, *Talking to the Moon* has lessons for living that still resonate. Faced with a spiritual crisis, Jo Mathews retreated from the business and industrial America he found so unsatisfying and built a new life for himself on the ridges of the prairie. Mathews's life at The Blackjacks was unique because it embraced a particular patch of land—the Osage prairie—from a particular point of view—that of a twentieth-century Indigenous man finding solace and meaning in Osage traditions. The result was a book of wide-ranging and deeply felt observations about nature and the human condition that transcended Mathews's experiences and spoke to a larger world. From his small sandstone house, Mathews watched the days unfold. He contemplated the cycles of the seasons, nature's patterns of birth, decay, and rebirth. On his best days at The Blackjacks, Mathews had no schedule, no obligations, and few expectations. He needed only *to be*—to live as he pleased among the blackjacks on his beloved bit of earth. Active and alert in his small cabin set against the

[66] Schweninger, "Irony," 42.

fullness of the prairie, Mathews found a way to renew his body and his spirit, to live the essential life he needed to thrive as a man and a writer.

References

Frazier, Ian. *Great Plains*. New York: Farrar, Straus and Giroux, 1989.

Higgs, Richard. "Searching for John Joseph Mathews," *This Land* 4, no. 16 (August 15, 2013) 12–13.

Krutch, Joseph Wood. *Thoreau: Walden and Other Writings*. New York: Bantam Books, 1962.

Mathews, John Joseph. *Talking to the Moon*. Norman: University of Oklahoma Press, 1981.

Ruoff, A. LaVonne Brown. "John Joseph Mathews's *Talking to the Moon*: Literary and Osage Contexts." In *Multicultural Autobiography: American Lives*, 1–31. Edited by James Robert Payne. Knoxville: University of Tennessee Press, 1992.

Schweninger, Lee. "Irony and the 'Balance of Nature on the Ridges' in Matthews's *Talking to the Moon*," *Studies in American Indian Literature* 9, no. 2 (Summer 1997), 41–56.

Synder, Michael. *John Joseph Mathews: Life of an Osage Writer*. Norman: University of Oklahoma Press, 2017.

Warrior, Robert Allen. "Intellectual Sovereignty and the Struggle for an American Indian Future," *Wicazo Sa Review* 8, no. 1, (Spring 1992). 1–20.

Warrior, Robert Allen. *Tribal Secrets: Recovering American Indian Intellectual Traditions*. Minneapolis: University of Minnesota Press, 2005.

Edible Armageddon: Insect Superheroes and Other Villains

Tessa Ditner Amorosi

Insects, not human beings, are the great survivors of evolution, originating at the same time as the earliest terrestrial plants about 500 million years ago.[1] These insect ancestors were sometimes alarmingly large. For instance, ancient carnivorous dragonfly relatives had slashing mandibles and a wingspan of 60–70 cm.[2] Global news on insects tends to be fantastical in style. Rio reported "Mutant mosquitoes, designed to combat Dengue and Zika, were released in Brazil."[3] The UK went for "Mosquitoes 'put off biting' by human diet drugs."[4] The US explored "Death by tsetse

[1] Yeates, David, "Insects are the great survivors in evolution: new study," *The Conversation*, (November 4, 2014).

[2] Nel, André et al., "Palaeozoic giant dragonflies were hawker predators," *NCBI*, (August 14, 2018).

[3] Ying, Xiu, "Mutant Mosquitoes, Designed to Combat Dengue and Zika, Were Released in Brazil," *The Rio Times*, (September 26, 2019).

[4] *BBC Health*, "Mosquitoes 'put off biting' by human diet drugs," (February 7, 2019).

T. D. Amorosi (✉)
Central Saint Martins, University of the Arts London, London, UK
e-mail: t.amorosi@csm.arts.ac.uk

P. Calvi (ed.), *The Literary Journalist as a Naturalist*, Palgrave Studies in Literary Journalism,
https://doi.org/10.1007/978-3-031-56634-9_4

fly. Monsters inside me."[5] South Africa announced: "Biological control of mosquitoes to eliminate malaria."[6] Even when insects are harmless to people in a direct way, unappetizing and apocalyptic angles abound, such as: "'Many will starve': Locusts devour crops and livelihoods in Pakistan,"[7] and "Bedbugs can drink three times their own body weight in blood in one sitting,"[8] or "Blowfly maggots can consume 60 percent of a human corpse within a week."[9]

Literary journalists that have carried this theme into longer form include Pulitzer Prize winner Kathryn Schulz in her 2018 *The New Yorker* piece "When Twenty-Six Thousand Stinkbugs Invade Your Home," also entitled "Home Invasion" in the print version. The piece looked at the emotional burden of stinkbugs in the home, their impact on agriculture, their worldwide migration, and their olfactory "super powers."

Then in December of that same year, writer Brooke Jarvis reversed the narrative with "The Insect Apocalypse Is Here. What does this mean for the rest of life on Earth?" In the piece, Jarvis presented insects not as grotesque pests but as vital allies, working and dying as they attempt to rescue ecosystems from collapse. So what had changed between March and December 2018?

Historically, agricultural narratives have tended to view insects as pests,[10] yet even agricultural news stories were shifting their perspective.[11] This was driven by scientific data trickling into news channels but often wearing a family-friendly veil: a focus on bees, for instance,[12] the impact of

[5] *Animal Planet*, "Death by Tsetse Fly, Monsters Inside Me," (July 27, 2009).

[6] *SABC News*, "SA launches biological control of mosquitoes to eliminate malaria," (October 18, 2018).

[7] Ellis-Petersen, Hannah and Shah Meer Baloch, "Many will starve': locusts devour crops and livelihoods in Pakistan," *The Guardian*, (May 25, 2020).

[8] Gill, Victoria, "Bedbugs survived the dinosaur extinction event," *BBC*, (May 16, 2019).

[9] Kolbert, Elizabeth, "Where have all the insects gone?" *National Geographic*, (April 23, 2020).

[10] Looy et al., "How Then Shall We Eat? Insect-Eating Attitudes and Sustainable Foodways," *Agriculture and Human Values 31*, (March 1, 2014): 132.

[11] Foucart, Stephane, "En trente ans, près de 80% des insectes auraient disparu en Europe," *Le Monde*, (18 October, 2017).

[12] *The Economist*, "The decline of bees," (September 7, 2015).

losing pollinators on strawberries,[13] or discussing insects in the context of food and feed security.[14] Jarvis's strategy was to invert the villain role traditionally held by insects, while keeping the polarity, and handing the role of villain to humans. She also chose to write for a wide audience, in the style of a blockbuster.

The most frequently quoted insectageddon study is the 1989 to 2016 survey by Germany's Entomological Society of Krefeld, which collected insects from 63 protected areas. The study concluded that in 27 years, insect biomass declined by 76%.[15] Other small and large studies across the globe seemed to echo these findings: 80% beetle loss in New Hampshire's protected forests; butterflies missing in Mexico;[16] formerly abundant prairie grasshoppers gone from Kansas;[17] insect numbers in a dry forest in Mexico dropped 80% since the 1980s; scientists returning to the Luquillo rainforest in Puerto Rico after 35 years found 98% ground insects gone and 80% from the leafy canopy affecting bird survival;[18] a 20-year study in the Netherlands saw 70% beetles and 60% moths lost,[19] meanwhile academics combed through the data and concluded that over 40% of insect species were threatened with extinction.[20]

Ecologists and biologists noted that insect numbers had plummeted, but equally, it was not obvious that apocalyptic-sounding statements, the kind that Jarvis would select, were appropriate for a piece of literary journalism. After all, the topic was bugs. As British broadcaster and farmer,

[13] Compagnon, Sébastian, "Pourquoi les abeilles tombent comme des mouches?" *Le Parisien*, (June 7, 2018).

[14] Van Huis et al., "Edible insects: future prospects for food and feed security," (*FAO forestry paper, No. 171*. Rome: FAO UN, 2013); Durst et al., *"Forest insects as food: humans bite back,"* (Bangkok, Thailand: FAO UN, 2012).

[15] Hallmann et al., "More than 75 percent decline over 27 years in total flying insect biomass in protected areas," *PloS ONE*, Vol. 12. No. 10, (2017): 1–21.

[16] Kolbert, 2020.

[17] Carrington, Damian, "Insect numbers down 25% since 1990, global study finds," *The Guardian*, (April 23, 2020).

[18] Carrington, Damian, "Insect collapse: 'We are destroying our life support systems'" *The Guardian*, (January 15, 2019).

[19] Campbell, Eric, Tomas Ybarra, and ABC, "Insectageddon," *People and Power, Aljazeera*, (Jan 23, 2020).

[20] Sánchez-Bayo, Francisco and Kris A.G. Wyckhuys, "Worldwide decline of the entomofauna: A review of its drivers," *Biological Conservation*, Vol. 232, (2019): 8–27.

Jeremy Clarkson, pointed out: "Even a hardcore peace'n'love vegan will smash a wasp over the head with a copy of the *Socialist Worker* if it's being annoying."[21] When Dr. Sánchez-Bayo was quizzed about his strong vocabulary, he explained: "Even if we do not have enough data to prove it statistically, we know this is happening. It is better to do it now than ten years later when we have a more serious problem."[22]

This is where literary journalism has the advantage over traditional news: it can report on probable futures presenting scientists as characters. This softer approach is also referred to as humanizing the black box of science.[23] In her insect apocalypse piece, Jarvis begins with an ordinary person's point of view: Riis an "everyman" riding his bicycle in the countryside. Riis realizes something is missing: he is not eating any bugs. This technique is also used by Schulz, who opens with an "everywoman" character: Pam. Pam is watching television at home when she remembers she left a door open. When she goes upstairs, stinkbugs have climbed in and invaded her bedroom.

The reader's first contact with both pieces is not only these carefully selected characters, but also carefully selected visuals. Schulz's stinkbug invasion piece opens with an artwork that borrows from hieroglyphics. It has a whiff of *Return of the Mummy* about it. Jarvis's apocalypse piece in the *New York Times Magazine* opens with a photograph of an atomic bomb destroying a hornet. Design director Gail Bichler explains her use of an extreme visual by saying they were looking to capture the absence of something while also conveying its importance.[24]

The artwork is a far cry from the stock imagery used in everyday insect-related journalism. For instance, "Insect numbers down 25% since 1990, global study finds"[25] is accompanied by an uninspired photograph of a stonefly which looks like a mosquito. Equally, a stock photograph of a squashed bug accompanies France Culture's "Why are there no insects on

[21] Clarkson, Jeremy, "When it comes to farmyard pest control, you're either a man or a mouse," *The Sunday Times Magazine*, (1 May 2022): 47.

[22] Alexander, Ruth, "Behind the stats: Insectageddon," (*More or less*, BBC4, March 2, 2019).

[23] Passos et al., "The Chudnovsky Case: How Literary Journalism Can Open the 'Black Box' of Science'," *Literary Journalism Studies*, Vol. 2, No. 2, (2010): 27–46.

[24] Bichler, Gail, "Behind the Cover: The Insect Apocalypse," *The New York Times Magazine*, (Nov 30, 2018).

[25] Carrington, 2020.

your windscreen?"[26] and a series of stock images illustrates "Global insect decline may see 'plague of pests'."[27] The *Return of the Mummy* style and the atomic bomb art serve as a framing function, the role is to reel audiences in, just like a film-marketing poster. The visuals also play a role in explaining the genre of the story, which is horror and fantasy in the case of Schultz' piece. This genre is then fine-tuned through the careful selection of language such as: "he heard her scream," "crawling," "horrifying," "edged steadily closer," "stinkbugs scuttle and crawl and amass like enemy armies," "prehistoric look and a post-mortem smell." By contrast, the atomic bomb visual is followed by dystopian terminology: "vaguely," "melancholy," "eerie," "something about the experience was amiss," "he couldn't recall," "why hadn't he noticed?," "the most disquieting thing," "he could not shake a feeling."

Both authors use a strategy well-loved by fiction storytellers when they move from the small and personal (easy to picture in the mind's eye) to the national and international arena. In the invasion story we are presented with an extensive list of regions while the insects are discussed as international invaders: "That was the beginning of the grand American journey of the brown marmorated stinkbug." In the apocalypse narrative, people across the world are noticing something is not quite right: "Riis was not alone in noticing," "an entire world going blind and deaf," "a world of people become stupid," "the base of food webs everywhere," "we've dropped the ball in some giant collective way."

It may appear that writing nonfiction within the framework of a genre lacks sobriety, that genre is frivolous and we ought to express ourselves as scientists detached from the emotions experienced by concerned entomologists.[28] Yet sobriety could also be viewed as a specific kind of sales pitch. "Explicit, partisan, vulgar, jingoistic, informal" writing has been viewed as objectively factual, with sobre reports failing to be.[29] A different definition of journalistic objectivity is one where writers sympathize with

[26] Ropert, Pierre, "Pourquoi n'y a-t-il plus d'insectes sur vos pare-brises?" *France Culture*, (February 13, 2019).

[27] McGrath, Matt, "Global insect decline may see 'plague of pests'" *BBC*, (February 11, 2019).

[28] Hallmann et al., "Insect biomass decline scaled to species diversity: General patterns derived from a hoverfly community," *PNA*, (January 11, 2021).

[29] Winston, Brian, and Matthew Winston, *The Roots of Fake News: Objecting to Objective Journalism*, (London, New York: Routledge, Taylor & Francis Group, 2021): 9–11.

underprivileged or marginalized participants.[30] *The Washington Post*, for instance, uses emotional terms when it reports on tackling $1 trillion in food waste using adjectives like "calamity" to discuss the 1.4 billion tons of food annually ending in landfills while 800 million people go hungry every day.[31]

When writers take on the topic of insectageddon, they have the challenge of picturing the concept "global insect population," a vast and varied amount of living creatures. Ecological activist Timothy Morton defines these kinds of concepts, like global warming or the sum total of all plastic bags on earth, as "hyperobject." These fundamentally unknowable things can only be experienced as pieces at any one time. Since we fail to see the end, these feel uncanny and remain intrinsically tricky to engage with.[32] While reporters may hone in on small details and side with a beautiful butterfly or a hard-working bee, defining hyperobjects such as the global insect population is more challenging. One method seems to be the artful creation of a villain.

A villain in storytelling is a mechanism for holding the reader's attention. It remains effective so long as it is a viable threat to the hero. Creating a gripping villain out of a small bug relies on two aspects. Firstly, the association of insects with what psychologists call "core disgust" due to association with feces, vomit, and rotten food. Core disgust occurs when insects or animals are seen as vectors of dirt and disease.[33] These are not helped by ecocritical approaches that mesh together nature and culture. While considered useful in ecological narratives,[34] this meshing may enhance this disgust in the case of insect narratives due to a lack of separation between human beings and the insect kingdom. A reaction to being reminded of our interconnectedness with bugs could result in insisting on difference. For instance, through our perceptions of insects as "other." The system that "allows us to comprehend one aspect of a concept in

[30] Hearns-Branaman, Jesse Owen, *Journalism and the Philosophy of Truth: Beyond Objectivity and Balance*, (New York and London: Routledge, Taylor & Francis Group, 2016): 15.

[31] Shanker, Deena, "Big money joins fight against $1 trillion in wasted food," *The Washington Post*, (February 23, 2019).

[32] Morton, Timothy, *Hyperobjects, philosophy and ecology after the end of the world*. (Minneapolis, London: University of Minnesota Press, 2013): 1–4, 55, 70.

[33] Deroy, Ophelia, Ben Reade and Charles Spence, "The insectivore's dilemma, and how to take the West out of it," *Food Quality and Preference* 44, (2015): 47.

[34] von Mossner, Alexa Weik, *Affective ecologies: empathy, emotion, and environmental narrative*, (Columbus: The Ohio State University Press, 2017): 10.

terms of another will necessarily hide other aspects of the concept." To associate insects with scuttling aliens, for instance, over other potential metaphors such as "fragile" or "mother," fits in with culturally familiar norms. This means we may not notice when we are reducing a living being to a metaphor.[35]

The second aspect that helps turn an insect into a gripping villain, is the already fantastical (and sometimes grotesquely comic) representation of insects and the micro-world in fiction and pop culture. Films like *The Fly*[36] or books like *Micro* that feature shrunken scientists having to survive the insect kingdom capture our fears in visual form.[37] Visual storytelling, conflation, comparison, and exaggeration in "Home invasion" include: a stinkbug in someone's granola, one surviving a hot spin in the washing machine, suggestions that what is needed is a silver bullet, and a cat vomiting up two. Another section is dedicated to putting the stinkbug in league with skilled olfactory supervillains:

> Wolverines, anteaters, and polecats all have scent glands that produce an odor rivalling that of a skunk; bombardier beetles, when threatened, emit a foul-smelling chemical hot enough to burn human skin; vultures keep predators at bay by vomiting up the most recent bit of carrion they ate; honey badgers achieve the same effect by turning their anal pouch inside out. All these creatures produce a smell worse than the stinkbug's, but none of them do so in your home.[38]

By aligning vibrant but contrasting imageries of olfactory-skilled creatures versus the clean private home, the invasion metaphor builds. The overlaid images purposefully clash. This polarity generates struggle and suspense.[39] The reader may start to imagine an overlay of a vulture vomiting on a clean sofa, or a badger rubbing his bottom on a freshly washed pillowcase. The fact, of course, that stinkbugs are not in some villain stink school honing their craft is beside the point. The image has been placed in

[35] Lakoff, George and Mark Johnson, *Metaphors we live by*, (Chicago: University of Chicago Press, 1980 [2003]): 10–11.

[36] *The Fly*, Directed by David Cronemberg, (Los Angeles: Brooksfilms, 20 Century Fox, 1986.)

[37] Crichton, Michael, and Richard Preston, *Micro*, (New York: Harper Collins, 2011).

[38] Schulz, Kathryn, "When twenty-six thousand stinkbugs invade your home," *The New Yorker*, (March 5, 2018).

[39] Vogler, Christopher, *The Writer's Journey, mythical structures for writers, 3rd edition*, (CA: Michael Wiese Productions. 2007): 318.

the reader's mind, it is fun and it works. Schultz also uses comparison to her advantage; for instance, a stinkbug's brown body is described as excrement-colored. It could just as easily be compared to a cocoa bean. We are also reminded of the negative traits associated with insects, even though stinkbugs have none of these: "They bite, sting, transmit disease, or gnaw through foundations, scuttle and crawl and amass like enemy armies." The polarity is masterfully maintained by painting people as frail and vulnerable, outnumbered, and living protected by permeable walls. The project carries on with a long list of food stinkbugs eat, to show they are competitors for human food sources. They are compared to a medieval princeling to give the impression that these foods are being eaten by a single, gargantuan creature whose daily shopping list includes figs, quinces, cherry trees, sugar maples, vine maples, horse chestnuts, black walnuts, sycamores, elms, broccoli, cucumbers, pecans, hazelnuts, grapes, apples, jalapeños, pears, etc. For extra comedy, the list ends with: "This is just the hors d'oeuvres."

Humor has the advantage of enabling journalists to make data or scientific and technical information more palatable.[40] It can also have the advantage of making a saturated topic feel fresh. For instance, comedic attempts to write about obesity and lockdown food supply shortages in the UK used the headlines: "The real lockdown mystery: where have Wimbledon's strawberries gone?"[41] and "Porky Brits now fatter than pigs."[42] Buried under Schultz's comedic visuals is the more serious angle about biodiversity loss. As Conservation Scientist Andrew Balmford explains: "Many of the species on the planet are highly specialised for very particular sets of conditions, they like things just so."[43] Stinkbugs have this survival advantage of being easy-going eaters, potentially making them the pigeon or rat of the insect world.

Literary journalists remain journalists however, and so as not to slip into what could feel like pure entertainment, Shultz inserts facts and data to convey the realities of crop damage. She explains that a US Department of Agriculture task force was formed during a stinkbug outbreak to

[40] Passos et al., 29.

[41] Das, Shanti, "The real lockdown mystery: where have Wimbledon's strawberries gone?" *The Sunday Times* UK, (June 14, 2020): 1.

[42] Leake, Jonathan, "Porky Brits now fatter than pigs," *The Sunday Times*, (June 21, 2020): 1.

[43] *Cambridge University*, "Eureka Moments: Professor Andrew Balmford," (Nov 22, 2016).

establish just how serious it was. The Mid-Atlantic apple growers' losses come to 37 million dollars. It was wise to highlight data related to apple loss over, for instance, Brussels sprouts which are less loved. Apples are semiotically associated with pies, childhood, and national identity; as painter Cezanne said of their humble ordinariness: "With this apple, I will astonish Paris."[44]

In Jarvis' apocalypse narrative, the villain-hero dynamic is reversed. In her story, humans are the villains. Jarvis could have played it safe and elevated insects to the role of endangered hero without being specific about a villain. This is the technique used by Elizabeth Kolbert in "Where have all the insects gone?" In the piece, she soberly explains biodiversity loss across all animal groups as reflective of "seven billion people on the planet and the process of feeding, clothing, housing, and transporting them is altering the planet in fundamental ways."[45] Note the lack of an active verb or provocative language, "the process of feeding," instead of "people are eating," or the more provocative: "Fatties lack the willpower to stop eating";[46] also "the process of clothing," rather than the more accusatory tone of "Why fashion brands destroy billions' worth of their own merchandise every year" that discusses the burning of unsold clothing to maintain scarcity and exclusivity.[47] Kolbert's technique relies on what she calls the "imaginative exercise" of looking at facts through "different eyes—or even without eyes, since so many of our fellow-creatures lack them." There is empathy for the human experience, which she describes as anxious and limited. She describes human beings as unable to see colors in the dark, and unaware that showing up on new continents is rare in nature, only occurring when there is a tsunami. When writing about other species, Kolbert slips into the quaint tone of children's author and natural scientist Beatrix Potter. Her *Peter Rabbit* series features tales and watercolors of well-dressed rabbits, frogs, and foxes. In this vein, Kolbert calls a tree frog "a very charming fellow," and the reader is stirred by referencing the

[44] Stamberg, Susan, "For Paul Cezanne, an apple a day kept obscurity away," *NPR*, WAMU, (July 10, 2014).

[45] Kolbert (2020).

[46] Platell, Amanda, "Platell's people: Obesity a sickness? No, it's just plain greed," *The Daily Mail*, (August 19, 2017).

[47] Lieber, Chavie, ""Why fashion brands destroy billions' worth of their own merchandise every year," *Vox*, (September 17, 2018).

extinction of visually appealing species such as "the most beautiful bird in the world."[48]

Jarvis, on the other hand, aims for drama and conflict. This is created using dystopia, a genre which typically has a hero entrenched in states or institutions that filters their perceptions through ideological control.[49] Jarvis borrows from the Orwellian tradition that unpicks the institutions in which we find ourselves. Control and tyranny are not only about guns and restrictions on free speech, but a soothing of the citizenry through sophisticated entertainment, empty news reports, and exciting new gadgets.[50] Dystopia is also a genre that is notably devoid of animal life, this absence is often a marker of social or environmental loss in a world in which humanity has reached a point of functional extinction.[51] Jarvis asks why we do not notice the death of insects nor take seriously the work of our "vital pollinators and recyclers of ecosystems and the base of food webs everywhere," hence allow birds to starve and our ecosystem to collapse. The human being in Jarvis' piece is not anxious, not threatened in an overt way, but mistakenly believes that insects are just pests and data is boring.[52] This echoes Foucault's concept of "normalised power," which describes how human beings internalize norms and support the existing narrative structures, in effect, policing each other through online clicks and purchases and having opinions that reaffirm dominant narratives.[53] The presence of this effect may be observed in the wide use of the word "insect," suitable for all manner of flying, aquatic, stinging, armored, or soft-shelled creatures. The word is effectively used in any context from food to famine, medicine, and disease, farming to pesticides. While human beings are biologically well-equipped to notice insects, there is limited demand on these perceptual powers because "culture and environment largely determine which of the senses are favored."[54] The world is chopped

[48] Kolbert, Elizabeth, "The fate of earth," *The New Yorker,* (October 12, 2017); McCrum, Robert, "The 100 best nonfiction books: No 1 – The Sixth Extinction by Elizabeth Kolbert (2014)," *The Guardian*, (February 1, 2016).

[49] Atkinson, Michael, "The road to dystopia," *The Village Voice*, (March 7, 2006).

[50] *The School of Life,* "Literature: George Orwell," (November 25, 2016).

[51] Norledge, Jessica, "Experiencing Dystopia Through Umwelt: Modelling the Nonhuman Animal in Hollow Kingdom," *English Studies*, (March 29, 2022).

[52] Jarvis, Brooke,"The insect apocalypse is here, what does it mean for the rest of life on Earth?" *New York Times*, (November 27, 2018).

[53] West, Stephen, "Michel Foucault pt. 1" *Philosophize this!* 121, (August 15, 2018).

[54] Tuan, Yi-Fu, *Topophilia:A Study of Environmental Perceptions, Attitudes, and Values*, (New York: Columbia University Press, 1990): 245.

up in such a way that everyday uses of the concept of "insect" are vague, have no time or place, and can operate like an old coin, all rubbed away of specificity.[55] The word "insect" while appearing to refer to something, acts like a mirror, reflecting our own needs and cultural and conceptual associations. An example is the references to leaf-cutter ants. On the one hand, *Science* describes them as polluters: "Leaf-cutter ant 'compost piles' produce potent greenhouse gases";[56] on the other, their usefulness is considered: "Leaf-cutter ants in wound closure"[57] and also "Leaf-cutter ants aid to the establishment success of Tapirira velutinifolia seedlings."[58] While some might argue these are just glimpses, not an attempt to provide the 360-degree picture or to define the hyperobject, others would argue this incomplete picture is a "most total (though by no means innocent) illogicality."[59] Being quantified, defined, explored, and particular aspects identified does not amount to understanding;[60] with the result that ants are worthy of affection and protection only when useful or when reduced to the status of innocence and childishness.[61]

Jarvis reminds us that we were not always in this visceral disconnect from insects. She uses Riis on his bicycle like a Proustian moment to place us back in time.

> Riis was on a bike ride … enjoying the sun slanting over the fields and woodlands near their home … Riis was transported to his childhood … summer bike rides meant closing his mouth to cruise through thick clouds of insects … he remembered, the car's windshield was frequently so smeared with insect carcasses that you almost couldn't see through it.[62]

[55] Derrida, Jacques, *Marges de la Philosophie*, (Paris: Les Editions de Minuit, 1972): 250.

[56] Perkins, Sid, "Leafcutter Ant 'compost Piles' Produce Potent Greenhouse Gases," *Science*, (January 1, 2019).

[57] Davis, Hillary E., "Leaf-cutter ants in wound closure," *Wilderness & Environmental Medicine* Vol. 30, No. 4, (Elsevier, December, 2019): 469–470.

[58] Brener, Alejandro G. Farji, and J. F. Silva, "Leaf-Cutter Ants' (Atta Laevigata) Aid to the Establishment Success of Tapirira Velutinifolia (Anacardiaceae) Seedlings in a Parkland Savanna," *Journal of Tropical Ecology* 12, No. 1 (1996): 163–68.

[59] Baudrillard, Jean, *The consumer society, myths and structures*, London: Sage Publications. (1970, [1998]): 41.

[60] Lakoff and Johnson, (1980 [2003]): 26.

[61] Baudrillard, Jean, *Simulacra and Simulation*, (Michigan: The University of Michigan Press, 1981 [1994]): 134.

[62] Jarvis (2018).

French author Proust, in his 1909 book *In Search of Lost Time*, described a madeleine cake dipped in hot tea that, years on, could vividly transport him to his childhood. The power of smell and taste to conjure up memories is used here in an original way: being young and riding a bicycle is a memory on par perhaps with a taste memory thanks to it engaging our sixth sense of balance. It forms a bridge with any reader with similar childhood experiences.

It may seem that a move towards neutral, genre-free, impartial journalism is needed to serve a balanced view. However, attempts at deflating and climbing down from the more dramatic tone have lacked actual neutrality. News updates like the *BBC's* "'Insect apocalypse' more complicated than thought"[63] or announcements from *The Economist* that "Insectageddon is not imminent" and "The insect apocalypse is not here"[64] direct our attention to a lack of global proof. They also refer to a study that demonstrates ecological collapse occurs at 90% insect pollinator deaths, not merely at 75%.[65]

To provide a balance of data that includes the work insects do for human beings (not just the damage they cause), entomologists have attempted to reach a global figure. For the USA, the figure of $57 billion a year was reached to include dung burial, pest control, pollination, and wildlife nutrition.[66] An Australian dung beetle import project valued the work on soil, water, and pasture health at many millions of dollars a year.[67] Another method is to look at replacement value; in China's Maoxian Valley, studies showed that shortages of insect pollinators led farmers to hire human workers at up to $19 per human worker per day.[68] Radboud University in the Netherlands calculated that "eighty percent of crops depend on insects for pollination. Eighty percent of the wild plant species

[63] McGrath, Matt, "Nature crisis: 'Insect apocalypse' more complicated than thought" *BBC*, (April 23, 2020).

[64] *The Economist* (a) "Plague without locusts. Insectageddon is not imminent. But the decline of insect species is still a concern," (March 23, 2019):14. *The Economist* (b) "The insect apocalypse is not here. But there are reasons for alarm about the long-term health of many species," (March 23, 2019): 71–73.

[65] *The Economist* (a) (2019):14.

[66] Kolbert (2020).

[67] *The Queensland Dung Beetle Project*, (2002): 54.

[68] Partap, Uma and Tang Ya (2012) "The Human Pollinators of Fruit Crops in Maoxian County, Sichuan, China," *BioOne Complet. Mountain Research and Development*, Vol. 32. No. 2 (2012).

as well."[69] The value for annual global insect pollination has generated the figure $500 billion.[70] However, trying to turn insect pollination into data may result in officially classifying insects as livestock, which results in bees being hired out like "large-scale machines" to the "toxic chemical soup" that is the Californian almond industry. The insertion of the attribute "commercial" in front of "honeybee" provides some logic to the idea of "fattening bees on corn syrup."[71]

In their bid to write compelling narratives both Jarvis and Schultz have left out large areas of knowledge. They also both speak to a Western perspective on insects that has been accused of underestimating the value of insects.[72] *The Japan Times* reflects on a past when musical insects were bought at night fairs and kept as pets for their sonorous qualities.[73] Equally by China's Sung dynasty, crickets were a popular topic of discussion at teashops and in markets: "Fashionable people were never without chirping crickets concealed under their robes."[74] Globally, over 1200 varieties of insects are considered ingredients for food with taste and texture diversity that goes from the foie gras to the potato chips.[75] However the West traditionally has seen insects as "emblematic of the alien" and eating them as "a threat to our psychological and cultural identity."[76] *The Sunday Times* bemoans the UK's squeamishness, quoting *The Great British Bake Off* judge Pru Leith, who predicts a move towards using insect flour. Leith thinks there is no logic to our distaste for eating insects since we eat prawns, which she also finds insect-like.[77] While experimental insect eating may be on the rise, traditional foraging for insects is on the decline due to clashes with technology. In Sanambele, Mali, parents discourage their children from eating foraged grasshoppers concerned they may be contaminated by chemicals from cotton farming.[78] Meanwhile in Tokyo, a special

[69] Campbell (2020).

[70] Jarvis (2018).

[71] McGivney, Annette, "'Like sending bees to war': the deadly truth behind your almond milk obsession," *The Guardian* (2020).

[72] Looy et al., 132.

[73] Flanagan, Damian, "Insect Literature," *The Japan Times* (January 23, 2016).

[74] Ryan, Lisa Gail, "Insect Musicians & Cricket Champions, a cultural history of singing insects in China and Japan," (San Francisco: China Books & periodicals, 1996): XIII.

[75] Deroy, Reade and Spence (2015): 52.

[76] Looy et al., 131.

[77] Eccles, Louise, "Waiter, there's a fly in my soup, it looks delicious," *The Sunday Times* (August 8, 2021).

[78] Looy et al., 131.

event saw 100 bowls of insect noodles sold in four hours with feedback from customers including: "It's deep-fried, so it's really crispy."[79] In Australia, at Melbourne Museum, a playful menu for children was served including honey described as "bee's vomit" and tips to roast tarantulas to remove the hairs.[80] While in Madagascar, a cricket farm successfully produces powdered protein for famine relief which is baked into sweet cakes (which local school children report tastes delicious).[81] From Thailand to Paris, from street food to haute cuisine, as protein shakes, in cookery books, in chocolate boutiques, and whipped into mousse, insects are seen as an ecologically savvy protein source for the human diet.[82] Though some bugs (such as roasted, hairless tarantulas) would make even the most ardent insectivores hesitate, analysts might argue that a repulsion for eating insects stems from our "colonized imaginations."[83] Companies like France's *Ynsect* factory, the UK's *Eat Grub* and *Bug Farm Foods*, and South Africa's *Gourmet Grubb* that sell peanut butter black soldier fly ice cream, among other flavors, could all be seen as taking steps to decolonize not just our food, but also our minds.

Discussions around insects as food are largely ignored in insect apocalypse narratives because they disrupt the clean lines of the hero/villain dynamic. Insectaggedon narratives that explore insect apocalypses are concerned with a "complex risk scenario"[84] that would be undermined by the realities of our current use of insects for the production of silk, lacquer, dye, fish bait, honey, pet food, and as livestock for pollination work. Equally, insect invasions in the home may appear less horrifying if accompanied by a recipe for stinkbug bonbons, high-protein cookies, or lactose-free ice cream. Yet even with the exclusion of some aspects, the impact of these literary journalism insectageddon narratives has trickled into other kinds of writing. Reporters comfortably add historical or cultural richness in short news reports, for instance by specifying which insects are considered halal in what could be a straight piece of journalism that warns against

[79] Ha, Kwiyeon, "Ramen flavored with insects attracts thrillseekers in Tokyo," *The Japan Times* (April 11, 2017).

[80] Deroy, Reade and Spence (2015): 51.

[81] Filou, Emilie, "Could you be convinced to swap a steak for a plate of tasty crickets?" *From Our Own Correspondent*, BBC 4, podcast (September 28, 2019).

[82] Topham, Nurdin, "Moth Mousse," *Nordic Food Lab blog*, (July 2, 2013).

[83] Looy et al., 132.

[84] von Mossner, 147.

ingesting pesticide-treated locusts.[85] The cliched term "plague of locusts" is still used, but with the emphasis that farmers are looking for support and the blame is shifted from the insects themselves to governments who are described as being "in deep slumber."[86] In addition, accompanying visuals are considered part of the narrative power and "effective vessels for persuasion."[87] *National Geographic's* close-up portrait of a potentially extinct blue Xerces butterfly is able to combine "attention, imagery, and emotion" that draws readers into a narrative world.[88] These highly detailed insect photographs are considered so aesthetically powerful they have become narrative journeys in their own right. A photography exhibition of insects was curated with the aim of attracting young people to careers in science.[89]

There is, however, a temptation when writing about our hidden world to engage the aesthetic ideal. This would imply that exposure to the insect kingdom gives some a sensitivity that provides access to a domain of aesthetic value.[90] This shift to the aesthetic and value-laden is used by both Schulz and Jarvis in their conclusions: "The most troubling thing about the natural world today"[91] and "What gave a creature value? (…) is it grace?"[92] To end with morals can give a piece of writing a fairy tale quality, but it may also avoid the hard questions. Firstly by doing what might seem like sound-bite journalism,[93] secondly because moralizing may fail to change people's structures and social institutions,[94] and finally because moralizing reinforces a naive understanding of the relationship between man and nature. In such a landscape, insects only have a license to live when docile and with all their defects removed. As Tuan puts it in his book

[85] Van Itterbeeck, Joost, "Swarming locusts: people used to eat them, but shouldn't anymore," *The Conversation*, (April 24, 2020).

[86] Ellis-Petersen, Hannah and Shah Meer Baloch (2020).

[87] Mazzocco et al., "This Story Is Not for Everyone: Transportability and Narrative Persuasion," *Social Psychological and Personality Science*, Vol. 1. No. 4 (2010): 361.

[88] Kolbert (2020).

[89] Leeds Museum and Galleries, *Dead Inspiring: Insect collections and empowering young scientists:* https://museumsandgalleries.leeds.gov.uk/virtual-visit/dead-inspiring

[90] Riggle, Nick, "On the Aesthetic Ideal," *The British Journal of Aesthetics*, Vol. 55, No. 4, (October, 2015): 433–447.

[91] Schulz (2018).

[92] Jarvis (2018).

[93] Landa, Amy Snow, "How Literary Journalism Can Inform Bioethics," in: *Literary Journalism Studies* Vol. 2. No. 2 (2010): 48.

[94] Geuss, Raymond, "Marxism lecture 2/8" Kotti Everdene (January 29, 2014).

Topophilia,[95] this fantasy of nature features "plants and animals useful and friendly to man abound."

Instead, some call for ethics to be practical, such as the creation of "Zoöps." These roles would be a formal legal entity that gives voice to non-human life forms,[96] since "nonhuman beings are responsible for the next moment of human history."[97] Other writers try to have an impact by providing "a bundle of sensations," yet challenges continue with attempts to put ourselves in the shoes of insects by applying "critical anthropomorphism."[98] Perhaps the value of these narratives is more than translating science, practical ethics, and entertainment. Perhaps with them comes a new curiosity that chips away at cultural stereotypes by highlighting where metaphors have calcified into norms. Mycologist Paul Stamets happily describes how insect observation was one of his starting points. He observed bees deriving medicinal properties from virus-fighting fungi.[99] He has since collaborated with the US National Institute of Health and Department of Defense to search for compounds in those wood-rotting fungi that could fight biological warfare.[100] Natural sciences writer Janine Benyus also warns against only elevating and admiring humans as problem solvers while "there is a vast array of knowledge from the natural world." For instance, a wingless midge survives the cold by coating its eggs in antifreeze, or the sleeping chironomid fly has larvae that shrink to desiccated flakes in dry times, entering a suspended animation from which they can recover after 15 years. Engineers hope to develop a coating for buildings that extracts water from air, based on studying Namibian desert beetles who drink water out of fog through bumps on their wings.[101]

To conclude, while insect declines come as no surprise to entomologists[102] it is the imbalanced, discussion-generating, filmic writing of liter-

[95] Tuan, 247.

[96] Sudiono, Tonya, "Let's start a Zoöp! For more biodiversity," *Waag Future Lab*, (June 15, 2021).

[97] Morton (2013): 201.

[98] von Mossner, 21, 113.

[99] Stamets et al. "Extracts of Polypore Mushroom Mycelia Reduce Viruses in Honey Bees," *Scientific Report* 8 (2018).

[100] Sheldrake, Merlin, *Entangled Life*, (London: Penguin Random House, 2021): 221.

[101] Benyus, Janine, "Biomimicry's surprising lessons from nature's engineers," *Ted Talks*, (2005).

[102] Thomas, Chris D., T. Hefin Jones, and Sue E. Hartley, "'Insectageddon': A call for more robust data and rigorous analyses," *Global Change Biology*, Vol. 25, No. 6, (March 1, 2019).

ary journalism that has acted as a springboard to bring the micro to a wide readership. As emotions are engaged and interest grows, so does a readership's appreciation for the complexities and intricacies involved in our relationship with insects.

There are fascinating moral and ethical questions to be asked but also fascinating practical and technical ones whose solutions may require us to recalibrate our thinking across countries, industries, and time. Literary journalism, with its vivid images and compelling narratives, can help the science community turn hyperobjects into public discussion, thanks to its handling of compelling heroes and terrible villains that capture the imagination. While our most used concepts, such as "insect" and "bug" serve us well, they also play their part in creating a detachment, erasing the detail, and reaffirming dominant narratives that can allow policies and practices to become outdated.[103] Until we are insect-literate, drama can showcase the data from science-in-the-making.

References

Alexander, Ruth, "Behind the stats: Insectageddon," *More or less*, BBC4, (March 2, 2019). Accessed 29 June 2020 https://www.bbc.co.uk/programmes/p072c44x

Animal Planet, "Death by Tsetse Fly, Monsters Inside Me," (July 27, 2009). Accessed 29 June 2020 https://www.youtube.com/watch?v=4aVUrGO97Zg

Atkinson, Michael, "The road to dystopia," *The Village Voice*, (March 7, 2006). Accessed 28 September 2023 https://www.villagevoice.com/road-to-dystopia

Baudrillard, Jean, *The consumer society, myths and structures*, London: Sage Publications. 1970 [1998].

Baudrillard, Jean, *Simulacra and Simulation*, Michigan: The University of Michigan Press, 1981 [1994].

BBC Health, "Mosquitoes 'put off biting' by human diet drugs," (February 7, 2019). Accessed 7 July 2020 https://www.bbc.co.uk/news/health-47157017

Benyus, Janine, "Biomimicry's surprising lessons from nature's engineers," *Ted Talks*, (2005). Accessed 29 June 2020 https://www.ted.com/talks/janine_benyus_biomimicry_s_surprising_lessons_from_nature_s_engineers?language=en

Bichler, Gail, "Behind the Cover: The Insect Apocalypse," *The New York Times Magazine*, (Nov 30, 2018). Accessed 30 August 2021 https://www.nytimes.com/2018/11/30/magazine/behind-the-cover-the-insect-apocalypse.html

[103] Jarvis (2018).

Brener, Alejandro G. Farji, and J. F. Silva, "Leaf-Cutter Ants' (Atta Laevigata) Aid to the Establishment Success of Tapirira Velutinifolia (Anacardiaceae) Seedlings in a Parkland Savanna," *Journal of Tropical Ecology* 12, No. 1 (1996): 163–68. https://doi.org/10.1017/S0266467400009378.

Cambridge University, "Eureka Moments: Professor Andrew Balmford," *Zoology 150 Years 1866–2016*. (Nov 22, 2016). Accessed 29 June 2020 https://www.youtube.com/watch?v=zTVC32Bfl-U#action=share

Campbell, Eric, Tomas Ybarra, and ABC, "*Insectageddon*," *People and Power*, *Aljazeera*, (Jan 23, 2020). Accessed 5 September 2021 https://www.youtube.com/watch?v=-Hb9DTxo98g

Carrington, Damian, "Insect collapse: 'We are destroying our life support systems'" *The Guardian*, (January 15, 2019). Accessed 1 September 2021 https://www.theguardian.com/environment/2019/jan/15/insect-collapse-we-are-destroying-our-life-support-systems

Carrington, Damian, "Insect numbers down 25% since 1990, global study finds," *The Guardian*, (April 23, 2020). Accessed 29 June 2020 https://www.theguardian.com/environment/2020/apr/23/insect-numbers-down-25-since-1990-global-study-finds

Crichton, Michael, and Richard Preston, *Micro*, New York: Harper Collins. 2011.

Clarkson, Jeremy, "When it comes to farmyard pest control, you're either a man or a mouse," *The Sunday Times Magazine*, (1 May 2022).

Compagnon, Sébastian, "Pourquoi les abeilles tombent comme des mouches?" *Le Parisien*, (7 June, 2018). Accessed 1 September 2021 https://www.leparisien.fr/societe/pourquoi-les-abeilles-tombent-comme-des-mouches-07-06-2018-7758600.php

Das, Shanti, "The real lockdown mystery: where have Wimbledon's strawberries gone? The two million berries won't go to waste – they could be jam tomorrow," *The Sunday Times* UK, (June 14, 2020).

Davis, Hillary E., "Leaf-cutter ants in wound closure," *Wilderness & Environmental Medicine* 30: 4, Elsevier, December (2019): 469–470.

Deroy, Ophelia, Ben Reade and Charles Spence, "The insectivore's dilemma, and how to take the West out of it," *Food Quality and Preference* 44 (2015): 44–55. https://doi.org/10.1016/j.foodqual.2015.02.007.

Derrida, Jacques, *Marges de la Philosophie*, Paris: Les Editions de Minuit, 1972.

Durst, Patrick B., Dennis V. Johnson, Robin N. Leslie and Kenichi Shono, ed., "Forest insects as food: humans bite back," Bangkok, Thailand: Food and Agriculture Organization of the United Nations Regional Office for Asia and The Pacific, (2012). Accessed 26 August 2021 http://www.fao.org/3/i1380e/i1380e.pdf

Eccles, Louise, "Waiter, there's a fly in my soup, it looks delicious. Food brands are so sure crickets and beetles will become kitchen staples they now offer courses on how to cook them," *The Sunday Times*, (August 8, 2021). Accessed 29

August 2021 https://www.thetimes.co.uk/article/waiter-theres-a-fly-in-my-soup-it-looks-delicious-z8xd6sh8g

Ellis-Petersen, Hannah and Shah Meer Baloch, "Many will starve': locusts devour crops and livelihoods in Pakistan," *The Guardian*, (May 25, 2020). Accessed 29 June 2020 https://www.theguardian.com/world/2020/may/25/many-will-starve-locusts-devour-crops-and-livelihoods-in-pakistan

Filou, Emilie, "Could you be convinced to swap a steak for a plate of tasty crickets? Emilie Filou visits a noisy farm in Madagascar, to find out how one company wants to put 'cricket powder' in everyone's kitchen cabinet," in "Can Afghanistan find peace?" *From Our Own Correspondent*, BBC Radio 4, (September 28, 2019). Accessed 29 June 2020 https://www.bbc.co.uk/programmes/m0008x9f

Flanagan, Damian, "Insect Literature," *The Japan Times*, (January 23, 2016). Accessed 29 June 2020 https://www.japantimes.co.jp/culture/2016/01/23/books/insect-literature

Foucart, Stephane, "En trente ans, près de 80% des insectes auraient disparu en Europe, Ce déclin catastrophique est dû à l'intensification des pratiques agricoles et au recours aux pesticides. Il menace la chaîne alimentaire," *Le Monde* (18 October, 2017). Accessed 1 September 2021 https://www.lemonde.fr/biodiversite/article/2017/10/18/en-trente-ans-pres-de-80-des-insectes-auraient-disparu-en-europe_5202939_1652692.html

Geuss, Raymond, "Marxism lecture 2/8," Kotti Everdene, (January 29, 2014). Accessed 2 August 2020 https://www.youtube.com/watch?v=25aYrfvOGUs

Gill, Victoria, "Bedbugs survived the dinosaur extinction event," *BBC*, (May 16, 2019). Accessed 25 August 2021 https://www.bbc.co.uk/news/science-environment-48274090

Hallmann, Caspar A., Martin Sorg, Eelke Jongejans, Henk Siepel, Nick Hofland, Heinz Schwan, Wener Stenmans, Andreas Muller, Hubert Sumser, Thomas Horren, Dave Goulson and Hans de Kroon, "More than 75 percent decline over 27 years in total flying insect biomass in protected areas," *PloS ONE*, Vol. 12. No. 10 (2017): 1–21. https://doi.org/10.1371/journal.pone.0185809.

Hallmann, Caspar A., Axel Ssymank, Martin Sorg and Eelke Jongejans, "Insect biomass decline scaled to species diversity: General patterns derived from a hoverfly community," *Proceedings of the National Academy of Science of the United States of America*, (January 11, 2021). Accessed 28 September 2023. https://doi.org/10.1073/pnas.2002554117.

Ha, Kwiyeon, "Ramen flavored with insects attracts thrillseekers in Tokyo," *The Japan Times*, (April 11, 2017). Accessed 25 August 2021 https://www.japantimes.co.jp/life/2017/04/11/food/ramen-flavored-with-insects-attracts-thrillseekers-in-tokyo

Hearns-Branaman, Jesse Owen, *Journalism and the Philosophy of Truth: Beyond Objectivity and Balance*, New York and London: Routledge, Taylor & Francis Group, 2016.

Jarvis, Brooke,"The insect apocalypse is here, what does it mean for the rest of life on Earth?" *New York Times*, (November 27, 2018). Accessed 29 June 2020 https://www.nytimes.com/2018/11/27/magazine/insect-apocalypse.html

Kolbert, Elizabeth, "The fate of earth," *The New Yorker*, October 12, 2017. Accessed 1 September 2021 https://www.newyorker.com/tech/annals-of-technology/the-fate-of-earth

Kolbert, Elizabeth, "Where have all the insects gone?" *National Geographic*, April 23, 2020. Accessed 29 June 2020 https://www.nationalgeographic.com/magazine/2020/05/where-have-all-the-insects-gone-feature

Lakoff, George and Mark Johnson, *Metaphors we live by*, Chicago: University of Chicago Press, 1980 (2003).

Landa, Amy Snow, "How Literary Journalism Can Inform Bioethics," in: *Literary Journalism Studies*, Vol. 2. No. 2 (2010): 47–61. Accessed 25 August 2021 https://ialjs.org/wp-content/uploads/2014/12/049-063-LJS_v2n2.pdf

Leake, Jonathan, "Porky Brits now fatter than pigs. While livestock becomes leaner; humans have been piling on excess weight," *The Sunday Times* (June 21, 2020): 1. Accessed 25 August 2021 https://www.thetimes.co.uk/article/porky-brits-now-fatter-than-pigs-hvzrnxgp9

Leeds Museum and Galleries, *Dead Inspiring: Insect collections and empowering young scientists.* Accessed 3 September 2021 https://museumsandgalleries.leeds.gov.uk/virtual-visit/dead-inspiring

Lieber, Chavie, "'Why fashion brands destroy billions' worth of their own merchandise every year," *Vox*, (September 17, 2018). Accessed 28 June 2021 https://www.vox.com/the-goods/2018/9/17/17852294/fashion-brands-burning-merchandise-burberry-nike-h-and-m

Looy, Heather, Florence V. Dunkel, and John R. Wood, "How Then Shall We Eat? Insect-Eating Attitudes and Sustainable Foodways," *Agriculture and Human Values 31*, (March 1, 2014): 131–141. https://doi.org/10.1007/s10460-013-9450-x.

Mazzocco, Philip J., Melanie C. Green, Jo A. Sasota and Norman W. Jones, "This Story Is Not for Everyone: Transportability and Narrative Persuasion," *Social Psychological and Personality Science*, Vol. 1. No. 4 (2010): 361–368. https://doi.org/10.1177/1948550610376600.

McCrum, Robert, "The 100 best nonfiction books: No 1 – The Sixth Extinction by Elizabeth Kolbert (2014)," *The Guardian*, (February 1, 2016). Accessed 30 August 2021 https://www.theguardian.com/books/2016/feb/01/100-best-nonfiction-books-of-all-time-the-sixth-extinction-elizabeth-kolbert

McGivney, Annette, "'Like sending bees to war': the deadly truth behind your almond milk obsession," *The Guardian*, (2020). Accessed on 28 June 2021 https://www.theguardian.com/environment/2020/jan/07/honeybees-deaths-almonds-hives-aoe

McGrath, Matt, "Global insect decline may see 'plague of pests'" *BBC*, (Feb 11, 2019). Accessed 29 June 2020 https://www.bbc.co.uk/news/science-environment-47198576

McGrath, Matt, "Nature crisis: 'Insect apocalypse' more complicated than thought" *BBC*, (April 23, 2020). Accessed 29 June 2020 https://www.bbc.co.uk/news/science-environment-52399373

Morton, Timothy, *Hyperobjects, philosophy and ecology after the end of the world*. Minneapolis, London: University of Minnesota Press. 2013.

Nel, André, Jakub Prokop, Martina Pecharová, Michael S. Engel and Romain Garrouste, "Palaeozoic giant dragonflies were hawker predators," *NCBI*, (August 14, 2018). Accessed 29 June 2020 https://www.ncbi.nlm.nih.gov/pmc/articles/PMC6092361

Norledge, Jessica, "Experiencing Dystopia Through Umwelt: Modelling the Nonhuman Animal in Hollow Kingdom," *English Studies*, (March 29, 2022). https://doi.org/10.1080/0013838X.2022.2043003.

Partap, Uma and Tang Ya (2012) "The Human Pollinators of Fruit Crops in Maoxian County, Sichuan, China," *BioOne Complet. Mountain Research and Development*, Vol. 32. No. 2 (2012). Accessed 6 August, 2020 https://bioone.org/journals/mountain-research-and-development/volume-32/issue-2/MRD-JOURNAL-D-11-00108.1/The-Human-Pollinators-of-Fruit-Crops-in-Maoxian-County-Sichuan/10.1659/MRD-JOURNAL-D-11-00108.1.full

Passos, Mateus Yuri, Erica Masiero Nering, and Juliano Mauricio de Carvalho (2010) "The Chudnovsky Case: How Literary Journalism Can Open the 'Black Box' of Science'". In: Hartsock, J. C (ed) *Literary Journalism Studies* Vol. 2, No. 2 (2010): 27–46.

Perkins, Sid, "Leafcutter Ant 'compost Piles' Produce Potent Greenhouse Gases," *Science*, (January 1, 2019). Accessed 22 May 2022 www.science.org/content/article/leafcutter-ant-compost-piles-produce-potent-greenhouse-gases

Platell, Amanda, "Platell's people: Obesity a sickness? No, it's just plain greed," *The Daily Mail*, (August 19, 2017). Accessed 1 September 2021 https://www.dailymail.co.uk/debate/article-4804330/PLATELL-S-PEOPLE-Obesity-sickness-No-s-just-greed.html

Proust, Marcel, *A la recherche du temps perdu: Du côté de chez Swann*. Paris: Flammarion. (1913 [2019]).

Riggle, Nick, "On the Aesthetic Ideal", *The British Journal of Aesthetics*, Vol. 55, No. 4, (October, 2015): 433–447. https://doi.org/10.1093/aesthj/ayv026.

Ropert, Pierre, "Pourquoi n'y a-t-il plus d'insectes sur vos pare-brises?" *France Culture*, (February 13, 2019). Accessed 29 June 2020 https://www.franceculture.fr/ecologie-et-environnement/pourquoi-ny-a-t-il-plus-dinsectes-sur-vos-pare-brises

Ryan, Lisa Gail, "Insect Musicians & Cricket Champions, a cultural history of singing insects in China and Japan," San Francisco: China Books & periodicals. 1996.

SABC News, "SA launches biological control of mosquitoes to eliminate malaria" (October 18, 2018). Accessed 29 June 2020 https://www.sabcnews.com/sabcnews/sa-launches-biological-control-of-mosquitoes-to-eliminate-malaria

Sánchez-Bayo, Francisco and Kris A.G. Wyckhuys, "Worldwide decline of the entomofauna: A review of its drivers," *Biological Conservation,* Vol. 232 (2019): 8–27. https://doi.org/10.1016/j.biocon.2019.01.020.

Schulz, Kathryn, "When twenty-six thousand stinkbugs invade your home," *The New Yorker*, (March 5, 2018). Accessed 29 June 2020 https://www.newyorker.com/magazine/2018/03/12/when-twenty-six-thousand-stinkbugs-invade-your-home (Also titled "Home Invasion" in print, March 12, 2018).

Shanker, Deena, "Big money joins fight against $1 trillion in wasted food," *The Washington Post*, (February 23, 2019). Accessed 10 August 2020 https://www.washingtonpost.com/business/economy/companies-fighting-the-1-trillion-in-food-waste-are-raking-in-the-dough/2019/02/23/67633e9e-37cc-11e9-a400-e481bf264fdc_story.html

Sheldrake, Merlin, *Entangled Life*, London: Penguin Random House. 2021.

Stamberg, Susan, "For Paul Cezanne, an apple a day kept obscurity away," *NPR*, WAMU 88.5 Morning Edition, (July 10, 2014). Accessed 1 September 2021 https://www.npr.org/2014/07/10/330159592/for-paul-cezanne-an-apple-a-day-kept-obscurity-away?t=1630922682728

Stamets, Paul E., Nicholas L. Naeger, Jay D. Evans, Jennifer O. Han, Brandon K. Hopkins, Dawn Lopez, Henry M. Moershel, Regan Nally, David Sumerlin, Alex W. Taylor, Lori M. Carris and Walter S. Sheppard, "Extracts of Polypore Mushroom Mycelia Reduce Viruses in Honey Bees," *Scientific Report* 8 (2018). https://doi.org/10.1038/s41598-018-32194-8

Sudiono, Tonya, "Let's start a Zoöp! For more biodiversity," *Waag Future Lab*, (June 15, 2021). Accessed 28 April 2022 https://waag.org/en/article/lets-start-zoop-more-biodiversity

The Economist, "The decline of bees," *The Economist Explains,* (September 7, 2015), Accessed 1 September 2021 https://www.economist.com/the-economist-explains/2015/09/06/the-decline-of-bees

The Economist (a) "Plague without locusts. Insectageddon is not imminent. But the decline of insect species is still a concern," *Leaders*, (March 23, 2019a):14.

The Economist (b) "The insect apocalypse is not here. But there are reasons for alarm about the long-term health of many species," *Science and Technology*, (March 23, 2019b): 71–73.

The Fly, Directed by David Cronemberg, Los Angeles: Brooksfilms, 20 Century Fox, 1986.

The Queensland Dung Beetle Project, "Improving sustainable land management systems in Queensland using dung beetles. Final Report of the 2001–2002," *Queensland Dung Beetle Project*, (2002). Accessed 6 August 2020 http://www.dungbeetle.com.au/queenslandreport.pdf

The School of Life, "Literature: George Orwell," (Nov 25, 2016). Accessed 29 June 2020 https://www.youtube.com/watch?v=kvXU3vzHq8E

Thomas, Chris D., T. Hefin Jones, and Sue E. Hartley, "'Insectageddon': A call for more robust data and rigorous analyses," *Global Change Biology*, Vol. 25, No. 6, (March 1, 2019). https://doi.org/10.1111/gcb.14608

Topham, Nurdin, "Moth Mousse, recipe development for our Pestival menu, by Nurdin Topham – now Head Chef at NUR in Hong Kong," *Nordic Food Lab blog*, (July 2, 2013). Accessed 30 August 2021 https://nordicfoodlab.org/blog/2013/07/moth-mousse

Tuan, Yi-Fu, *Topophilia:A Study of Environmental Perceptions, Attitudes, and Values*, New York: Columbia University Press. 1990.

Van Huis, Arnold, Joost Van Itterbeeck, Harmke Klunder, Esther Mertens, Afton Halloran, Giulia Muir, Paul Vantomme "Edible insects: future prospects for food and feed security". *FAO forestry paper, No. 171.* Rome: Food and Agriculture Organization of the United Nations, 2013.

Van Itterbeeck, Joost, "Swarming locusts: people used to eat them, but shouldn't anymore". *The Conversation*, (April 24, 2020). Accessed 29 June 2021 https://theconversation.com/swarming-locusts-people-used-to-eat-them-but-shouldnt-anymore-135058

Vogler, Christopher, *The Writer's Journey, mythical structures for writers, 3rd edition*, CA: Michael Wiese Productions. 2007.

von Mossner, Alexa Weik, *Affective ecologies: empathy, emotion, and environmental narrative*, Columbus: The Ohio State University Press. 2017.

West, Stephen, "Michel Foucault pt. 1" *Philosophize this!* 121, (August 15, 2018). Accessed 1 September 2021 https://www.youtube.com/watch?v=HG_Q9WnOuxU

Winston, Brian, and Matthew Winston, *The Roots of Fake News: Objecting to Objective Journalism.* London, New York: Routledge, Taylor & Francis Group. 2021.

Yeates, David, "Insects are the great survivors in evolution: new study," *The Conversation*, (November 4, 2014). Accessed 29 June 2021 https://theconversation.com/insects-are-the-great-survivors-in-evolution-new-study-33786

Ying, Xiu, "Mutant Mosquitoes, Designed to Combat Dengue and Zika, Were Released in Brazil. Genetic engineering made it possible: for mutant mosquitoes to be released in Brazil. Their offspring should not be able to survive," *The Rio Times*, (September 26, 2019). Accessed 29 June 2020 https://riotimesonline.com/brazil-news/miscellaneous/mutant-mosquitoes-spreading-uncontrollably-in-brazil

Toward an Activist Aesthetic of Environmental Literary Journalism: Deep and Social Ecology in Thoreau, Carson, and Jenkins

David O. Dowling

Social and political activism has played a crucial role in environmental literary journalism, one that functions in tandem with the genre's appreciation of nature as a window into the universe and the human soul. The activist tradition of environmental literature concerned with human uses and impact on nature leverages political philosophy in a way epitomized by Rachel Carson's *Silent Spring*.[1] Deep ecology, the foundation of such activist writing, emphasizes the value of nature in and of itself, as told through narratives of self-sufficiency in the wild by authors such as Wendell

[1] Rachel Carson, *Silent Spring* (Houghton Mifflin: Boston, 1962).

D. O. Dowling (✉)
University of Iowa, Iowa City, IA, USA
e-mail: david-dowling@uiowa.edu

P. Calvi (ed.), *The Literary Journalist as a Naturalist*, Palgrave Studies in Literary Journalism,
https://doi.org/10.1007/978-3-031-56634-9_5

Berry, Gary Snider, and Edward Abbey.[2] Influenced by Martin Heidegger, Norwegian environmentalist Arne Naess portrays deep ecology in "The Shallow and The Deep"—"probably the most influential [brief] article of the last half century" on the subject[3]—as the pursuit of meaning in nature through "the realization of the self that encompasses both the individual and the cosmos."[4] From this tradition arose social ecology spearheaded by Murray Bookchin,[5] the radical anticapitalist whose vision for renewable energy inspired the Green New Deal.

The politics of environmental writing are closely linked to the Green New Deal, a series of legislative efforts to promote policy to counter climate change in ways that create jobs and reduce economic inequality. Although the name derives from US President Franklin Delano Roosevelt's New Deal, which was a series of public works projects for social and economic reform during the Great Depression, the European Green New Deal was passed by the European Council and the European Parliament in 2020. In the United States, the attempt to pass legislation to support the Green New Deal by Congressional Representatives Alexandria Ocasio-Cortez and Ed Markey was voted down by conservatives in the US Senate in 2019. The Green New Deal builds on Roosevelt's original concern for resource efficiency and renewable energy to address the social and economic impacts of contemporary climate change. Following Green Party pioneer Howie Hawkins's 2010 bid to become governor of New York based on a Green New Deal platform, Green Party candidate Jill Stein ran on a Green New Deal platform in the 2012 and 2016 presidential elections.[6]

Social ecology, which also aligns with these political principles, sees environmental degradation as a consequence of economic inequality and

[2] Ursula K. Heise, "The Hitchhiker's Guide to Ecocriticism," *Publication of the Modern Language Association* Vol. 121, No. 2, March 2006, 507.

[3] Timothy D. Quick, "American Transcendentalism and Deep Ecology in the History of Ideas," M.A. Thesis, School of Environmental Studies, 2004, University of Victoria, 84 .

[4] Heise, "Hitchhiker's Guide," 511; Arne Naess, "The Shallow and the Deep, Long-range Ecology Movement. A Summary," *Inquiry*, Vol. 16, No. 1-4, May 1973: 95-100.

[5] Murray Bookchin, "Social Ecology vs. Deep Ecology: A Challenge to the Ecology Movement," *Green Perspectives: Newsletter of the Green Program Project, 1987* No. 4-5 (1987).

[6] Ann Pettifor, *The Case for the Green New Deal* (New York: Verso, 2020); Yanis Varoufakis and David Adler, "It's Time for Nations to Unite around an International Green New Deal," *The Guardian*, April 23, 2019. Retrieved from: https://www.theguardian.com/commentisfree/2019/apr/23/international-green-new-deal-climate-change-global-response

social injustice, thus affirming that "there will be no solution to the ecological crisis without a solution to the problem of social domination."[7] Although more biocentric, activist institutional reform also informs the deep ecology grassroots movement that developed during the decades following Carson's influential book, whose philosophical differences from social ecology have been largely overestimated.[8] The expansion of industry for its own sake, Bookchin emphasized, is also a grave concern embedded in the principles of deep ecology.

The environmental literary journalism of Henry David Thoreau and Rachel Carson and the contemporary eco-activist writings of McKay Jenkins exhibit core principles of both Naess' deep ecology and Bookchin's social ecology. Due to literary journalism's unique affordances as a genre, enabling both fact-based reportage and subjective emotional expression, these writers resolve tensions between the two movements, especially biocentrism versus anticapitalism and aesthetic appreciation of nature versus social and political activism.[9] Rather than pitting these traditions in a zero-sum game, environmental literary journalism both proselytizes on behalf of nature's transcendent beauty and advocates on behalf of compromised ecosystems and human populations. Deep ecology's appreciation of the intrinsic value of nature inheres in social ecology's platform of political advocacy. Major concerns of both movements appear mutually reinforcing rather than at odds in environmental literary journalism's characteristic rhetorical structures rooted in outrage and hope, horror and ecstasy. The ensuing critical analysis demonstrates that aesthetic appreciation of nature's value is always already politicized in Thoreau's *Walden; or, Life in the Woods*,[10] Carson's *Silent Spring*,[11] and Jenkins's *ContamiNation: My Quest to Survive in a Toxic World*[12] and *Poison Spring: The Secret History of Pollution and the EPA*.[13] Drawing on tensions in Thoreau's deep ecology that vacillates between exultation and activism—ecstasy in nature's beauty

[7] John P. Clark, "A Dialogue with Arne Naess on Social Ecology and Deep Ecology (1988-1997)," *The Trumpeter* Vol. 26, No. 2, June 2010), 34.

[8] Clark, "A Dialogue."

[9] Bookchin, "Social Ecology"; Clarke, "A Dialogue."

[10] Henry D. Thoreau, *Walden* (Bill McKibbon ed. Boston: Beacon Press, 1854 [2017]).

[11] Carson, *Silent Spring*.

[12] McKay Jenkins, *ContamiNation: My Quest to Survive in a Toxic World* (New York: Avery, Random House, 2011).

[13] Evaggelos Vallianatos and McKay Jenkins, *Poison Spring: The Secret History of Pollution and the EPA* (New York: Bloomsbury, 2014).

and horror in its industrial abuse—each leverages its own signature journalistic storytelling through subjective narration and felt detail dramatizing the human experience missing from abstract scientific discourse on the environment.

The Anthropocene, the systematic destruction of nature by humans that has accelerated since the industrial revolution, is a process which Naess' deep ecology movement implicitly forecasted and social ecology explicitly politicized. Environmental literary journalism combines strands of both movements, notably in casting nature simultaneously as a source of human inspiration and as a force vulnerable to human impact. Through critical analysis of the environmental literary journalism of Thoreau, Carson, and Jenkins, this chapter draws on the seminal writings of deep and social ecology to highlight how the two movements are not mutually exclusive but converge in an activist aesthetic. Urgency in these environmental narratives intensifies insofar as nature functions as a channel for self-discovery and sacred force vulnerable to consumerist and industrial logics that justify its destruction.

The major works of Thoreau, Carson, and Jenkins were selected for the current study because they exhibit environmental literary journalism's development from Thoreau's *Walden*, recognized as the foundation of many core principles of the environmental movement. Thoreau's criticism of commercial enterprise and its influence on the environment informs his experiment in deliberate living at Walden Pond in the 1850s. Thoreau's concern was not merely for his own spiritual awakening and transcendence through nature but to critically interrogate systematic separation of humans from the environment via specialized work and divided labor. An intimate relation with nature, he argued, was necessary for leading an intellectually engaged life. Thoreau's *Walden* aimed at documenting "the changing relationship between people and the natural environment that has shaped the land and continues to determine where it is heading."[14]

Just as Thoreau was an acute observer and critic of the people of New England and their impact on the natural world, Rachel Carson was deeply concerned with human impact on the environment. Carson was writing in the mid-twentieth century during a time when pesticide use was on the rise in the wake of synergies between corporate chemical industries and the suburbanization of the United States. The rampant use of pesticides in

[14] David Foster, *Thoreau's Country: Journey Through a Transformed Landscape* (Cambridge: Harvard University Press, 1999), 2.

concentrations, now widely recognized as highly dangerous to both humans and natural ecosystems, is meticulously traced in *Silent Spring*, which builds on Thoreau's legacy that calls into question the advance of industry and commercial prosperity at the expense of the environment's myriad habitats. McKay Jenkins then draws on both Thoreau's self-experimental method and Carson's scientific analysis in his investigation into the effect of toxins used in everyday life on the human body. Thoreau's concern for preserving the mind-body-nature connection forms the basis of Jenkins's alarming revelation of the levels of toxicity that routinely enter the body. Prompted by the discovery of a tumor in his abdomen the size of an orange, which was followed by questions from researchers asking about his levels of chemical exposure, Jenkins launched an investigation into toxic exposure in the body, the home, drinking water, the lawn, and hardware stores. Consumerism is a major focal point of his critique of the causes of this normalization of industrial production leading to exposure to such contaminants as asbestos dust, vinyl chlorine, and pesticides. Such toxins accumulate within the so-called "healthy" levels of exposure.

To frame the critical analysis of these three texts, an overview is provided in the following section addressing the main tenets of deep ecology and social ecology with respect to advocacy journalism. It details how the human right to non-toxic environments forms the foundation of environmental activism via the Biophilia Hypothesis.[15] Interpretive critical analysis is methodologically deployed in these case studies as historically-informed close reading, a type of textual analysis that, "beyond the manifest content of the media, focuses on the underlying ideological and cultural assumptions of the text."[16] Thus "context is an important part of interpretive analysis."[17] particularly as applied to media texts selected strategically and parsed to build analysis in support of the broader argument.[18]

[15] Edward O. Wilson, "Afterword." In *Silent Spring*, by Rachel Carson, 357-363, (Boston: Houghton Mifflin, 2002); Richard Louv, "Outdoors For All: A Nascent Global Movement Proclaims that Access to Nature is a Human Right," *Sierra*, Vol. 7, May 2019. Retrieved from https://www.sierraclub.org/sierra/2019-3-may-june/feature/outdoors-for-all-nature-is-a-human-right

[16] Elfriede Fürsich, "In Defense of Textual Analysis: Restoring a Challenged Method for Journalism and Media Studies," *Journalism Studies* No. 10, Vol. 2, April 2009, 240.

[17] Bonnie Brennan, *Qualitative Research Methods for Media Studies* (New York: Routledge, 2017), 22.

[18] Fürsich, "In Defense," 240.

Foundations of Deep and Social Ecology in Advocacy Journalism

The foundational definition of advocacy journalism alludes to journalists who adopt the role of "active interpreters and participants who 'speak on behalf' of central groups," particularly those who are denied "powerful spokes[people]."[19] The "reformist impulse" intrinsic to such advocacy journalism derives from the urge "to promote perspectives that are typically under or misrepresented by the media."[20] In other contexts, advocacy journalism can consist of "organized groups that influence reporting, and ultimately affect public policy."[21] Historically, such efforts have been spearheaded by figures such as Rebecca Harding Davis, tracing back to the nineteenth century, and other authors who leveraged both the popular and elite press to shape learned debate and advocate for legislative change. This was apparent in the asylum reform of Davis,[22] wage labor advocacy of Orestes Brownson, and other transcendentalist followers of Ralph Waldo Emerson who were active in the nineteenth-century press, particularly Horace Greeley's *New York Tribune*.[23]

This reformist impulse of advocacy journalism is particularly well suited to the two schools of environmental writing stemming from deep and social ecology. Both strands are intertwined in the writings of Thoreau, Carson, and Jenkins. Deep ecology's contemplative approach toward the power of nature as a tonic to modernization and industrialization marked by increasingly routinized and segmented daily life that robs people of their connection to the wilderness establishes and defines the value of nature to humans as an asset greater than the sum of its constituent parts.[24] Deep ecology's individualism is embodied by Thoreau's sojourn into the wilderness where he seeks solitude, peace, and in many cases ecstatic

[19] Morris Janowitz, "Professional Models in Journalism: The Gatekeeper and the Advocate," *Journalism Quarterly*, Vol. 52, No. 4, December 1975, 619.

[20] Silvo Waisbord, "Advocacy Journalism in a Global Context," In *The Handbook of Journalism Studies* (Karin Wahl-Jorgenson and Thomas Hanitzsch eds.) (New York: Routledge, 2009), 371.

[21] Waisbord, "Advocacy," 371.

[22] David Dowling, "Davis, Inc.: The Business of Asylum Reform in the Periodical Press," *American Periodicals* Vol. 20, No. 1, January 2010, 23-45.

[23] David Dowling, "Emerson's Newspaperman: Horace Greeley and Radical Intellectual Culture, 1836-1872," *Journalism & Communication Monographs* Vol. 19, No. 1, March 2017, 7-74.

[24] Clark, "A Dialogue."

reveries in the kinetic aesthetic energy of nature. Social ecology, by contrast, focuses less on the immersion of the individual in nature seeking spiritual sustenance but instead represents a more activist bent, typically aimed at investigating and exposing damage to the natural environment, often caused by industrialization and, in Carson's time, corporate capitalism.[25] Social ecology aims to expose the social impact of humans on nature, operating in a role akin to watchdog journalists on behalf of not only the environment but the humans who rely on it for their health and substance, especially in the forms of clean air and water. According to the Biophilia Hypothesis,[26] access to clean air and water as well as wilderness for recreation is not a luxury but a human right, one worth advocating for through published investigations produced as politically efficacious narrative journalism.

The Biophilia Hypothesis

Edwin H. Ford's *A Bibliography of Literary Journalism in America*[27] names Thoreau as one of six representative writers who blend literature and journalism in their work. Thoreau's critique of modernity diagnosed the cultural malaise of the Industrial Revolution as the root cause of human alienation from the natural environment. Nancy L. Roberts described *Walden* as a "literary journalistic call to action for environmental preservation" and "the first literary journalism of environmental advocacy," one that calls attention to consumer culture's increasing detachment from the natural world that abets its destruction.[28] A century after *Walden*, Carson's *Silent Spring* led a new wave of narrative nonfiction, fusing what Ford calls the "interpretive cast of literature" with "the contemporary interest of journalism" to expose the impact of government and industrial interests on the environment and human communities.[29]

[25] John P. Clark, "How Wide is Deep Ecology?" *Inquiry* Vol. 39, June 1996, 189-201.

[26] Edward O. Wilson, "Biophilia and the Conservation Ethic," In *Evolutionary Perspectives on Environmental Problems*, (Dustin J. Penn and Iver Myerstud eds.) (New York: Routledge, 2017), 263-272.

[27] Edwin Hopkins Ford, "Foreword," *A Bibliography of Literary Journalism in America* (Minneapolis, MN: Burgess, 1937), 1-168.

[28] Nancy L. Roberts, "Literary Journalism and Social Activism," *The Routledge Companion to American Literary Journalism* (William E. Dow and Roberta S. Maguire eds.) (New York: Routledge, 2020), 264.

[29] Ford, "Foreword." 1.

We now see environmental literary journalists of the twenty-first century such as Jenkins waging campaigns to protect the environment and human health from industrial and consumer toxins.

This activist literary journalism is premised in a rights-based claim to access to a healthy natural environment. Nooshin Razani, a physician and clinical scientist, argues that universal and equitable access to nature is fundamental to our humanity and future of life of earth. Anchored in ideals of justice and fairness, Edward O. Wilson's Biophilia Hypothesis suggests that humans are genetically programmed to have an affiliation with the rest of nature.[30] David Orr, leader in green urban design, claims that human connection to a healthy natural environment is "the ultimate human right, upon which all other rights depend."[31] The natural environment is deemed vital because, according to Carson, "Those who contemplate the beauty of the earth will find reserves of strength that will last as long as life lasts."[32] The Biophilia Hypothesis thus takes a definitive stand against ecocide, such as the US government's attempted eradication of the imported fire ant in the South during the 1960s. This "Vietnam of Entomology," as Wilson aptly called it, was another unwinnable and self-destructive war hastily and myopically waged on an enemy that "bounded back after the pesticide carpet bombing."[33] The Biophilia Hypothesis acknowledges nature's value as a precondition for defending against such destructive campaigns that laid waste to the ecosystem of vegetation, mammals, and aquatic life well beyond the red fire ant. Just as the prohibition of DDT pesticide use led by Carson brought the American eagle back from near extinction, the elimination of neonicotinoids (common pesticides) can also spark species recovery. But larger systemic change must occur to slow the alarming rate at which the bird population is dying due to climate change throwing the ecosystem into disarray, as seen in the 29% decline of birds—more than three billion—in North America since 1970.[34]

[30] Wilson, "Biophilia."

[31] Louv, "Outdoors for All."

[32] Carson, *Silent Spring*, 64.

[33] Wilson, "Afterword," 360.

[34] Elizabeth Pennisi, "Three Billion North American Birds Have Vanished Since 1970, Surveys Show," *Science*, September 2019), Retrieved from https://www.sciencemag.org/news/2019/09/three-billion-north-american-birds-have-vanished-1970-surveys-show

Within the dialectic of deep and social ecology, the Biophilia Hypothesis is perhaps best captured by Laura Dassow Walls's apt observation that "[Thoreau] lived not in a pristine wilderness but on the outskirts of Boston, on land that had been cut over, farmed, and abused for generations."[35] This point is crucial in understanding deep ecology's appreciation of nature's present, compromised condition, rather than only its idealized pristine state. "His ability to see beauty in that land—to make it stand, symbolically, for the principle of wild nature—can help twenty-first-century generations to recover the desperately abused land they will inherit from us, and to see in it beauty, hope, regeneration."[36] Seeing "such beauty in the everyday environment of a Boston suburb," according to Walls, transforms the Thoreauvian aesthetic encounter of nature from mere escape into a "call for social justice."[37] The wild in the twenty-first century is in a decidedly post-pristine state, one which nonetheless contains the seeds for regeneration.

In *Silent Spring*, Carson finds such hope in the stretches of roadside vegetation spared from chemical spraying, oases teeming with wildflowers, butterflies, and caterpillars that birds seek for sustenance. "In such places my spirt lifted," she wrote, "to the sight of the drifts of clover or the clouds of purple vetch with here and there the flaming cup of a wood lily."[38] These "oases of beauty in the midst of austere and regimented control...made the desecration of the greater part of the road unbearable,"[39] thus intensifying the activist impulse. *The Sense of Wonder*,[40] Carson's aptly titled book that followed *Silent Spring*, epitomizes the core principle of the Biophilia Hypothesis that posits connection and access to nature as a basic human right. The apocalyptic gloom of *Silent Spring* and its disquieting images of eggs with no young birds within sounded the clarion call to action in defense of nature's "*equal right to live and blossom*," whose restriction to humans is anthropocentrism "with detrimental effects on humans themselves," as Naess pointed out.[41]

[35] Laura D. Walls, "Thoreau's *Walden* in the Twenty-First Century," *The Concord Saunterer* Vol. 12, No. 13 (2004), 15.
[36] Walls, "Thoreau's *Walden*," 16.
[37] Walls, "Thoreau's *Walden*," 16.
[38] Carson, *Silent Spring*, 71.
[39] Carson, *Silent Spring*, 71.
[40] Rachel Carson, *The Sense of Wonder* (New York: Harper and Rowe, 1965).
[41] Naess, "The Shallow and the Deep," 96.

Deep and Social Ecology in *Silent Spring* and *Walden*

Silent Spring's haunting portrait of unregulated chemical pollution was the catalyst for the modern environmental movement and, along with *Walden*, a major inspiration for the deep ecology movement. Armed with scientific evidence and powerful narrative reportage, the movement posed a direct threat to the chemical industry, which responded by impugning Carson's character and dismissing her work as hyperbole. "If facts are seeds that later produce knowledge and wisdom," she urged, "then the emotions and the impressions of the senses are the fertile soil in which the seeds must grow."[42] Carson's expertise in biology supported literary techniques such as the composite vignette of the book's opening. It describes a town's loss of insects, birds, fish, and roadside vegetation to a silent killer, invisible except for "a white granular powder," which had "fallen like snow upon the roofs and the lawns, the fields and the streams."[43] She allows that although "this town does not actually exist," and that she knows of "no community that has experienced all the misfortunes I describe," there are "a thousand counterparts in America and elsewhere in the world." More pointedly, "every one of these disasters has actually happened somewhere, and many real communities have already suffered a substantial number of them."[44] The ensuing chapters then provide overwhelming evidence of such destruction through anecdotes that animate scientific evidence of the chemicals' deadly effects.

Thoreau's lyrical narrativity was similarly grounded in science, particularly the fields of botany and civil engineering in his case. The precision demanded by the practice of land surveying directly informed his observations of the Walden habitat.[45] Environmental journalism's closeness to the natural sciences predisposes it to fact-based reportage. In the hands of a masterful stylist and compelling narrator, those facts can be delivered through rhetorical devices commonly associated with fiction and poetry. Thoreau, Carson, and Jenkins thus challenge journalistic discourse limited to what Pauly calls the binary of either the "empire of facts," especially

42 Carson, *Wonder*, 45.

43 Carson, *Silent Spring*, 3.

44 Carson, *Silent Spring*, 3.

45 Patrick Chura, *Thoreau the Land Surveyor* (Gainesville, FL: University Press of Florida, 2011).

sober hard news accounts, or literature's "garden of the imagination" detached from social reality.[46]

Such rhetorical versatility in environmental narrative journalism is consonant with deep ecology's biocentric disposition, an alternative to the traditional anthropocentric "man-in-nature" paradigm. Shallow ecology is "a *limited* science, which makes *use* of scientific methods."[47] Deep ecology, on the other hand, emphasizes "ecological harmony or equilibrium" through enlightened "policy wisdom" with a broader understanding of "priority announcements *and* hypotheses concerning the state of affairs in our universe." By framing scientific findings within the context of "value priorities," deep ecology thus stands as a corrective against narrow "scientific description and prediction,"[48] an aim conducive to the expressive affordances of literary journalism.

Deep ecology thus connects scientific ecology with a broader understanding of the value systems behind its application, in the process fostering more intimate connection and awareness of humanity's place in the ecosystem. McKay Jenkins notes that "the degree to which we have become abstracted from our natural surroundings might tell us a lot about the degree to which we are willing—consciously or not—to live with synthetic chemicals."[49] That degree of abstraction, crucially, is only visible insofar as ecstatic witness to nature's power—even in its impure, post-pristine state—can be experienced. Thoreau's famous contact with nature in *The Maine Woods* is perhaps the most iconic of his ecstatic reveries reflecting his deep ecology. "Talk of mysteries! Think of our life in nature,—daily to be shown matter, to come in contact with it,—rocks, trees, wind on our cheeks!" The impetus for writing is etched into his desire to meet life face to face in such moments where he discovers "the *solid* earth! The *actual* world! the *common* sense! *Contact! Contact! Who are we? Where are we?*"[50] This ecstatic witness testifies to nature's interconnectedness, one in which the body is part of—and not separate from—the wilderness.

[46] John Pauly, "The Politics of the New Journalism," *Literary Journalism in the Twentieth Century* (Norman Sims ed.) (Oxford: Oxford University Press, 1990), 123.

[47] Naess, "The Shallow and the Deep," 99.

[48] Naess, "The Shallow and the Deep," 99.

[49] Jenkins, *ContamiNation*, 165

[50] Henry D. Thoreau, *The Maine Woods* (Joseph J. Moldenhauer ed.) (Princeton: Princeton University Press, 1972), 69-72.

Thoreau's literary journalism is characterized by both a visceral nearness to nature and an understanding that universal insight is contingent rather than permanent. Thoreau's records of the environmental impact of rising seasonal temperatures were among the earliest to denote a pattern we now recognize as climate change.[51] To merely record such data in an abstract ledger was not enough, however. Thoreau was dedicated to the expression of his own subjective moment, especially the ways his deepest sensibility might be altered by natural phenomena. He understood the myriad contingencies, provisions—and even temporal lifespan—of any fact. "There is no such thing as pure objectivity," he observed in his journal, noting its alienating effect on both the naturalist reporter and reader. "Your observation to be interesting, i.e., significant, must be *subjective*."[52] The subject must not only be involved in the observation for it to have meaning but must also acknowledge their role in the observation. As Carson, Naess, and Jenkins would concur, detachment from one's subject practiced by "the man of science makes this mistake," Thoreau explained, "and the mass of mankind along with him: that you should cooly give your chief attention to the phenomenon which excites you as something independent of you, and not as it is related to you"[53] Instead he focused on the "important fact" and its "effect on me," not as solipsistic anthropocentrism but with a distinctly ecocentric sensibility. Thoreau's concern is not with objects in themselves, "(with which men of science deal); the point of interest is somewhere between them and me (i.e., the objects),"[54] thus closing the gap between subject and object according to the literary journalist's relation to the phenomenological world.

Thoreau's resistance to rigid approaches to nature via narrowly practiced scientific objectivity opens space not only for his narrative journalistic method that can exalt in the interrelatedness of the universe. It also allows for his anti-capitalistic register. His writing anticipates both Carson's and Naess' critique of profit motives, particularly the results of mindless economic growth according to a "grow or die" mentality,[55] and the

[51] Richard B. Primack, *Walden Warming: Climate Change Comes to Thoreau's Woods* (Chicago: University of Chicago Press, 2014).

[52] Henry D. Thoreau, *The Journal of Henry David Thoreau, vol. 6*, (Bradford Torrey and Francis Allen, eds.) (Boston: Houghton Mifflin, 1949), 236-7.

[53] Thoreau, *Journal 6*, 236-7.

[54] Thoreau, *Journal 6*, 236-7.

[55] Bookchin, "Social Ecology."

subordination of environmental sustainability to consumer lifestyle choices. In this vein, Carson revealed "the very limited awareness of the nature of the threat" of mass use of toxic chemicals such as DDT as herbicides and insecticides "with little or no advance investigation of their effect on soil, water, wildlife, and man himself."[56] Carson's activism is policy-oriented unlike Jenkins' promotion of Thoreauvian self-sufficiency via DIY (do it yourself) culture. "We don't need the government, we don't need to keep our fingers crossed that conservation organizations will do the work for us. We can do it ourselves," Jenkins suggests, offering a variety of alternative consumer choices for home, lawn, and water use.[57]

Carson addresses a more vulnerable consumer in 1962 with far fewer organic and sustainable alternatives. She thus targets "tranquilizing pills of half-truths" delivered by government officials and industrial apologists designed to placate protest. The other cause of limited public awareness of the effects of regulatory entomology via chemical application she ascribes to hyperspecialization of increasingly narrow scientific approaches to the environment. "This is an era of specialists," she said of her historical moment in 1962, "each of whom sees his own problem and is unaware of or intolerant of the larger frame into which it fits." Such specialists cater to the demands of industry, whose entitlement to "the right to make a dollar at whatever cost is seldom challenged."[58] Carson's activist rhetoric dons the unmistakable language of advocacy journalism serving public interest. "Who has decided—who has the *right* to decide—for the countless legions of people who were not consulted that the supreme value is a world without insects, even though it be also a sterile world ungraced by the curving wing of a bird in flight?" she asks, noting that "the authoritarian temporarily entrusted with power" has the final word. Such decisions, she urges, are inattentive to "the millions to whom beauty and the ordered world of nature still have a meaning that is deep and imperative."[59]

Naess' articulation of Carson's principles is visible in his sense that shallow ecology is aligned with the dominant values and attitudes of modernity, which treats environmental issues as problems to be solved through chemical and technological innovation and scientific resource management specialists, whose anthropocentric bias is aimed at increasing

[56] Carson, *Silent Spring*, 13.
[57] Jenkins, *ContamiNation*, 196.
[58] Carson, *Silent Spring*, 13.
[59] Carson, *Silent Spring*, 127.

agricultural profits (for industrial farming) and appeasing the rising demands of bourgeois consumer culture (for middle-class lifestyles including lawn maintenance). Deep ecology bears both Thoreau's distrust of the principles and aims of scientific authority as well as Carson's call for policy change. For example, Naess touted self-realization in conjunction with biocentric equality he termed "biospherical egalitarianism," which promotes "diversity and symbiosis" through an "anti-class posture." The exploiter who abets the use of products known to damage the environment "lives differently from the exploited, but both are adversely affected in the potentialities and self-realization."[60] Like Thoreau and Carson, Naess argued the need to "forsake the meretricious attractions of 'a higher standard of living' in favor of the deeper satisfactions of 'situations of inherent value,'" as Clark describes.[61] Such deep satisfactions are epitomized by the tableau of Thoreau in his self-constructed hut at Walden Pond marveling at the sweetness of the rain in "Sounds," the liberating feel of soil beneath his bare feet while cultivating the food that will sustain him in "The Bean Field," and how, in "Solitude," "there can be no very black melancholy to him who lives in the midst of Nature and has his senses still."[62]

Such appreciation for nature's intrinsic value is premised in peace and social justice. Social ecology brings particular emphasis to economic and political decentralization (found in the "Economy" chapter of *Walden*), alternative technologies (used in surveying the pond and heating his cabin), and grassroots community (in the "Visitors" chapter). Far from the detached, apolitical New Age mystic his detractors portrayed him as, Naess lauded social ecology's emphasis on how "the human attempt to dominate nature stems from human domination of other humans," calling it an "excellent frame" he hoped would "inspire an *increasing* number of people."[63] Thoreau's activist politics on behalf of social justice not only saturate *Walden*—which names the outcasts and runaway slaves who seek asylum at the pond—but also inform his famous treatise on civil disobedience and radical defense of John Brown's anti-slavery insurrection. Human domination of nature is visible in the industrial designs on Walden Pond: "Since I left those shores the woodchoppers have still further laid them

[60] Naess, "The Shallow and the Deep," 96.
[61] Clark, "A Dialogue," 4.
[62] Thoreau, *Walden*, 124.
[63] Quoted in Clark, "A Dialogue," 26.

waste, and the wood, with occasional vistas through which you can see the water."[64] Anticipating Carson's controlling metaphor of lost songbirds, he dreads his Muse of nature might become "silent henceforth." "How can you expect the birds to sing when their groves are cut down?" he asks. Further destroying nature's music is "that devilish Iron Horse"—the train—"whose ear-rending neigh is heard throughout the town, browsing off all the woods on Walden shore" and scarring the landscape.[65]

The Right to Know

Jenkins's revelation of industrialization's threat to the ecosystem and human health builds on Thoreau and Carson by re-situating the human body in nature. Following Carson's probe into the government's failure to protect citizens from harmful toxins, Jenkins exposes profound malfeasance in his as-told-to narrative journalistic exposé *Poison Spring: The Secret History of Pollution and the EPA*. In it, the testimony of twenty-five-year EPA veteran E.G. Vallianatos demonstrates that the EPA "is infected with political appointees, a great many of them with strong ties to the very industries they have been hired to regulate." Regulations are routinely molded by EPA staff "to satisfy their former (and often future) industry employers."[66] Graft is apparent when "former government officials are able to persuade their former colleagues to be more lenient in their scrutiny of data provided by industry," a process that "ensures that new and more dangerous pesticides continue to be 'registered' and to enter the market."[67]

Jenkins shares Carson's "belief that the public had a fundamental 'right to know'" and "should be mobilized to act to improve the system."[68] He draws on Carson to emphasize how recent body burden studies based on hair, nail, and blood samples reveal toxic chemicals are not only dangerous, but are in all of our bodies at alarmingly high levels, a point with rational, affective, and ethical appeal. Jenkins builds on David Duncan's *National Geographic* story, "The Pollution Within," which Rob Alexander

[64] Henry D. Thoreau, *The Writings of Henry David Thoreau: Journal 4, 1851-1852*, (Robert Sattelmeyer ed.) (Princeton, NJ: Princeton University Press, 1992), 273.

[65] Thoreau, *Walden*, 125.

[66] Vallianatos and Jenkins, *Poison Spring*, 5.

[67] Vallianatos and Jenkins, *Poison Spring*, 6.

[68] Priscilla Murphy, *What a Book Can Do: The Publication and Reception of* Silent Spring (Amherst, MA: University of Massachusetts Press, 2007), 13.

calls a "work recognizable as literary journalism."[69] Duncan reports the results of his own chemical burden, tracing each to the time in his life when he was exposed to it. In this case the news, as Alexander astutely observes, "is not simply someplace 'out there': it is breaking inside his own body."[70] This distillation of journalistic subject into the reporter's body extends to the source of the contaminants in a canoe trip down the Brandywine River that flows from southeastern Pennsylvania into northern Delaware. Accompanied by a toxicologist on the journey, Jenkins learns that the bacteria giardia appeared in 90% of samples from the river, and that an antiquated century-old sewage treatment plant—bearing a striking architectural resemblance to "a fifteenth-century Hungarian bathhouse" with its "unimaginably huge underground caverns"[71]—overflowed 60% of the time it rained, dumping thousands of gallons of untreated sewage into the waterways.[72]

The urgency of social ecology's call to action through such images thus depends on the value of nature established by deep ecology, two separate yet reciprocally influential traditions traceable from Thoreau's antebellum sensibility to Carson's mid-twentieth century lament and the contemporary toxic discourse of Jenkins. The aesthetic of environmental journalism exults in the intrinsic beauty and value of the environment while combating a condition in which "We are all oysters," like those dying in the Chesapeake Bay, exposed to toxins we only partially filter out.[73] This sort of argument by analogy common to Jenkins's narrative journalism offers a sharp critique via allegories found in nature, a technique commonly used by Thoreau. As Foster points out, Thoreau found "remarkable local analogues to a wide range of human experience and emotion," applying them to geopolitical conflict as when he aligned the Homeric epics he was reading with the black and red ants "that engaged in anxious, single-minded battle in his cabin dooryard at Walden."[74] Jenkins brings Thoreau's method of social and cultural critique through analogues in nature into

[69] Robert Alexander, "Literary Journalism and Ecocriticism," *The Routledge Companion to American Literary Journalism*, (William E. Dow and Roberta S. Maguire eds.) (New York: Routledge, 2020), 492.

[70] Alexander, "Literary Journalism," 492.

[71] Jenkins, *ContamiNation*, 150.

[72] Jenkins, *ContamiNation*, 146.

[73] Jenkins, *ContamiNation*, 26.

[74] David Foster, *Thoreau's Country: Journey Through a Transformed Landscape* (Cambridge: Harvard University Press, 1999), 2-3.

our contemporary world, meshing his analysis with the type of scientific environmental analysis offered by Carson.

Taken together, all three texts support a powerful claim for the importance of the philosophical foundations of deep and social ecology for environmental literary journalism. All three had a passionate concern for what Nixon describes as the "complicity of the military-industrial complex in disguising toxicity, both physically and rhetorically."[75] In accord with the watchdog function of the press, these texts hold to account government and industrial power that has chronically compromised the well-being of its citizens. Environmental writing confined to conservationism and nature appreciation in deep ecology becomes distinctly more political and socially engaged when wedded with social ecology. The two strands have provided the essential elements of the formation of environmental justice movements, which rely directly on the power of felt detail in longform narrative journalism. Critically, such literary journalism offers a corrective for what Nixon calls "the dubious funding of partitioned knowledge" on toxic waste and ecological destruction that yield "baleful public health implications."[76] Together, the environmental activism of these works calls into question the industrial logics and capitalist values that generate toxic waste and nuclear weapons, which rank among the most significant threats facing the planet.

References

Alexander, Robert. "Literary Journalism and Ecocriticism." In *The Routledge Companion to American Literary Journalism*, edited by William E. Dow and Roberta S. Maguire, 282–497. New York: Routledge, 2020.

Bookchin, Murray. "Social Ecology vs. Deep Ecology: A Challenge to the Ecology Movement." *Green Perspectives: Newsletter of the Green Program Project, 1987* No. 4–5 (1987).

Brennan, Bonnie. *Qualitative Research Methods for Media Studies*. New York: Routledge, 2017.

Carson, Rachel. *Silent Spring*. Boston: Houghton Mifflin, 1962.

Carson, Rachel. *The Sense of Wonder*. New York: Harper and Rowe, 1965.

Chura, Patrick. *Thoreau the Land Surveyor*. Gainesville, FL: University Press of Florida, 2011.

[75] Rob Nixon. *Slow Violence and the Environmentalism of the Poor* (Cambridge, MA: Harvard University Press, 2011), xi.

[76] Nixon, *Slow Violence*, xi.

Clark, John P. A dialogue with Arne Naess on social ecology and deep ecology (1988–1997). *The Trumpeter* Vol. 26, No. 2 (June 2010): 20–39.
Clark, John P. "How Wide is Deep Ecology?" *Inquiry* Vol. 39 (June 1996): 189–201.
Dowling, David. "Davis, Inc.: The Business of Asylum Reform in the Periodical Press," *American Periodicals* Vol. 20, No. 1 (January 2010): 23–45.
Dowling, David. "Emerson's Newspaperman: Horace Greeley and Radical Intellectual Culture, 1836–1872," *Journalism & Communication Monographs* Vol. 19, No. 1 (March 2017): 7–74.
Ford, Edwin Hopkins. "Foreword." *A Bibliography of Literary Journalism in America*. Minneapolis, MN: Burgess, pp. 1–68, 1937.
Foster, David. *Thoreau's Country: Journey Through a Transformed Landscape*. Cambridge: Harvard University Press, 1999.
Fürsich, Elfriede. "In Defense of Textual Analysis: Restoring a Challenged Method for Journalism and Media Studies," *Journalism Studies* No. 10, Vol. 2 (April 2009): 238–252.
Heise, Ursula K. "The Hitchhiker's Guide to Ecocriticism." *Publication of the Modern Language Association* Vol. 121, No. 2 (March 2006): 503–516.
Janowitz, Morris. Professional models in journalism: The gatekeeper and the advocate. *Journalism Quarterly*, Vol. 52, No. 4 (December 1975): 618–626.
Jenkins, McKay. *ContamiNation: My Quest to Survive in a Toxic World*. New York: Avery, Random House, 2011.
Louv, Richard. "Outdoors For All: A Nascent Global Movement Proclaims that Access to Nature is a Human Right," *Sierra*, Vol. 7 (May 2019). Retrieved from https://www.sierraclub.org/sierra/2019-3-may-june/feature/outdoors-for-all-nature-is-a-human-right
Murphy, Priscilla. *What a Book Can Do: The Publication and Reception of* Silent Spring. Amherst, MA: University of Massachusetts Press, 2007.
Naess, Arne. "The Shallow and the Deep, Long-range Ecology Movement. A Summary," *Inquiry*, Vol. 16, No. 1–4 (May 1973): 95–100.
Nixon, Rob. *Slow Violence and the Environmentalism of the Poor*. Cambridge, MA: Harvard University Press, 2011.
Pauly, John. "The Politics of the New Journalism." In *Literary Journalism in the Twentieth Century*, edited by Norman Sims, 110–132. Oxford: Oxford University Press, 1990.
Pennisi, Elizabeth. "Three Billion North American Birds Have Vanished Since 1970, Surveys Show," *Science* (September 2019). Retrieved from https://www.sciencemag.org/news/2019/09/three-billion-north-american-birds-have-vanished-1970-surveys-show
Pettifor, Ann. *The Case for the Green New Deal*. New York: Verso, 2020.
Primack, Richard B. *Walden Warming: Climate Change Comes to Thoreau's Woods*. Chicago: University of Chicago Press, 2014.

Quick, Timothy D. "American Transcendentalism and Deep Ecology in the History of Ideas," M.A. Thesis, School of Environmental Studies, 2004, University of Victoria.

Roberts, Nancy L. "Literary Journalism and Social Activism." In *The Routledge Companion to American Literary Journalism*, edited by William E. Dow and Roberta S. Maguire, 256–268. New York: Routledge, 2020.

Thoreau, Henry D. *The Journal of Henry David Thoreau, vol. 6*, edited by Bradford Torrey and Francis Allen. Boston: Houghton Mifflin, 1949.

Thoreau, Henry D. *The Writings of Henry David Thoreau: Journal 4, 1851–1852*, edited by Robert Sattelmeyer. Princeton, NJ: Princeton University Press, 1992.

Thoreau, Henry D. *The Maine Woods*, edited by Joseph J. Moldenhauer. Princeton: Princeton University Press, 1972.

Thoreau, Henry D. *Walden*, edited by Bill McKibbon. Boston: Beacon Press, 1854 [2017].

Vallianatos, Evaggelos and McKay Jenkins. *Poison Spring: The Secret History of Pollution and the EPA*. New York: Bloomsbury, 2014.

Varoufakis, Yanis and David Adler. "It's Time For Nations to Unite around an International Green New Deal." *The Guardian*, 23 April 2019. Retrieved from https://www.theguardian.com/commentisfree/2019/apr/23/international-green-new-deal-climate-change-global-response

Waisbord, Silvo. "Advocacy Journalism in a Global Context." In *The Handbook of Journalism Studies*, edited by Karin Wahl-Jorgenson and Thomas Hanitzsch, 371–385). New York: Routledge, 2009.

Walls, Laura D. "Thoreau's *Walden* in the Twenty-First Century." *The Concord Saunterer* Vol. 12, No. 13 (2004), 15–16.

Wilson, Edward O. "Biophilia and the Conservation Ethic." In *Evolutionary Perspectives on Environmental Problems*, edited by Dustin J. Penn and Iver Myerstud, 263–272. New York: Routledge, 2017.

Wilson, Edward O. "Afterword." In *Silent Spring*, by Rachel Carson, 357–363. Boston: Houghton Mifflin, 2002.

"The Bitter Taste of Extinction": Writing the Environmental Crisis Through Food

L. Sasha Gora

Introduction

Jonathan Gold—the first food writer to have won a Pulitzer Prize—started his career in the 1980s covering music, but it was when he turned his attention from his ears to his mouth and started reviewing restaurants that he found his voice. In 2011 he penned a *Los Angeles Times* op-ed. Gold's name appeared regularly in newspaper and magazine food sections, but not in the op-eds. He begins the story recounting the last time he ate shark fin—a delicacy in China—fifteen years ago, at a now-defunct "seafood palace." Remembering the specialities exhibited in glass cases, "where they had the stark, ghostly presence of museum displays," the scene Gold

L. S. Gora (✉)
The University of Augsburg, Augsburg, Germany
e-mail: Sasha.gora@uni-a.de

P. Calvi (ed.), *The Literary Journalist as a Naturalist*, Palgrave Studies in Literary Journalism,
https://doi.org/10.1007/978-3-031-56634-9_6

sets feels out of time.[1] His memory transitions to a description of why shark fin is prized—its "ethereal, gelatinous texture"—before delivering a reality check: "there is no sustainable source." This adds new meaning to its "ghostly presence," and leaves a haunting aftertaste. Gold drifts between memories and reports. Attaining a balanced respect for both traditional practices and fishing regulations, he channels cultural sensitivity while taking an environmental stance. Although Gold confesses shark fin soup is "delicious," he concludes: "It is hard to work up an appetite for the bitter taste of extinction."

Gold's brother Mark—an environmental scientist and California's Deputy Secretary for Oceans and Coastal Policy—encouraged him to write this op-ed. Two assemblymen had introduced a bill in California that would ban the sale of shark fin, only for it to meet resistance in the state Senate out of concern that, as an inadvertent side effect, it would discriminate against Chinese Americans, a point Gold addresses. "It's true," he says—the bill would require them, as opposed to other California communities, to give up a food seasoned with tradition. But he argues that Chinese culinary culture is resilient, it adapts. Shark numbers, however, are not. "There is no third way with shark's fin," Gold explains, "—we either stop eating it because we choose to preserve the species, or we stop eating it because soon there will be none left to eat." Food today, a culinary ghost tomorrow. Following Jonathan's 2018 death, his brother claimed this piece was central in California's decision to ban shark fin sales.[2]

Although his tattered Dodge Ram 1500 was green, Gold's lifestyle was far from eco-friendly. His culinary research across Los Angeles clocked in at 20,000 miles a year. "I'm an L.A. guy," he said. "I drive. I am my truck. My truck is me."[3] Yet his op-ed had such a positive environmental impact that it begs the question: how is food writing a means for social and environmental activism?

Shark fin only begins to reveal how human appetites contribute to the environmental crisis. Food systems account for more than one-third of

[1] Jonathan Gold, "Shark Fin Soup: The taste of extinction," *Los Angeles Times*, 7 August 7, 2011, https://www.latimes.com/opinion/la-xpm-2011-aug-07-la-oe-gold-shark-fin-20110807-story.html.

[2] Mark Gold, "He Bridged Cultural Divides and Saved the Sharks: Jonathan Gold's Brother Remembers his Impact," *Los Angeles Times*, July 24, 2018, https://www.latimes.com/food/la-fo-jonathan-gold-sharks-20180724-story.html.

[3] Laura Gabbert, *City of Gold* (New York: IFC Films, 2015).

global greenhouse gas emissions.[4] To eat is, thus, to harm the world.[5] Writer and farmer Wendell Berry describes eating as an agricultural act.[6] However, it is an act that goes beyond agriculture alone. Eating is an environmental act. In addition to the plants and livestock that humans cultivate, eating practices involve water and air, from aquaculture to irrigation and from transportation to waste management. In addition to greenhouse gas emissions, food systems contribute to deforestation, biodiversity loss, and water pollution.[7] Simply put, the environmental crisis is also a food production crisis. As the climate continues to change, current global food systems—which rely "on just about twelve plant and five animal species to provide 76% of the world's food"—become more and more vulnerable.[8] In fact, Sarah Bridle, the author of *Food and Climate Change Without the Hot Air*, quotes climate change policy researcher Tim Gore, who believes that "[the] main way that most people will experience climate change is through the impact on food: the food they eat, the price they pay for it, and the availability and choice they have."[9]

Gold was neither the first nor the last writer to make the connection between eating and the environment. This environmental consciousness motivates academics, journalists, and recipe writers alike to promote what scientist Dave Reay calls "climate-smart food."[10] For example, in 2019 Jonathan Safran Foer published *We Are the Weather*, whose subtitle reveals food's role in countering climate change: *Saving the Planet Begins at Breakfast*. A best-selling novelist's decision to address this topic represents an increasing awareness of how our culinary decisions affect our planet just as much as our bodies. From veganism and environmental vegetarianism to locavorism and flexitarianism, since there has been an environmental

[4] United Nations, "Food Systems Account for Over One-third of Global Greenhouse Gas Emissions," *UN News*. March 9, 2021, https://news.un.org/en/story/2021/03/1086822.

[5] Krishnendu Ray, *The Ethnic Restaurateur* (London and New York: Bloomsbury, 2016), xx.

[6] Wendell Berry, "The Pleasures of Eating," in *American Food Writing: An Anthology with Classic Recipes*, ed. Molly O'Neill (New York: Literary Classics of the United States, 2007), 552.

[7] FAO, *Sustainable Healthy Diets: Guiding Principles* (Rome: WHO, 2019), 8.

[8] Michael P. Hoffmann, Carrie Koplinka-Loehr, and Danielle L. Eiseman, *Our Changing Menu: Climate Change and the Foods We Love and Need* (Ithaca, New York: Cornell University Press, 2021), 45.

[9] Sarah L. Bridle, *Food and Climate Change Without the Hot Air* (Cambridge: UIT, 2020), 18.

[10] Dave Reay, *Climate-Smart Food* (London: Palgrave Macmillan, 2019).

movement, there have been environmentally conscious doctrines about what to eat—and what not to. Like Gold's, different articles rally different causes. Beyond voicing environmental concerns, others document and mourn foods that have already been lost.

Surveying food writing's growing environmental consciousness, this essay considers how American food writing represents, activates, or seeks to preserve the environment. Unfolding in three parts, this study fleshes out food writing as a genre. Departing from the seemingly simple question of what food writing is, it first demonstrates that to write about hunger is to write about the world at large. Then, it considers what I call the "environmental turn" in American food writing. Building off the previous section, it argues that culinary practices double as environmental manifestos. Together these sections flesh out how food writing is a means with which to uphold or, alternatively, to resist and challenge the social and environmental status quo. To conclude, the paper returns to discuss the relationship between food and extinction and how contemporary food writing doubles as a form of activism.

Wordy Appetites: Drafting the Culinary Genre

What is food writing? From describing ingredients in soil to food on plates, there is writing about planting and harvesting, writing about cooking, and writing about eating. "Most food writing is about cooking," claimed John Seabrook in *The New Yorker* in 2002, "—it's less about the ingredients than about the rendering of those ingredients, and the consuming of them in communal settings."[11] However, as the genre has developed, more and more food writers have ventured out of the kitchen, trekking through factory fields of corn, serving as witnesses in slaughterhouses, and adopting a fish—or shark's—eye view.

An elastic term, food writing spans cookbooks and recipes to reported features. It captures a meal's many life cycles, from production and distribution to consumption and waste. The literature of food pops up in publications centred on edibles alone—like *Bon Appétit* and its late Condé Nast sibling *Gourmet*—but also travels to other contexts, weaving together history, journalism, and biography. Although food writing has typically

[11] John Seabrook, "The Fruit Detective," in *Secret Ingredients: The New Yorker Book of Food and Drink*, ed. David Remnick (New York: The Modern Library, 2008 [2002]), 239.

been assigned to the lifestyle section of newspapers and magazines, as a genre it has begun to push against these boundaries.

Writing about food is a means of documenting a time and a place and profiling its inhabitants, human and more-than-human. There is a large body of historical writing that has featured food—from travel accounts and ethnographic studies to journals, letters, poems, and cookery manuals. Out of these documental forms emerged a genre. The work of M.F.K. Fisher (1908–1992), who, after spending years in France learning how to prepare snails and finish a bottle of wine by herself, returned to California and established a blueprint for American food writing, is a prominent example. Fisher's work inspired the likes of Craig Claiborne and James Beard and also established a model for the American version of what literary critic Rosalia Baena terms a "gastro-graphy."[12]

In 1937, Fisher published *Serve it Forth*, the first of many books that marry history, recipes, and cultural musings. "There are two kinds of books about eating," she begins, "those that try to imitate Brillat-Savarin's, and those that try not to."[13] She assumes the reader's familiarity with the French gastronome (and lawyer) Jean Anthelme Brillat-Savarin (1755–1826). Considered the gastronomic essay's founder, it is impossible to read about food without encountering his *Physiologie du Goût*. This 1825 publication marks the beginning of "the genre of gastronomic writing."[14] Over the years, Brillat-Savarin's aphorism "*Dis-moi ce que tu manges, je te dirai ce que tu es*" has melted down to "you are what you eat." Fisher later contributed her own translation of his seminal book, but his influence on her eating and writing was clear from the start.

Just like how this quote stands in for Brillat-Savarin's work, the first lines in Fisher's *The Gastronomical Me* (1943) play again and again in all sorts of food writing. A collection of essays penned between 1912 and 1941, she opens with: "People ask me: Why do you write about food, and eating and drinking? Why don't you write about the struggle for power and security, and about love, the way others do?"[15] The shortest answer—hunger—is the easiest, but she confesses there is more. "It seems to me

[12] Rosalia Baena, "Gastro-Graphy: Food as Metaphor in Fred Wah's *Diamond Grill* and Austin Clarke's *Pig Tails 'N Breadfruit*." *Canadian Ethnic Studies* 38, no. 1 (2006): 105–16.

[13] M.F.K. Fisher, *Serve It Forth* (San Francisco: North Print Press, 1989 [1937]), 3.

[14] Arlene Avakian, "Cooking Up Lives: Feminist Food Memoirs," *Feminist Studies* 40, no. 2 (2014): 277–303.

[15] M.F.K. Fisher, *The Gastronomical Me* (San Francisco: North Print Press, 1989 [1943]), ix.

that our three basic needs, for food and security and love, are so mixed and mingled and entwined that we cannot straightly think of one without the others. So it happens that when I write of hunger, I am really writing about love and the hunger for it."[16] This answer yields two revelations: the first is that food writing requires an explanation. As food studies scholar Meryl S. Rosofsky points out: "Implicit in this question is a subtle indictment: How can so talented a writer squander her enormous gift on something so trivial, even frivolous, as food?"[17] The second revelation is that food writing is not always about food.

More than half a century after Fisher's treaty, Molly O'Neill (1952–2019), a *New York Times* food columnist, edited *American Food Writing: An Anthology with Classic Recipes* (2007). In her introduction to 250 years of culinary history, she calls food writing "a relatively recent-almost a contemporary—invention" but draws from its predecessors, miscellaneous literary works documenting affairs of the stomach and the mouth.[18] Historically, those who have been able to find space to write about food reflects hierarchies of gender, class, race, and ethnicity. Fisher represents an early breakthrough—a woman writing about the pleasure of eating as opposed to cooking as domestic labour. But she was a privileged white woman who had grown up with a family cook.

American Food Writing employs a "voice of God" style of narration. O'Neill's commentary is present but her own contributions are not. However, in her obituary, the *New York Times* credits her with "[ushering in] an era of food writing in the 1990s that was as much about journalism as deliciousness." As *American Food Writing* began to reveal, food writing had expanded. Moving beyond a venue that traditionally upheld the status quo, it increasingly began to host voices that challenged it, critically

[16] Ibid., ix.

[17] Meryl S. Rosofsky, "Writing the Wolf Away: Food Meaning & Memories from World War II," *Food, Culture & Society* 7 no. 1 (2004): 48

[18] Molly O'Neill, ed., *American Food Writing: An Anthology with Classic Recipes* (New York: Literary Classics of the United States, 2007), xx.

reflecting on the relationship between food and power and who is able to claim American cultural citizenship through their foodways.[19]

This concern with reporting—and not just culinary critique—continues. Its timing also overlaps with the emergence of a new field of academic study. Although there were earlier scholars studying edibles, such as the anthropologists Sidney Mintz and Mary Douglas, food studies scholar Marion Nestle points out that "it was only in 1996 that this collective term began to describe a legitimate field of academic study."[20] The growth of food studies meant a growth in critical writing about food and its role in the human experience.

In an era of social media and online reviews, contemporary food writing is less about how pillowy a restaurant's tortillas are and more about how the restaurant treats its employees and sources its ingredients. More reporting today focuses on food's entanglement with larger social issues, as exemplified by Soleil Ho, who became the *San Francisco Chronicle*'s restaurant critic in 2019, and Alicia Kennedy, who pens a newsletter that is as equally popular as it is critical of mainstream food systems. Stories of sexual harassment and unethical labour practices are difficult to swallow, no matter how fiery a homemade, lacto-fermented hot sauce is or how fresh the bread.[21] So, if cooking is an act of transforming plants and animals into meals—a transformation that is equally about culture as it is about raw materials—how do food writers represent the environment?

[19] I address the relationship between food writing, memory, immigration, and assimilation in L. Sasha Gora, "Secret Ingredient: Food, Knowledge, and Migration," *Arts of the Working Class* 16 (2021): 14. See also: Rebecca J. Kinney, "Riding Shotgun with an LA Son: Narratives of Race, Place, and Mobility in Roy Choi's Culinary Autobiography," *Food, Culture & Society* 20, no. 1 (2017): 59–75. Rosalia Baena, "Gastro-Graphy: Food as Metaphor in Fred Wah's *Diamond Grill* and Austin Clarke's *Pig Tails 'N Breadfruit*," *Canadian Ethnic Studies* 38, no. 1 (2006): 105–16.

[20] Marion Nestle and W. Alex McIntosh, "Writing the Food Studies Movement," *Food, Culture & Society* 13 no. 2 (2010): 160. For the difference between food studies and food writing see: Andrew Smith, Jeffrey M. Pilcher, and Darra Goldstein, "Food Scholarship and Food Writing," *Food, Culture & Society* 13 no. 3 (2010): 319–329.

[21] Theodore Gioia, "The Midlife Crisis of the American Restaurant Review," *Los Angeles Review of Books*, December 31, 2019, https://lareviewofbooks.org/article/midlife-crisis-american-restaurant-review/.

Green Gourmets: An Environmental Turn

Early American food writing was concerned with experience and taste—applauding that bistro that serves the perfect trout amandine or unmasking a secret ingredient. The environmental turn started in the 1960s. Around that time, both an environmental and a culinary consciousness awoke. Food writing grants a vantage point from which to view how the two have met, crossed, and merged. The term *turn* is a cliché and yet I like that it stipulates a direction something takes without being descriptive of how—a sharp turn maybe, or perhaps a sudden one that sneaks up on you, or maybe even a long creeping one.

But one could also consider a culinary turn in environmental writing. In 2008 Bill McKibben edited *American Earth: Environmental Writing Since Thoreau*, which is the environmental equivalent of O'Neill's food volume. In the anthology, McKibben defines environmental writing as something that "takes as its subject the collision between people and the rest of the world, and asks searching questions about this collision…"[22] He finds environmental writing in political speeches, song lyrics—like "Mercy Mercy Me (The Ecology)" by Marvin Gaye—cartoons, fiction, and essays.

In his summary of the genre's ebbs and flows, McKibben includes food. He begins with Wendell Berry, who started writing in the 1960s and 1970s, and became "the father of the farmers' market" and "the voice of the new agrarianism."[23] McKibben reports that a surprising amount of writing "follows Berry's lead and concentrates on food—just in the past year or so, Michael Pollan and Barbara Kingsolver have written best sellers about relocalising our eating, about escaping the republic of irresponsibility."[24] This brings McKibben to the conclusion, in 2008, that "the farmer's market is as potent an emblem as the wilderness area, the smell of fresh-baked local bread as tonic as the tang of mountain air."[25] In this 1047-page anthology, there are texts about preparing food—heating dehydrated peaches and milk while backpacking in Colin Fletcher's "A Sample Day in the Kitchen" (1968)—growing it—Berry's reflection on healing land and himself in "The Making of a Marginal Farm" (1981)

[22] Bill McKibben, ed., *American Earth: Environmental Writing Since Thoreau* (New York: Library of America, 2008), xxii.

[23] Ibid., 504.

[24] Ibid., xxix.

[25] Ibid., xxviii–xxx.

—and contaminated food as a health hazard in Sandra Steingraber's "Having Faith" (2001) and César Chávez's "Wrath of Grapes Boycott Speech" (1986). Food also appears in discussions of ecosystems—as demonstrated by an excerpt from Rachel Carson's *Silent Spring* (1962)—as well as through the many plants and animals humans eat, from beans to bison and from fish to maple sap.

However, most prominently, food appears in the context of raising industrial cows with cheap field corn and petroleum: an excerpt from Pollan's 2006 bestseller *The Omnivore's Dilemma: A Natural History of Four Meals*. In this passage, Pollan (born in 1955) follows steer 534 from the ranch to the feedlot to profile a production system that casts cows as protein machines capable of converting "thirty-two pounds of feed into four pounds of gain."[26] An equation that turns corn into hamburgers. Unlike Fisher (who, when asked why she writes about food, replied that she was, first, hungry), Pollan admits hunger was the last thing he felt.[27] But, as many can attest, the stomach plays tricks on the eyes, nose, and mind. "Yet I'm sure that after enough time goes by, and the stink of this place is gone from my nostrils," writes Pollan, "I will eat feedlot beef again. Eating industrial meat takes an almost heroic act of not knowing or, now, forgetting."[28] An example of food writing as environmental exposé.

It is telling that *American Food Writing*—which contains over 150 texts, with the oldest dating to 1753–61—ends with an excerpt from Pollan's *The Omnivore's Dilemma*. His "gastro-opus"[29] begins with the question: "What should we have for dinner?"[30] *American Food Writing* reprints a section where Pollan reckons with organic as an ideal and as an industry, complicating the belief that it is a panacea.

[26] Michael Pollan, "From The Omnivore's Dilemma," in *American Earth: Environmental Writing Since Thoreau*, ed. Bill McKibben (New York: Library of America, 2008 [2006]), 956.

[27] Ibid., 959.

[28] Ibid., 959.

[29] Robert Ji-Song Ku, *Dubious Gastronomy: The Cultural Politics of Eating Asian in the USA* (Honolulu: University of Hawai'i Press, 2013), 36.

[30] Michael Pollan, *The Omnivore's Dilemma: A Natural History of Four Meals* (New York: The Penguin Press, 2006), 1. Two years later, Pollan published *In Defense of Food: An Eater's Manifesto*, which provides an answer: "Eat food. Not too much. Mostly plants." Michael Pollan, *In Defense of Food: An Eater's Manifesto* (New York: The Penguin Press, 2008), 1. He also gave this away in "Unhappy Meals." Michael Pollan, "Unhappy Meals," *New York Times Magazine*, January 28, 2007. https://www.nytimes.com/2007/01/28/magazine/28nutritionism.t.html.

Predating *The Omnivore's Dilemma*, Nestle credits journalist Eric Schlosser's 2001 bestseller *Fast Food Nation* for igniting a general public interest in food issues. "His account," Nestle writes, "turned masses of readers into food advocates eager to transform the current food system into something healthier for people, food animals and farm workers, as well as for the environment."[31] Sociologist W. Alex McIntosh agrees, but he also points out that earlier ethnographies made similar points about fast food and labour.[32] He then says the same about Pollan, asking himself, "Haven't I read this elsewhere?"[33] Like geographer Julie Guthman, McIntosh points out that Pollan's work erases the names of the scholars whose work he is in dialogue with.[34] Nonetheless, McIntosh recognises that Pollan, like Schlosser, "should be credited with making a broader audience aware of changes in agriculture and their relationship to changes in food consumption…"[35] Pollan's influence is undeniable.

Pollan lives in Berkeley, the epicentre of California's—and America's—farm-to-table movement, as exemplified by Alice Waters' seminal restaurant Chez Panisse. With the exception of Schlosser's success, comparing Pollan's work to what came before—penned by back-to-the-landers or health aficionados—is on par with comparing a small organic co-op to the organic aisle in a big box grocery store. Pollan reaches a mainstream audience. This, too, is what Waters does and is what the historian Paul Freedman calls the Chez Panisse paradoxes. He names one he considers particularly revolutionary: "viewing naturalness as something that enhances sensual pleasure, not as a form of penance. …it's not enough to have environmentally virtuous food if the food is boring."[36] A prestigious restaurant—once considered the nation's best—serving gourmet dishes seasoned with ideas for radical agricultural change. Green gourmets.

Just like Fisher, Pollan's name is now synonymous with American food writing. When historian Neil Prendergast polls students in his Food and Environment course about their motivation for enrolling, many mention

[31] Nestle and McIntosh, "Writing the Food Studies Movement," 163.

[32] Ibid., 172.

[33] Ibid., 172.

[34] Julie Guthman, "Commentary on Teaching Food: Why I am Fed Up with Michael Pollan et al," *Agriculture and Human Values* 24 (2007): 261–264.

[35] Nestle and McIntosh, "Writing the Food Studies Movement," 172.

[36] Paul Freedman, *Ten Restaurants that Changed America* (New York and London: Liveright Publishing Corporation, 2016), 369.

The Omnivore's Dilemma.[37] Many college courses also assign Pollan's texts. A 2013 study found his book did indeed influence undergraduate students. However, their change in attitude generally vanished after a year, a reminder that culinary habits die hard.[38]

Short-lived change connects to Guthman's critique of "how to eat" books.[39] She questions their ability to "move toward a more just and ecological way of producing and consuming food."[40] Instead of pioneering systemic change, the views that such books deal in often encourage conscious eaters to simply "opt out." Alternatives like farmers markets and organic and heritage livestock production provide alternatives to corporate grocery stores and factory farming; they do not reform mainstream food systems. Although Meghan Lynch and Audrey Giles, public health scholars, recognise that Pollan dominates "the field of sustainable eating in North America,"[41] they assert that his advice, like other mainstream media prescriptions, omits a "*critical* examination of the underlying relations of power that inform discourses about sustainable eating initiatives" and overlooks the complexities of everyday life that determine food decisions.[42] For these reasons, books like Pollan's give readers an easier, a lighter, a more nostalgic version of environmentalism.

In addition, interdisciplinary scholar Emily Lind Johnston draws attention to how Pollan "invokes a pastoral agrarian imaginary" that reconstructs "an agrarian past that never was."[43] A celebration of small-scale farms operated by those who own them—the Berry model of new agrarianism—on soil that is now American overlooks agrarianism's dirty laundry: how it is the outcome of colonisation, Indigenous removal, slavery, and gendered and racialised labour.[44] Pollan's writing, like Berry's, is sea-

[37] Neil Prendergast, "Food and the Environment," in *The Routledge History of American Foodways*, eds. Michael D. Wise and Jennifer Jensen Wallach (New York and London: Routledge, 2016), 261.

[38] Julia M. Hormes, Paul Rozin, Melanie C. Green, and Katrina Fincher, "Reading A Book Can Change Your Mind, But Only Some Changes Last For a Year: Food Attitude Changes in Readers of *The Omnivore's Dilemma*," *Frontiers in Psychology* 4 (2013): 6.

[39] Guthman, "Commentary on Teaching Food," 261.

[40] Ibid., 263.

[41] Meghan Lynch and Audrey Giles, "Let Them Eat Organic Cake: Discourses in Sustainable Food Initiatives," *Food, Culture & Society* 16 no. 3 (2013): 482.

[42] Ibid., 480, 485.

[43] Emily Lind Johnston, "Agrarian Dreams and Neoliberal Futures in Life Writing of The Alternative Food Movement," *Food and Foodways* 24 no. 1–2 (2016): 10–11.

[44] Ibid., 15.

soned with nostalgia that overlooks who in the United States has historically been allowed to own land and, therefore, participate in agrarianism.[45] Heidi Zimmerman, a media studies scholar, makes it clear that what she calls the "Pollan phenomenon" is "profoundly classed."[46] This, she argues, is not reason to dismiss his influence. Instead, it gives context with which to better understand class formation and "its relationship to food, environmentalism and politics."[47] Pollan as a cultural phenomenon, Zimmerman writes, "is part of a broader upswing in 'ethical consumerism'" and one of the reasons to which she attributes Pollan's appeal is: "it also speaks to longings for a different, less environmentally destructive, more equitable kind of world."[48] In short, he speaks to nostalgia.

Pollan inherited decades of work that connected eating with environmental concerns. Nine years after *Silent Spring*, one year after the first Earth Day (1970), and the same year Waters opened Chez Panisse, Frances Moore Lappé published *Diet for a Small Planet* (1971). Besides *Silent Spring*, journalist Jonathan Kauffman claims, "no single book had a greater impact on the way her generation thought about food."[49] *Diet for a Small Planet*—Lappé's first of many books—also departed from hunger. Instead of a search for pleasure, she prioritised politics. In the twentieth anniversary edition, she summarises her original mission: "to awaken people to this simple fact: Hunger is human made."[50] This mission came out of having learned: "half of our harvested acreage went to feed livestock" and that cattle—America's preferred meat—require 16 pounds of grain and soybeans for 1 pound of beef—a match to Pollan's description of steer 534.[51] Part detailed report and part recipe collection, the book critiques "the machine model of nature."[52] The recipes encourage people not to eat

[45] Sarah D. Wald, "Visible Farmers/Invisible Workers: Locating Immigrant Labor in Food Studies," *Food, Culture & Society* 14 no. 4 (2011): 567–586. Janet Fisko, "Unsettling Ecocriticism: Rethinking Agrarianism, Place, and Citizenship," *American Literature* 84, no. 2 (2012): 301–325.

[46] Heidi Zimmerman, "Caring for the Middle Class Soul: Ambivalence, Ethical Eating and the Michael Pollan Phenomenon," *Food, Culture & Society* 18 no. 1 (2015): 32.

[47] Ibid., 33.

[48] Ibid., 35, 33.

[49] Jonathan Kauffman, *Hippie Food: How Back-to-the-Landers, Longhairs, and Revolutionaries Changed the Way We Eat* (New York: William Morrow, 2018), 143.

[50] Frances Moore Lappé, *Diet for a Small Planet* (New York: Ballantine Books, 1991 [1971]), xvii.

[51] Ibid., 9.

[52] Ibid., xxviii.

meat, coaching them how. Kauffman considers the recipes quintessential "hippie food," an example of the relationship between food—a populist subject—and counterculture values. Cuisine of the counterculture, or what American studies scholar Warren Belasco calls countercuisine.[53] The belief that what people eat doubles as a manifesto of how they treat the earth. The objective to save the planet by eating ethically marked an early environmental turn in food writing, but one limited to the counterculture.

Silent Spring hit shelves on September 27, 1962, and that summer *The New Yorker* published a three-part serialisation, informing readers of Carson's pesticide research. The weekly magazine is also an important home for food writing. In 2008 its editor, David Remnick, assembled *Secret Ingredients: The New Yorker Book of Food and Drink*. The text provides an overview of how food writing has developed. Organised thematically, the first cluster of articles gather under "Dining Out." All of these pieces are by-lined by white men. The next section—"Eating In"—begins with three Fisher essays (including "The Secret Ingredient," from which the volume borrows its name)—a woman's voice at the table, but at a kitchen table. The following section—"Fishing and Foraging"—returns to an all-male choir but samples how food writing represents American environments. The most obvious example is John McPhee's "A Forager," a 1968 piece that would feel at home in McKibben's volume. A long-form profile of Euell Gibbons (1911–1975), wild plant expert and author of *Stalking the Wild Asparagus* (1962), McPhee persuades Gibbons to go on a six-day November trip to central Pennsylvania to live off the land. They begin by eating only what they forage and gradually introduce store-bought ingredients. While cooking over a Coleman camp stove, the two discuss the "sorry mistake" of framing nature as something to conquer. Gibbons remarks: "Nature has many, many balances, and we have to find a balance that includes man. If man accepts that he has to be a part of the balance, he must reject the idea of the conquest of nature."[54] To eat nature without crushing it.

[53] Warren J. Belasco, *Appetite for Change: How the Counterculture Took on the Food Industry* (Ithaca and London: Cornell University Press, 2007 [1989]), 4, 28.

[54] John McPhee, "A Forager," in *Secret Ingredients: The New Yorker Book of Food and Drink*, 188–227, ed. David Remnick (New York: The Modern Library, 2008 [1968]), 196.

Conclusion: Burns and Ghosts

As I write this, the American food media is on fire. Writers and editors are reaching for buckets of water, trying to save their careers, their field. There have been high-profile resignations and a much-needed conversation about labour, race, and industry gatekeepers. In the words of artist and chef Tunde Wey: “Can you renovate a burning house?”[55] The COVID-19 pandemic has revealed the many still-active ghosts of structural racism and social inequality. These ghosts loom large in food writing in a country where the history of slavery is a history of forced agriculture and culinary labour.[56] Publishing has prioritised the appetites of only a privileged few and although this has started to change, the pace of this change has been too slow.

Despite these limitations, food writing is an influential form of record-keeping—of documenting extinct and forgotten foods.[57] And as food writing’s environmental turn demonstrates, the genre is increasingly becoming a means of advocacy. Gold’s conclusion that although shark fin soup tastes wonderful, extinction does not, is an example of food writing as a means for encouraging environmental action. It documents all that can go extinct—animals and plants, environments and seasons, traditions and ways of life. It eulogises and commemorates while also warning about future loss. About culinary responsibility.

The environmental turn in American food writing demonstrates that food writers also play an important role in discussions about sustainable appetites and the future of planetary health. In 2021 the popular American recipe site *Epicurious* announced that it had “cut out beef,” meaning the nation’s favourite meat will no longer appear in new recipes or articles.[58]

[55] Helen Rosner, “The Case for Letting the Restaurant Industry Die,” *The New Yorker*, May 22, 2020, https://www.newyorker.com/culture/annals-of-gastronomy/the-case-for-letting-the-restaurant-industry-die.

[56] See: Michael W. Twitty, *The Cooking Gene: A Journey through African American Culinary History in the Old South* (New York: Amistad, 2017); Frederick Douglass Opie, *Hog & Hominy: Soul Food from Africa to America* (New York: Columbia University Press, 2008); and Psyche A. Williams-Forson, *Building Houses Out of Chicken Legs: Black Women, Food and Power* (Chapel Hill: The University of North Carolina Press, 2006).

[57] Lenore Newman, *Lost Feast: Culinary Extinction and the Future of Food* (Toronto: ECW Press, 2019).

[58] David Tamarkin and Maggie Hoffman, “The Planet on the Plate: Why Epicurious Left Beef Behind,” *Epicurious*, April 26, 2021, https://www.epicurious.com/expert-advice/why-epicurious-left-beef-behind-article.

Recognising the sensitivity of this decision in a "Red Meat Republic,"[59] *Epicurious* explained to its readers that this decision was not "anti-beef but rather pro-planet." In an interview, cookbook author and *New York Times* recipe writer Melissa Clark revealed that environmental concerns were now an essential ingredient in the recipes she develops. "I really think the environment is our number-one impending crisis," Clark explained, "And we can help. Food writers can help. We can make a difference. Maybe it's not a huge difference. But I know that the work I do can be important in that way."[60] A recipe codes what is—as well as what is not—acceptable to eat and, thus, demands social, cultural, and environmental responsibility.

Responsibility also guides Foer's *We Are the Weather*. Nearly too abstract to classify as a "what to eat book," it is instead rich in analogies. Stories about his childhood home and life as a father, about the guilt of craving—and even eating—hamburgers after his 2009 book *Eating Animals*, stories about the history of farming, the Frozen Ark Project that collects DNA from endangered animals, and a civil rights lawyer who set himself on fire as a "protest suicide" in Brooklyn's Prospect Park on a Saturday morning in April 2018 as a statement about protecting the environment. *We Are the Weather* begins with a different suicide note, the world's oldest—written in ancient Egypt around 2181 to 2040 BCE. This sets up Foer's thinking about history, death, and sacrifice, topics he threads together while leaving readers with this haunting climate change protest-cum suicide. As Foer makes clear: "We cannot keep the kinds of meals we have known and also keep the planet we have known. We must either let some eating habits go or let the planet go."[61] The world as we know it is endangered, he warns, and our appetites are largely to blame.

What leads a novelist to write about diet and suicide? Why write about food? O'Neill suggests one answer: "Perhaps because reading about food reminds us that we are humans and alive and a tangle of contradictions, some of which can be eased by a meal and some of which can be resolved by remembering and imagining what it feels like to bite off a piece of the world, to pull it across the tongue, and to swallow it."[62] More and more

[59] Joshua Specht, *Red Meat Republic: A Hoof-to-Table History of How Beef Changed America* (Princeton: Princeton University Press, 2019).

[60] Alicia Kennedy, "A Conversation with Melissa Clark," *From the Desk of Alicia Kennedy*. October 29, 2021, https://www.aliciakennedy.news/p/a-conversation-with-melissa-clark.

[61] Jonathan Safran Foer, *We Are the Weather: Saving the Planet Begins at Breakfast* (New York: Macmillan, 2019), 71.

[62] O'Neill, *American Food Writing*, xxiii.

pieces of the world are burning—so hot they singe the tongue—or are too dry to swallow. Other bits are swarmed by ghosts, just like wasps crashing a picnic. Food writing records these connections. It charts the contradictions between the foods people are hungry for and the environmental crisis they cultivate—or attempt to heal.

References

Avakian, Arlene. "Cooking Up Lives: Feminist Food Memoirs." *Feminist Studies* 40, no. 2 (2014): 277–303.

Baena, Rosalia. "Gastro-Graphy: Food as Metaphor in Fred Wah's *Diamond Grill* and Austin Clarke's *Pig Tails 'N Breadfruit*." *Canadian Ethnic Studies* 38, no. 1 (2006): 105–116.

Belasco, Warren J. *Appetite for Change: How the Counterculture Took on the Food Industry*. Ithaca and London: Cornell University Press, 2007 [1989].

Berry, Wendell. "The Pleasures of Eating." In *American Food Writing: An Anthology with Classic Recipes*, 551–8. Edited by Molly O'Neill. New York: Literary Classics of the United States, 2007.

Bridle, Sarah L. *Food and Climate Change Without the Hot Air*. Cambridge: UIT, 2020.

Brillat-Savarin, Jean Anthelme. *The Physiology of Taste*. Translated by Anne Drayton. London: Penguin Books, 1970.

Chang, Andrea. "Los Angeles Times Restaurant Critic Jonathan Gold Dies at 57." *Los Angeles Times*, July 21, 2018. https://www.latimes.com/local/obituaries/la-fo-jonathan-gold-obit-20180721-story.html.

FAO. *Sustainable Healthy Diets: Guiding Principles*. Rome: WHO, 2019.

Fisher, M.F.K. *Serve It Forth*. San Francisco: North Print Press, 1989a [1937].

Fisher, M.F.K. *The Gastronomical Me*. San Francisco: North Print Press, 1989b [1943].

Fisko, Janet. "Unsettling Ecocriticism: Rethinking Agrarianism, Place, and Citizenship." *American Literature* 84, no. 2 (2012): 301–325, https://doi.org/10.1215/00029831-1587359.

Foer, Jonathan Safran. *We Are the Weather: Saving the Planet Begins at Breakfast*. New York: Macmillian, 2019.

Foer, Jonathan Safran. *Eating Animals*. New York: Little, Brown and Company, 2009.

Freedman, Paul. *Ten Restaurants that Changed America*. New York and London: Liveright Publishing Corporation, 2016.

Gabbert, Laura. *City of Gold*. New York: IFC Films, 2015.

Gibbons, Euell. *Stalking the Wild Asparagus*. Chambersburg, PA: Alan C Hood & Co, 1962.

Gioia, Theodore. "The Midlife Crisis of the American Restaurant Review." *Los Angeles Review of Books*, December 31, 2019. https://lareviewofbooks.org/article/midlife-crisis-american-restaurant-review/.

Gold, Jonathan. "Shark Fin Soup: The taste of extinction." *Los Angeles Times*, 7 August 7, 2011. https://www.latimes.com/opinion/la-xpm-2011-aug-07-la-oe-gold-shark-fin-20110807-story.html.

Gold, Mark. "He Bridged Cultural Divides and Saved the Sharks: Jonathan Gold's Brother Remembers his Impact." *Los Angeles Times*, July 24, 2018. https://www.latimes.com/food/la-fo-jonathan-gold-sharks-20180724-story.html.

Gora, L. Sasha. "Secret Ingredient: Food, Knowledge, and Migration." *Arts of the Working Class* 16 (2021): 10–14.

Guthman, Julie. "Commentary on Teaching Food: Why I am Fed Up with Michael Pollan et al." *Agriculture and Human Values* 24 (2007): 261–264, https://doi.org/10.1007/s10460-006-9053-x.

Hormes, Julia M., Paul Rozin, Melanie C. Green, and Katrina Fincher. "Reading A Book Can Change Your Mind, But Only Some Changes Last For a Year: Food Attitude Changes in Readers of *The Omnivore's Dilemma*." *Frontiers in Psychology* 4 (2013): 1–8, https://doi.org/10.3389/fpsyg.2013.00778.

Hoffmann, Michael P., Carrie Koplinka-Loehr, and Danielle L. Eiseman. *Our Changing Menu: Climate Change and the Foods We Love and Need*. Ithaca, New York: Cornell University Press, 2021.

Johnston, Emily Lind. "Agrarian Dreams and Neoliberal Futures in Life Writing of The Alternative Food Movement." *Food and Foodways* 24 no. 1–2 (2016): 9–29, https://doi.org/10.1080/07409710.2016.1142779.

Kauffman, Jonathan. *Hippie Food: How Back-to-the-Landers, Longhairs, and Revolutionaries Changed the Way We Eat*. New York: William Morrow, 2018.

Kennedy, Alicia. "A Conversation with Melissa Clark." *From the Desk of Alicia Kennedy*. October 29, 2021. https://www.aliciakennedy.news/p/a-conversation-with-melissa-clark.

Kinney, Rebecca J. "Riding Shotgun with an LA Son: Narratives of Race, Place, and Mobility in Roy Choi's Culinary Autobiography." *Food, Culture & Society* 20, no. 1 (2017): 59–75, https://doi.org/10.1080/15528014.2016.1272202.

Ku, Robert Ji-Song. *Dubious Gastronomy: The Cultural Politics of Eating Asian in the USA*. Honolulu: University of Hawai'i Press, 2013.

Lappé, Frances Moore. *Diet for a Small Planet*. New York: Ballantine Books, 1991 [1971].

Lynch, Meghan, and Audrey Giles. "Let Them Eat Organic Cake: Discourses in Sustainable Food Initiatives." *Food, Culture & Society* 16 no. 3 (2013): 479–493, https://doi.org/10.2752/175174413X13673466711967.

McPhee, John. "A Forager." In *Secret Ingredients: The New Yorker Book of Food and Drink*, 188–227. Edited by David Remnick. New York: The Modern Library, 2008 [1968].

McKibben, Bill. ed. *American Earth: Environmental Writing Since Thoreau.* New York: Library of America, 2008.

Nestle, Marion, and W. Alex McIntosh. "Writing the Food Studies Movement." *Food, Culture & Society* 13 no. 2 (2010): 159–179, https://doi.org/10.2752/175174410X12633934462999.

Newman, Lenore. *Lost Feast: Culinary Extinction and the Future of Food.* Toronto: ECW Press, 2019.

O'Neill, Molly. ed. *American Food Writing: An Anthology with Classic Recipes.* New York: Literary Classics of the United States, 2007.

Opie, Frederick Douglass. *Hog & Hominy: Soul Food from Africa to America.* New York: Columbia University Press, 2008.

Prendergast, Neil. "Food and the Environment." In *The Routledge History of American Foodways*, 261–275. Edited by Michael D. Wise and Jennifer Jensen Wallach. New York and London: Routledge, 2016.

Pollan, Michael. *In Defense of Food: An Eater's Manifesto.* New York: The Penguin Press, 2008a.

Pollan, Michael. "My Organic Industrial Meal." In *American Food Writing: An Anthology with Classic Recipes*, 717–727. Edited by Molly O'Neill. New York: Literary Classics of the United States, 2007a [2006].

Pollan, Michael. "From The Omnivore's Dilemma." In: *American Earth: Environmental Writing Since Thoreau*, 948–960. Edited by Bill McKibben. New York: Library of America. 2008b [2006].

Pollan, Michael. *The Omnivore's Dilemma: A Natural History of Four Meals.* New York: The Penguin Press, 2006.

Pollan, Michael. "Unhappy Meals." *New York Times Magazine*, January 28, 2007b. https://www.nytimes.com/2007/01/28/magazine/28nutritionism.t.html.

Reay, Dave. *Climate-Smart Food.* London: Palgrave Macmillan, 2019.

Ray, Krishnendu. *The Ethnic Restaurateur.* London and New York: Bloomsbury, 2016.

Remnick, David. ed. *Secret Ingredients: The New Yorker Book of Food and Drink.* New York: The Modern Library, 2008.

Rosner, Helen. "The Case for Letting the Restaurant Industry Die." *The New Yorker*, May 22, 2020. https://www.newyorker.com/culture/annals-of-gastronomy/the-case-for-letting-the-restaurant-industry-die.

Rosofsky, Meryl S. "Writing the Wolf Away: Food Meaning & Memories from World War II." *Food, Culture & Society* 7 no. 1 (2004): 47–58, https://doi.org/10.2752/155280104786578120.

Schlosser, Eric. *Fast Food Nation: The Dark Side of the All-American Meal.* London: Penguin Press, 2011.

Seabrook, John. "The Fruit Detective." In *Secret Ingredients: The New Yorker Book of Food and Drink*, 229–240. Edited by David Remnick. New York: The Modern Library. 2008 [2002].

Severson, Kim, and Neil Genzlinger. "Molly O'Neill, Writer Who Explored and Celebrated Food, Is Dead at 66." *New York Times*, June 17, 2019. https://www.nytimes.com/2019/06/17/dining/molly-oneill-dead.html.

Smith, Andrew, Jeffrey M. Pilcher, and Darra Goldstein. "Food Scholarship and Food Writing." *Food, Culture & Society* 13 no. 3 (2010): 319–329, https://doi.org/10.2752/175174410X12699432700827s.

Specht, Joshua. *Red Meat Republic: A Hoof-to-Table History of How Beef Changed America*. Princeton: Princeton University Press, 2019.

Tamarkin, David, and Maggie Hoffman. "The Planet on the Plate: Why Epicurious Left Beef Behind." *Epicurious*. April 26, 2021. https://www.epicurious.com/expert-advice/why-epicurious-left-beef-behind-article.

Twitty, Michael W. *The Cooking Gene: A Journey through African American Culinary History in the Old South*. New York: Amistad, 2017.

United Nations. "Food Systems Account for Over One-third of Global Greenhouse Gas Emissions." *UN News*. March 9, 2021. https://news.un.org/en/story/2021/03/1086822.

Wald, Sarah. D. "Visible Farmers/Invisible Workers: Locating Immigrant Labor in Food Studies." *Food, Culture & Society* 14 no. 4 (2011): 567–586, https://doi.org/10.2752/175174411X13046092851479.

Williams-Forson, Psyche A. *Building Houses Out of Chicken Legs: Black Women, Food and Power*. Chapel Hill: The University of North Carolina Press, 2006.

Zimmerman, Heidi. "Caring for the Middle Class Soul: Ambivalence, Ethical Eating and the Michael Pollan Phenomenon." *Food, Culture & Society* 18 no. 1 (2015): 31–50, https://doi.org/10.2752/175174415X14101814953729.

Silent Spring: The Rise of the Environmental Movement

Sue Joseph

> *We live in an age of rising seas… In our own lifetime we are witnessing a startling alteration of climate.*[1]

Introduction

The term eco-feminism was coined in 1974 by French author Françoise d'Eaubonne in her book *Le Féminisme ou la Mort*.[2] So, technically speaking, tagging American scientist and writer Rachel Carson as an ecofeminist is inaccurate, as she had passed away a decade earlier. But, as Gaard writes: "What makes an action or analysis ecofeminist is its intersectional approach

[1] Carson in Lepore, Jill, "The Right Way to Remember Rachel Carson," *The New Yorker*, March 26, 2018.

[2] Françoise d'Eaubonne *Le Féminisme ou la Mort*, (P. Horay: Paris, 1974).

S. Joseph (✉)
University of South Australia, Adelaide, SA, Australia
e-mail: sue.joseph@unisa.edu.au

P. Calvi (ed.), *The Literary Journalist as a Naturalist*, Palgrave Studies in Literary Journalism,
https://doi.org/10.1007/978-3-031-56634-9_7

to understanding and acting to correct eco-social and environmental problems. Specifically, ecofeminism is unique for bridging human justice, interspecies justice, and human environmental justice…".[3] Mapping Carson's literary journalism work against this definition, this paper argues that her message is genealogically ecofeminist, rendering her in the vanguard of the '70s movement as one of its finest protagonists. As Birkeland writes: "Ecofeminist theory and analysis has only been developing since the 1970s, but the practice has been around for much longer, and has been growing in many parts of the world".[4]

Gates writes that "ecofeminism involves activism as well as ideology".[5] Carson had both. In claiming in her foundational long-form narrative text *Silent Spring* that the unfettered use of synthetic chemical pesticides (DDT, amongst others)[6] was destroying not just the insects they were designed to eradicate but also food chains, potentially decimating the ecosystem and poisoning humans, she shook the foundation of post-war/ Cold War American scientific and political communities. The public's interest was so piqued that as the scientific and chemical industries, and political communities pushed back on her claims in order to gag her, discrediting and ridiculing her as much for her science as her gender, public support erupted. As Stein writes of Carson: "She questioned the practices and belief systems of economic entomologists, pesticide manufacturers, agribusiness, government regulatory agencies and common citizens. Indeed, she challenged the very vision of American scientific, technological, moral, and political supremacy in the Cold War period".[7]

One of the most obvious long-form techniques Carson uses to write *Silent Spring* is unpacking complex scientific and technical jargon, making it accessible to the general public. Dowling writes that she: "recreates imagined scenarios rooted in sociological fact fraught with threatening

[3] Greta Gaard "Ecofeminism," Keywords for Environmental Studies (Joni Adamson, William A. Gleason & David N. Pellow eds.) (New York University Press: New York USA, 2016) 69.

[4] Janis Birkeland "Ecofeminism: Linking Theory and Practice," Ecofeminism (Greta Gaard ed.) (Temple University Press: Philadelphia, USA, 1993), 19.

[5] Barbara Gates "A Root of Ecofeminism," Ecofeminisme, ISLE, *Interdisciplinary Studies in Literature and Environment*, Vol. 3, No. 1, Summer, (1996): 7.

[6] Carson was concerned about DDT—but also about 2,4-D, 2,4-T, DDE, aldrin, dieldrin, malathion, and heptachlor.

[7] Karen Stein "Rachel Carson: challenging authors", Introduction, (Sense Publishers: Rotterdam, 2012), xiii.

dramatic tension pitting an unsuspecting public at the peril of an industry and government willing to compromise its safety for profit".[8] Gaard writes that she makes: "… interspecies connections among the environmental health of avians, humans, and ecosystem flows (air, water, food)" arguing for "an end to pesticides, publishing her findings in a voice both literary and scientific".[9] And Stein flags the predictive nature of her research, bearing current gravitas: "Her concept of the interconnectedness of natural systems is increasingly relevant to our time as globalization produces a more intricate and tightly woven fabric of connections across the planet".[10]

Its power was deep. Lockwood explains: "Much of the impact of *Silent Spring* … on the effects of pesticide use, is credited to its literary style and rhetorical force, its adaptations and amplifications of nuclear and Cold War fears contemporary to publication".[11] But it is its literary journalistic lyricism with its poetical lilt at its heart that makes it shine. Her publisher at the time describes Carson as: "… a trained scientist with a poet's wonder".[12] As Al Gore, American Vice President for eight years beside President Bill Clinton, writes in the Houghton Mifflin (1994) edition: "*Silent Spring* came as a cry in the wilderness, a deeply felt, thoroughly researched, and brilliantly written argument that changed the course of history".[13]

Before discussing the poetic fabric of *Silent Spring*, this paper gives an overview of Rachel Carson; then gives a brief synopsis of ecofeminism scholarship and how I argue Rachel Carson was indeed in the vanguard of the movement. I then textually evaluate key literary and poetic movements in the text to demonstrate her literary prowess, before concluding.

[8] Dowling, "Banned in Britain," 65.

[9] Gaard, "Ecofeminism", 68.

[10] Stein, "Rachel Carson", xv-xvi.

[11] Alex Lockwood "The Affective Legacy of Silent Spring," Environmental Humanities, Vol. 1, (2012): 123.

[12] Bayard Webster, "Rachel Carson – a scientist with a poet's wonder," *The New York Times*, May 25, 1982.

[13] Al Gore "Rachel Carson and Silent Spring," in Courage for the Earth (Peter Matthiessen ed.) (Houghton Mifflin: New York, 2007): 63.

RACHEL CARSON: *WITNESS FOR NATURE*[14] AND *THE GENTLE SUBVERSIVE*[15]

Rachel Carson was born May 27, 1907, in Springdale, Pennsylvania. The third of three children, she published her first writing at the age of 10 in the *St Nicholas League* magazine, where she won a competition with her story "A Battle in the Clouds".[16] By 14, she was selling her stories to magazines and periodicals. In 1954, Carson tells a group of female students in a speech:[17]

> I can remember no time even in earliest childhood, when I didn't assume I was going to be a writer. Also, I can remember no time when I wasn't interested in the out of doors and the whole world of nature.[18]

Melding these two loves, Carson produced four texts throughout her life, another published posthumously: *Under the Sea*,[19] *The Sea Around Us*,[20] *The Edge of the Sea*,[21] *Silent Spring*,[22] and *The Sense of Wonder*,[23] after her death.

While at Pennsylvania College for Women, Carson changed her major from English to Biology, subsequently winning a scholarship to undertake a Master's of Zoology in 1929 at Johns Hopkins University in Maryland, which she was awarded in 1932. Three years later she began the journalistic tranche of her career, employed as a part-time feature writer to create 52 short radio programs on marine life for the US Bureau of Fisheries in Washington; the following year, the Bureau hired her as a junior aquatic biologist. Her first book was published in 1941 and in 1944, she pitched

[14] Title of Carson biography by Linda Lear (1997).

[15] Title of Carson biography by Mark H, Lytle (2007).

[16] Ferrari, Rachel Carson, *American Experience*, PBS, May 28, 2019.

[17] 'The real world around us', speech, April 21, 1954.

[18] Linda Lear *Rachel Carson, Witness for Nature*, (Houghton Mifflin Harcourt: New York; Boston, 1997):

[19] Rachel Carson *Under the Sea* (Simon and Schuster: New York, 1941).

[20] Rachel Carson *The Sea Around Us* (Oxford University Press: UK, 1951).

[21] Rachel Carson *The Edge of the Sea* (Houghton Mifflin: New York, 1955).

[22] Carson, *Silent*.

[23] Rachel Carson *The Sense of Wonder* (Harper Collins: New York, 1965).

a story idea to *Reader's Digest;* an article on DDT. The periodical turned it down, claiming the topic was too "unpleasant".[24]

Dichlorodiphenyltrichloroethane, or DDT, was first developed in the nineteenth century but not used until "1939 when a Swiss chemist named Paul Müller discovered that it was a very potent insecticide and killed all kinds of bugs very readily".[25] During WW2, The US Army, in a bid to protect its soldiers from disease-carrying (typhus, malaria) insects in the tropics, began to use DDT around its soldiers. Within the next decade, Carson's major concern was not the unintended consequences of synthetic pesticides, but that so little was actually known about their deleterious effects, yet they were used ubiquitously: "Widely employed by government agencies to protect health and agriculture, as well as American interests abroad, synthetic pesticides also were sold directly to consumers—who, by 1957, could choose from an array of some 6000 different products".[26] And Carson argues, without impartial knowledge of the damage inherent in their use. This is, I argue, cultural ecofeminism at its purest, aligning with Buckingham: "Its strength was to demonstrate the possibility of a way of thinking and being which reversed the normal hierarchy in which men stood at the peak".[27]

Carson was diagnosed with breast cancer in 1950. She sold *The Sea Around Us* to Oxford University Press, and *The New Yorker* published nine of its chapters, in three parts. In 1960, ten years after her diagnosis, Carson underwent a radical mastectomy.

In December 1961, *The New Yorker* editor William Shawn told Carson he wished to serialise *Silent Spring*.[28] As Koehn writes: "The book, combined with *The New Yorker* serialization, created a sensation".[29] And, as Walker writes: "*Silent Spring* was not just a book. It was an event that kept happening, a debate that kept circulating, and a felt conversation that kept people talking".[30]

[24] Linda Lear *The Life and Legacy of Rachel Carson*, 2020 <https://www.rachelcarson.org/>

[25] Ferrari, "Rachel Carson".

[26] Ferrari, "Rachel Carson".

[27] Susan Buckingham "Ecofeminsim in the twenty-first century," *The Geographical Journal*, Vol. 170, No. 2, June, (2004): 147

[28] Her work appeared in *The New Yorker* on June 16, 23 and 30, 1962.

[29] Nancy Koehn "From Calm Leadership, Lasting Change," *The New Yorker*, October 27, 2012.

[30] Kenny Walker "Without Evidence, there is No Answer": Uncertainty and Scientific Ethos in the Silent Spring[s] of Rachel Carson, *Environmental Humanities*, 2, (2013): 101.

The following summer at the height of the Cold War, citing from her book, President John F. Kennedy appointed a committee to study the use of pesticide—her dream that her research would inform policy was evolving. On May 15, 1963, the President's Science Advisory Committee (PSAC) published its report "The Uses of Pesticide", substantiating Carson's claims in *Silent Spring* on the devastating effects of over-use of synthetic chemical pesticides; in August that year, Kennedy attributed the PSAC investigation to the publication of Carson's work in *The New Yorker*.

Her scientific journalism and Carson herself were viciously attacked by the science communities, the pharmacological communities and various political quarters. She "…was accused of being a Communist, of being a hysterical, female Luddite. The reaction was to attack the messenger".[31] Gore claims the attacks on Carson can be compared to the attacks on Charles Darwin when he published *The Origin of Species.* He writes: "By the time the book became widely available, the forces against her were formidable".[32] The scientific community was perhaps the most vehement. Lytle writes: "Carson, they charged, had practiced bad science and thereby misrepresented the value of DDT. More than that, the reverence for nature she encouraged, created formidable roadblocks to the rapid exploitation of natural resources they advocated".[33] Elder says: "… very often they objected to it as a feminine way of talking about what should be left to the good objective male scientists. It was not subtle … They wanted her to be quiet". Elder continues, describing Carson as the: "Voice of objection to the arrogant application of power".[34] But on April 3, 1963, the national TV program *CBS Reports with Eric Sevareid* went to air with "The Silent Spring of Rachel Carson". Carson tells Sevareid, and the nation:

> The balance of nature is built with a series of inter-relationships between living things and between living things and their environment. You can't just step in with some brute force and change one thing without changing many others. Now this does not mean of course that we must never inter-

[31] Oreskes in Ferrari, "Rachel Carson".

[32] Gore, "Rachel Carson", 64.

[33] Mark H. Lytle "Rachel Carson: Saint or Sinner?" *Human Ecology Review*, Volume 23, Number 2 (2017): 56.

[34] *Rachel Carson and the Romantics*, radio broadcast, Radio National, (February 9, 2013). <https://www.abc.net.au/radionational/programs/archived/poetica/2013-02-09/4467452>.

fere, that we must not attempt to tilt that balance of nature in our favour but unless we do bring these chemicals under better control, we are certainly heading for disaster… [chemicals] are the sinister and little-recognized partners of radiation in changing the very nature of the world—the very nature of its life.[35]

This programme, along with the *Silent Spring* serialisation in *The New Yorker* and, then the book, placed her squarely in people's living rooms. And the people were listening. On June 4, 1963, Rachel Carson testified in front of the US Senate Committee on Government Operations—Senator Abraham Ribicoff's special sub-committee on environmental hazards. Senator Ribicoff began proceedings with these words: "Miss Carson… we welcome you here. You are the lady who started all this.[36] Will you please proceed".[37] Lear continues:

Rachel Carson was ready… This was the moment she had hoped for: one final chance to translate her vision into policy, to make a difference, to change the way people looked at the natural world, to stop the warfare against it…She began: 'The problem you have chosen to explore is one that must be solved in our time. I feel strongly that beginning must be made on it now – in this session of Congress'. Her words marked an end and a beginning. As she spoke, Rachel Carson was dying of cancer.[38]

Two days later she stood before the Senate Committee on Commerce, continuing her plea. Ten months after that, on April 11, 1964, Rachel Carson died, aged 56 of her breast cancer, "an illness later found to have significant links to the synthetic chemicals she studied".[39]

In March 2020, *Time Magazine* collated its yearly *Women of the Year* into the 100 most influential women of the century. In 1963, Rachel Carson made the first list; in 2020, she made the second. *Time Magazine* writes:

[35] *Reports with Eric Sevareid*, Documentary, *CBS*, (April 3, 1963) New York.

[36] This is an echo of what Abraham Lincoln allegedly said, alluding to her book *Uncle Tom's Cabin; or Life Among the Lowly* (1852), on meeting the author Harriet Beecher Stowe in 1862: 'Is this the little woman who made this great war?' (Gordon-Reed 2011).

[37] Lear, *Witness for Nature*, 3.

[38] Lear, *Witness for Nature*, 3–4.

[39] Gaard, "Ecofeminism", 68.

> By the time of her death from breast cancer in 1964, at 56, she had set in motion a movement that produced Earth Day, the Environmental Protection Agency, a domestic ban on DDT and a transformation of how Americans see the world they inhabit.[40]

No less than four environmental acts passed through Congress, paving the way for the establishment of the Environmental Protection Agency in 1970. Quaratiello writes: "Carson introduced concepts relating to the environment and conservation that were virtually ignored by most people during the mid–twentieth century and brought terms such as 'interdependence' and 'the balance of nature' into common usage".[41] Her message was heard: "If we are ever to solve the basic problem of environmental contamination, we must begin to count the many hidden costs of what we are doing".[42] In 1971, the EPA banned the indiscriminate use of DDT. On its website, it claims:

> The publication in 1962 of Rachel Carson's *Silent Spring* stimulated widespread public concern over the dangers of improper pesticide use and the need for better pesticide controls ... based on its adverse environmental effects, such as those to wildlife, as well as its potential human health risks. Since then, studies have continued, and a relationship between DDT exposure and reproductive effects in humans is suspected, based on studies in animals. In addition, some animals exposed to DDT in studies developed liver tumors. As a result, today, DDT is classified as a probable human carcinogen by U.S. and international authorities.[43]

Her dream to translate her "vision into policy" was set in stone;[44] government policy, informed by her research, passion, and advocacy, is her legacy.

[40] William Souder "100 Women of the Year, 1963: Rachel Carson," *Time Magazine*, March 5, 2020 <https://time.com/5793611/rachel-carson-100-women-of-the-year/>.

[41] Arlene Quaratiello *Rachel Carson: a biography*, Greenwood Press: Westport USA (2004): xi.

[42] Ferrari, Rachel Carson.

[43] Environmental Protection Agency, *DDT*.

[44] Lear, *Rachel Carson*, 3.

Ecofeminism and Rachel Carson

As mentioned above, the term ecofeminism is attributed to the French activist and author Françoise d'Eaubonne in her book *Le Féminisme ou la Mort*.[45] So clearly, Carson was not aware of this branch of feminism; indeed, as most women of her generation, she denied any connection to the burgeoning 1960s' feminist movement. But the ecofeminist genealogy is more than evident in all her books and advocacy, particularly and more acutely, in *Silent Spring*. In her text, d'Eaubonne depicts Western patriarchal society causing irreversible environmental destruction and loss through domination, suppression, and exploitation. So, at a basic level, ecofeminism links endemic patriarchy to the destruction of the environment—that man's "right" to exploit and mistreat women parallels man's "right" to exploit and mistreat the environment for profit. Nightingale writes: "…ecofeminist thinking was premised on the idea that the domination of women was linked to environmental destruction and other problematic social inequalities such as racism".[46] This form of environmental feminism is mapped onto Carson's writing and advocacy often—retrofitted if you like. In 2016, Greta Gaard asks a question, arriving at the same conclusion: "Was Rachel Carson an ecofeminist? Technically, no—the term "ecofeminist" did not appear until a decade after her death—yet Carson's work exemplifies ecofeminist praxis (the inseparability of theory and practice)".[47] Sturgeon claims her "as an ecofeminist foremother".[48]

Ecofeminism grew out of a heightened awareness of women throughout the '60s and '70s—and I would argue throughout the centuries[49]—of an intersectionality in their lives and identities; an intersectionality including oppression. Gaard calls this "a deep recognition of inter-being that bridges socially constructed boundaries of class, race, species, sexuality, gender, age, ability, nation, and more… that brings both compassion and action to the task of alleviating conditions of eco-social injustice".[50]

[45] d'Eaubonne *Le Féminisme*

[46] Andrea Nightingale "The nature of gender: work, gender, and environment," *Environment and Planning D: Society and Space*, volume 24, (2006): 166.

[47] Gaard, *Ecofeminism*, 68.

[48] Noël Sturgeon "Ecofeminist Movements," *Ecology: Key Concepts in Critical Theory,* (ed. Carolyn Merchant) (Prometheus Books: Buffalo, 1994): 244.

[49] In 1892, Edith Ward critiques the text *Animal Rights* by Henry Salt, drawing attention to the similarities between mistreatment of women and animals.

[50] Gaard, *Ecofeminism*, 68.

Compassion here is an integral word in Gaard's work, its notion definitively found throughout Carson's—an ethical and empathetic compassion to each other and for the environment in which we move and live. And Carson's "action" or advocacy is indisputable, her praxis cogently ecofeminist.

Ecofeminists correlate their identification of the oppression of women throughout the centuries to the burgeoning destruction of nature, its creatures, and the environment—the literal oppression of nature. Birkeland claims "...ecofeminism offers the most comprehensive and incisive socio-political analysis to guide both self- and social transformation' linking personal and political power abuses to 'human oppression and environmental exploitation".[51] Hay writes simply: "... ecofeminism argues for an explanation from *androcentrism*: that the destruction of nature is consequent upon structures of exploitation embedded *within* human society. And the particular structure to which ecofeminist attention is directed is that based around gender".[52] Put more simply, both the historical oppression of women, and the exploitation of the environment for profit are results of a deeply entrenched societal patriarchy. Birkeland regards ecofeminism "as feminism taken to its logical conclusion, because it theorises the interrelations among self, societies, and nature"[53] Daly claims ecofeminists have four central tenets. They: "make visible the interconnections ... between the domination of women and the domination of nature", "seek to overcome the hierarchical dualisms and to subvert the logic of domination that supports them", "are radically communal and contextual", and "are grounded in praxis".[54]

Domination aspects of ecofeminism are central, and echo throughout Carson's life narrative—it is patriarchy that cogently suppressed women for centuries, and to variant degrees, still does in most cultures. And as such, patriarchal society in the mid-twentieth century privileged certain sciences over the natural world, exploiting for profit the environment as a triumph of human ingenuity. As a female scientist in the 1950s, Carson was one of few women in the field. According to Corbett, by 1938, women

[51] Janis Birkeland "Ecofeminism: Linking Theory and Practice," *Ecofeminism* (Greta Gaard ed.) (Temple University Press: Philadelphia, USA, 1993): 16.

[52] Peter Hay *ECOFEMINISM, A Companion to Environmental Thought*, (Edinburgh University Press: Edinburgh, Scotland, 2002):73.

[53] Birkeland, "Ecofeminism", 17–18.

[54] Lois K. Daly "Ecofeminisms and Ethics." *The Annual of the Society of Christian Ethics* 14, (1994): 285–6.

made up less than eight percent of "zoologists and naturalists working for the entire federal government", that the mid-twentieth century was "the nadir for female scientists, who effectively disappeared from American science by the 1950s".[55] When her research into the effects of synthetic pesticides on the environment disrupted the androcentric model of progress at any cost, she was ridiculed both for her science and her gender. But her vision of the interconnectedness of everything "informed her understanding of ecology and inspired new ways of thinking about natural systems and about the human impact on the world in which we live".[56] Her message was simple, strong, and, significantly, accessible to the mainstream. And the tone and heart of her arguments echo through Birkeland's definition, thirty years later. She writes:

> … the very essence of ecofeminism is its challenge to the presumed necessity of power relationships. It is about changing from a morality based on 'power over' to one based on reciprocity and responsibility ('power to'). Ecofeminists believe we cannot end the exploitation of nature without ending human oppression, and vice versa … values and action are inseparable: one cannot care without acting.[57]

Carson's deep concerns about not just DDT but other synthetic pesticides and their ability to not only eradicate biota but create health issues and cancers in humans, make both her message and her practice implicitly ecofeminist.

Silent Spring

When *Silent Spring* was serialised as long-form literary journalism in three issues of *The New Yorker* in June, 1962, three months before its publication as a book on September 27, it shattered a too readily accepted conception—that the indiscriminate use of pesticides on crops and people was good and harmless, that scientists knew exactly what they were doing, and that there were no risks attached to their use. It became an immediate bestseller. Stein writes:

[55] Julia Corbett "Women, Scientists, Agitators: Magazine Portrayal of Rachel Carson and Theo Colborn," *Journal of Communication,* December (2001): 36.

[56] Stein, "Rachel Carson", xv-xvi.

[57] Birkeland, "Ecofeminism", 19.

> The book was effective because it assembled scientific evidence and interpreted complex technical information so that it was accessible to a wide reading public. It resonated with mounting fears of the dangers of pollution, sparked by the knowledge of radioactive fallout, pesticide residues in food, and the tragedy of birth defects caused by thalidomide, a drug prescribed to pregnant women.[58]

What Carson's journalism actually did was question the risk-free aspect of the science around synthetic pesticide use. Walker interprets this as "uncertainty in the form of ignorance to destabilize the science and in the form of risk to actualize potential harm".[59] He claims that by treating uncertainty in this way, she invoked the public to listen. He writes: "the public injected their values and fears about the science into the public policy process" creating "the tactical construction of scientific ethos in public forums".[60] Her questions and research brought to light the imperfections found at the heart of "the inner circle of science and government, charged with protecting the public against health threats".[61]

Although science is critical in her text, it is perhaps the lyricism rendering it accessible that enabled its huge impact. Twidle claims that "the celebrated accessibility of her writing is in fact a carefully worked-for effect".[62]

Carson uses many literary techniques to present her arguments: parable, anecdote, figurative language, allusion, metaphor, poetic metre, foreshadowing, rhetorical question, and juxtaposition, to name a few, mingling with rigorously mined and accumulative statistics, reportage, facts, and figures, to engage the general public. She strongly believed in the public's right to know, now enshrined in journalism codes around the globe—its right to know, in order to make decisions about chemical safeties, rather than leaving it to the scientific chemical industry, which privileged the science and the American dollar above citizens' safety. And she enacted this belief with a literary journalism impulse, combining the science and facts with lyricism and rhetorical technique.

[58] Stein, "Rachel Carson", xiii.

[59] Walker, "Without Evidence," 104.

[60] Walker, "Without Evidence," 104.

[61] Corbett, "Women, Scientists, Agitators", 724.

[62] Hedley Twidle "Rachel Carson and the Perils of Simplicity: Reading Silent Spring from the Global South," *Ariel: a review of international English literature*, vol. 44 no. 4 (The Johns Hopkins University Press and the University of Calgary: Baltimore, 2014): 49.

In this paper, I focus on examples of literary journalism technique, then the poetic allusions Carson uses throughout her text which gives it fluidity and agility, lending it readability and widening its scope to attain a mass readership. Ecofeminism and art, and for the means of this paper, literary long-form technique and poetry or poetic affect, are often aligned as the mode in which women express analogous cruelties concerning themselves and nature, and plays a crucial component of growing ecology movements. Sandra Steingraber, writer, biologist, and cancer survivor, says that Carson consciously uses metrical technique "to create a sense of momentum or to stop the action … cinematic and dramatic … when you scan the lines you realise she is writing in Shakespearean iambic pentameter", claiming this demonstrates the "… power Carson has to get you entirely caught up in the biological stories she is weaving".[63] John Elder, in the same radio broadcast, says: "… she does have cadenced lyrical language and a strong command of metaphor".[64] He likens Carson to the English Romantic poets. He writes, of Keats and Wordsworth's nature poetry particularly: "For them, an affirmation of nature's wholeness and humanity's participation in it was inevitably connected to revolt against the oppressiveness of urbanization, class privilege and nationalistic warfare".[65] He says: "Wordsworth and Keats, and such successors as Frost shaped her thinking very strongly … she shared with their prophetic resistance to a mechanistic manipulative dominating approach both to the larger community of life and to the human community".[66] Like Carson, their works evoke rejection of dominant thoughts and actions—in Carson's text, the ecofeminist frame aligns with the Romantic poetry ethos, and is potent as she seemingly explodes on the page:

> Who has the right to decide for the countless legions of people who were not consulted that the supreme value is a world without insects, even though it also be a sterile world, ungraced by the curving wing of a bird in flight? The decision is that of the authoritarian temporarily entrusted with power;

[63] in *Rachel Carson and the Romantics*, radio broadcast, Radio National, February 9, 2013. <https://www.abc.net.au/radionational/programs/archived/poetica/2013-02-09/4467452>.

[64] in *Rachel Carson and the Romantics*.

[65] John Elder "Withered Sedge and Yellowed Wood" in *Courage for the Earth* (Peter Matthiessen ed.) (Houghton Mifflin: New York, 2007): 81.

[66] in *Rachel Carson and the Romantics*.

> he has made it during a moment of inattention by millions to whom beauty and the ordered world of nature still have a meaning that is deep and imperative.[67]

As Carson well knew, these decisions were made by men in business, science, and government. As Vance writes: "The literature and the history that purport to record the interactions of human consciousness with the nonhuman world are in fact records of male consciousness".[68] Carson's voice was one of disruption, and as most in the vanguard of a movement, the resultant attacks were overt and performed as an onslaught of ridicule of herself and her research, a meta demonstration of the poor treatment of women by men. Interesting to note here, Carson writes more as an ecologist in the field than a scientist in a laboratory, and using literary language and imbuing her rigorous research with emotion, she places herself in another vanguard—the use of narrative and personal voice within the sciences to explicate experience, a practice which is currently ubiquitous in the sciences and medicine, in particular in the twenty-first century. Elder says: "Disruptive is a good way to describe Carson's voice … she is saying we have to wake up we have to come to … because in fact we are living in a poisoned world".[69]

It took Carson six years to research and write *Silent Spring*. The minutiae and depth of her scientific evidence is staggering but here, I focus on the literary technique she uses to animate the science. Tabulated below are just some of these techniques, chapter by chapter, each chapter heading itself, suggestive and explicit (Table 1).

Rigorous scientific facts and examples, statistics, and figures nestle closely within this array of literary affect, elevated by the language around them; her technical toolkit similar to all exemplary literary journalists. In this next section, I focus on the literal, poetic influences of the text.

[67] Carson, *Silent*, 121.

[68] Linda Vance "Ecofeminism and the politics of reality," *Ecofeminism* (Greta Gaard ed.) (Temple University Press: Philadelphia, USA, 1993): 119.

[69] in *Rachel Carson and the Romantics.*

Table 1 Literary techniques

	Literary technique examples:
1. A Fable for Tomorrow	Parabolic images—an idyllic town "in the heart of America", full of beautiful trees, crawling with animals and insects, living side by side in harmony with humans. She writes: "Even in winter the roadsides were places of beauty, where countless birds came to feed on the berries and on the seed heads of the dried weeds rising above the snow" (21); nostalgia—the streams were full and clear: "So it had been from the days many years ago when the first settlers raised their houses, sank their wells, and built their barns" (21); growing tension, as in most parables; a lesson hard learnt (21); descriptive language and juxtaposition as "a strange blight crept over the area and everything began to change. Some evil spell had settled on the community ... everywhere was a shadow of death. In this town, not just the animals and birds and fish suffer and die, but the people, with mysterious illness, even among the children" (21); poetic allusion in the disappearance of the birds—"it was a spring without voices" (22); "epiphany – the villain of this horror is the people themselves; this is not a true story, but everything in it has happened somewhere in America" (22).
2. The Obligation to Endure	Personifies chemicals and conflict language to bolster her position: "The most alarming of all man's assaults upon the environment is the contamination of air, earth, rivers and seas with dangerous and even lethal materials" (23); "chemicals are the sinister and little-recognized partners of radiation in changing the very nature of the world"; "man's war against nature" (24).
3. Elixirs of Death	Ironic wordplay: elixir being a magical potion or cure-all; explains the impact of DDT on insects, working its way through the food chain to our tables, invading our bodies, causing chemical changes, illness and death; substitutes the word "poison" for the, at the time, harmless "chemicals" and "pesticides", debunking this so-called "safety".
4. Surface Waters and Underground Seas	Personifies chemicals; writes of residual chemicals in waterways and groundwater: "... these chemicals sometimes defy detection by methods in ordinary use by purification plants. Most of them are so stable that they cannot be broken down by ordinary processes. Often they cannot even be identified" (51); cites incidences and places where fish and other wildlife mysteriously simply died in numbers.

(*continued*)

Table 1 (continued)

	Literary technique examples:
5. Realms of the Soil	Brings soil to life, as part of the ecosystem housing integral ingredients for the food chain for example bacteria, fungi, algae, insects, earthworms etc: "The soil exists in a state of constant change, taking part in cycles that have no beginning and no end"; startling description of its genesis, involving volcanoes, fiery streams chiselling of frost and splitting and shattering of rock (61); uses personification "Chemical control of insects seems to have proceeded on the assumption that the soil would sustain any amount of insult via the introduction of poisons without striking back" (64).
6. Earth's Green Mantle	Discusses plant life as integral, with water and soil, to the existence of animal life, including humans; discusses the web of life in which "there are intimate and essential relations between plants and the earth, between plants and other plants, between plants and animals; states humans unthinkingly bludgeon the landscape" (69); metaphor: the natural landscape is "spread before us like the pages of an open book in which we can read why the land is what it is, and why we should preserve its integrity" (69–70); personifies sagebrush as an example, "the sage – low growing and shrubby – could hold its place on the mountain slopes and on the plains, and within its small grey leaves it could hold moisture enough to defy the thieving winds" (70).
7. Needless Havoc	Use of moral and binary questioning—who to believe: "on the one hand conservationists and many wildlife biologists assert that the losses have been severe and catastrophic. On the other hand the control agencies tend to deny flatly and categorically that such losses have occurred, or that they are of any importance if they have. Which view are we to accept?" (87); privileges witness testimony; conjures the evil protagonists or anti-heroes—"the man with the spray gun" and the "the control men in state and federal governments – and of course the chemical manufacturers" (87–88); evocative language describing the painful death from pesticide poisoning of birds, fish and insects; rhetorical question—"by acquiescing in an act that can cause such suffering to a living creature, who among us is not diminished as a human being?" (99).

(continued)

Table 1 (continued)

	Literary technique examples:
8. And No Birds Sing	Poetic allusion to Keats' ballad *La Belle Dame Sans Merci*:[a] "… spring now comes unheralded by the return of the birds, and the early mornings are strangely silent where once they were filled with the beauty of bird song… The sudden silencing of the song of birds, this obliteration of the color and beauty and interest they lend to our world have come about swiftly, insidiously, and unnoticed" (100); symbolism—her story of the robin (spring) and the eagle (the nation); rhetorical questioning: "Who has made the decision that sets in motion these chains of poisonings, this ever-widening wave of death that spreads out, like ripples when a pebble is dropped into a still pond? … Who has decided – who has the *right* to decide?" (121).
9. Rivers of Death	Vivid description; using the cycle of salmon as a storied example of her argument; invokes uncertainty; foreshadowing: "…a pattern that had made the Miramichi [river] one of the finest salmon streams in North America. But that year the pattern was to be broken" (122).
10. Indiscriminately from the Skies	Discusses symbolism of the skull and crossbones on chemicals, now "showered indiscriminately from the skies" (142) across the land; rhetorical question: "who is safe-guarding the consumer?" from the effects of poisonous chemicals in the food chain; use of conflict language: "the government's all-out war" (144).
11. Beyond the Dreams of the Borgias	Describing poison as if it has agency: "this spreading contamination" (158); invokes a villain at work: "Lulled by the soft sell and the hidden persuader"; conjures symbolism: "If a huge skull and crossbones were suspended above the insecticide department the customer might at least enter it with the respect normally accorded death-dealing materials" (158); conflict language: "death-dealing" (158); juxtaposition: "use of poison in the kitchen is made both attractive and easy" (159).
12. The Human Price	Personification of radiation: "Today we are concerned with a… hazard that lurks in our environment" (168); direct question to reader: "Can he escape a pollution that is now so thoroughly distributed throughout the world?" (169); ecofeminist musing: "this is a problem of ecology, of interrelationships, of interdependence" (169).
13. Through a Narrow Window	Analogy: "A living cell, like a flame, burns fuel to produce the energy on which life depends. The analogy is more poetic than precise … all these billions of gently burning little fires sparks the energy of life" (179); metaphor: "This process by which the cell functions as a chemical factory is one of the wonders of the living world" (179).

(*continued*)

Table 1 (continued)

	Literary technique examples:
14. One in Every Four	Conflict language: "the battle of living things against cancer" (191); analogous discussion: naturally occurring dangers "ultra-violet radiation causing malignancy" (193) to man-made "for man, alone of all life forms, can create cancer-producing substances…in medical terms called carcinogens" (193).
15. Nature Fights Back	Flashback: "I remember from student days the miracle that could be wrought in a jar …" (215); conflict language: "we have turned our artillery against our friends" (219); presumption of reader question: "…isn't this theoretical?" and answering: "it is happening, here and now" (219).
16. The Rumblings of an Avalanche	Invoking Darwin's theory of survival of the fittest: "the insect world would delight and astound him with its impressive verification of his theories" (227); hyperbole: "sometimes resistance develops so rapidly that the ink is scarcely dry on a report hailing successful control of a species with some specified chemical when an amended report has to be issued" (230).
17. The Other Road	Explicit poetic appropriation in Frost's *The Road Not Taken* (240);[b] metaphor: "as crude a weapon as the cave man's club, the chemical barrage has been hurled against the fabric of life" (256); conflict language.

[a]John Keats "La Belle Dame sans Merci: A Ballad," *Poetry Foundation*, (1819). https://www.poetryfoundation.org/poems/44475/la-belle-dame-sans-merci-a-ballad.

[b]Robert Frost "The Road Not Taken," *Poetry Foundation*, 1915. https://www.poetryfoundation.org/poems/44272/the-road-not-taken.

Poetry

Elder writes: "Poetry, for Carson, was essential to the process of looking deeply into the world and making brave choices amid the clamor of competing opinions and interests".[70] Indeed, when she received the US National Book Award in 1952 for her earlier text *The Sea Around Us*, she said: "If there is poetry in my book about the sea, it is not because I deliberately put it there, but because no one could write truthfully about the sea and leave out the poetry".[71] Poetic and lyrical language, for Carson, was her way to elucidate the beauty and power she perceived in nature.

[70] Elder, "Withered Sedge", 80.
[71] Ibid.

Carson derived the title of her text from the 1819 short ballad written by Keats called *La Belle Dame Sans Merci*. Keats repeats two lines in the first and final stanzas of his twelve-quatrain ballad and Carson uses these as one of three epigraphs at the beginning of her text: "the sedge is wither'd from the lake/And no bird sings".[72] An English major in her first years of college, well read from a young age, and a lover of poetry, it is not surprising she draws on the hopefulness of the Romantics in writing her alarming and cautionary text. As Elder says, Carson writes with a: "Vividly emotional side and highly documented scientific side, then offers possibility for hope. If we see the problem we can now change our behaviour and reverse at least much of the damage".[73]

In Chapter "Alive to the Slow Terror of the Megadams: The Literary Journalism of Jacques Leslie's *Deep Water*", Carson again draws on Keats in its title: *And no birds sing*. This repeated line of his ballad—'and no birds sing'—resonated with her. She invokes vividly a land covered in the fine dust of DDT, its entry into the trees, grasses, and crops, into insects and animals, and finally into humans—an injection of silent poison. And then in her final chapter, Carson appropriates Robert Frost's short but renowned poem *The Road Not Taken*.[74] In The Other Road, Chapter "The Big Picture and the Small Scene: Anna Tsing's Assemblages vs. Paul Engle's Workshop, and the Nature In Between", her plea becomes loud in her opening paragraph:

> We stand now where two roads diverge. But unlike the roads in Robert Frost's familiar poem, they are not equally fair. The road we have long been traveling is deceptively easy … The other fork of the road – 'the one less travelled' – offers our last, our only chance to reach a destination that assures the preservation of our earth.
>
> The choice, after all, is ours to make.[75]

Laying out the "senseless and frightening risks" within which the American people were mindlessly living, she implores: "We should no longer accept the counsel of those who tell us that we must fill our world with poisonous chemicals".[76] She then proceeds to explain myriad alternate

[72] Keats "La Belle Dame".
[73] in *Rachel Carson and the Romantics*.
[74] Frost "The Road".
[75] Carson, *Silent*, 240.
[76] Carson, *Silent*, 240.

paths—the other roads. And her final words: "It is our alarming misfortune that ... science has armed itself with the most modern and terrible weapons, and that in turning them against the insects it has also turned them against the earth".[77] This is her manifesto, underscored by an immense need to see her research transmogrify into policy. She could not know at the time of writing the huge impact her words would have; she perhaps gleaned a little at the 1963 US Senate Committee on Government Operations where she gave testimony, her words instigating the 1971 government ban of the indiscriminate use of DDT in America, which in its turn influenced other countries around the world.

Conclusion

Brooks tells us that at the time of Carson's research and writing, the word "ecology" was unknown to the general public and "environment" had few of its current implications.[78] This elevates the work of Rachel Carson to that of the English Romantics or "poet-prophets"[79]—she was literally a lone voice, and throughout six years of painstaking field research and drawing on her innate linguistic talents, she gifted the American people agency. Renner writes:

> Carson accomplished the feat of raising a public outcry against DDT not just with her research on its deleterious effects, but with the descriptive imagery, strong rhetoric, and poetic language that lift *Silent Spring* into the realm of other great works of American literature.[80]

The intersectionality implicit throughout her arguments positions them within the vanguard of the 1970s ecofeminist movement, challenging the unwitting acceptance of the almost entirely male conventional scientific voice, upheld by big business and government of the time. At the very least, her writing was to invoke caution, what she actually accomplished was to enable a public to begin to question convention and mute acceptance of partial and invested voices. Fifty years on, Stein writes:

[77] ibid., 257.

[78] Corbett, "Women", 724.

[79] Elder, Withered, 84.

[80] Rebecca Renner "Silent Spring is More than a Scientific Landmark: It's Literature, On the Underrated Poetry of Rachel Carson's Masterpiece," *Literary Hub*, April 20, 2018.

> Carson teaches us that the earth and the oceans are living entities, shaping and shaped by their interactions with the living creatures that inhabit them. Among these creatures, humans are having the greatest impact, and our alteration of the environment is turning out to be greater than Carson imagined possible.[81]

With the rigour of the scientist, the skill of a literary journalist, the eye of the poet, and the heart and passion of the advocate, championing her cause, Rachel Carson embodies the 1970s ecofeminists, her vision, alarm, and command of language and poetry extending across the next 60 years, ringing warning for us still. Bending her love of words and poetry into dialogue with science, her greatest desire was to be heard, and heard in time, and to effect policy. Her literary journalistic text sits timelessly at the cornerstone of ecological policy in the USA, and beyond, still pertinent today.

References

Birkeland, Janis, "Ecofeminism: Linking Theory and Practice," *Ecofeminism* (Greta Gaard ed.) Philadelphia, USA: Temple University Press, 1993.

Buckingham, Susan. "Ecofeminism in the twenty-first century," *The Geographical Journal,* Vol. 170, No. 2 (June 2004): 146–154, https://doi.org/10.1111/j.0016-7398.2004.00116.x.

Carson, Rachel *Silent Spring*, Boston: Houghton Mifflin, 1962 [1994].

Corbett, Julia, "Women, Scientists, Agitators: Magazine Portrayal of Rachel Carson and Theo Colborn," *Journal of Communication* (December 2001): 720–749, https://doi.org/10.1093/joc/51.4.720.

Daly, Lois K., "Ecofeminisms and Ethics." *The Annual of the Society of Christian Ethics,* 14 (1994): 285–90. Accessed September 17, 2020. http://www.jstor.org/stable/23559643.

d'Eaubonne, Françoise, *Le Féminisme ou la Mort*, Paris: P. Horay, 1974

Dowling, David, "Banned in Britain: Marilynne Robinson's Environmental Literary Journalism," *Literary Journalism Studies* Vol. 11, No. 2 (December 2019): 56–87, http://www.ialjs.org/

Elder, John, Withered Sedge and Yellowed Wood in *Courage for the Earth* (Peter Matthiessen ed.) New York: Houghton Mifflin, 2007.

Environmental Protection Agency 2017, *DDT – A Brief History and Status,* accessed 17 September 2020, <https://www.epa.gov/ingredients-used-pesticide-products/ddt-brief-history-and-status>

[81] Stein, Rachel Carson, xvi.

Ferrari, Michelle (director/writer), Rachel Carson, she set out to save a species…, Documentary, *American Experience* (January 18, 2019) Boston: PBS. https://www.pbs.org/wgbh/americanexperience/films/rachel-carson/#cast_and_crew.

Frost, Robert, "The Road Not Taken," *Poetry Foundation*, 1916, accessed October 28 at: https://www.poetryfoundation.org/poems/44272/the-road-not-taken.

Gaard, Greta, "Ecofeminism," *Keywords for Environmental Studies* (Joni Adamson, William A. Gleason & David N. Pellow eds.) New York: New York University Press, 2016.

Gates, Barbara, "A Root of Ecofeminism," *Ecoféminisme, ISLE, Interdisciplinary Studies in Literature and Environment*, Vol. 3, No. 1 (Summer 1996): 7–16, ISSN: 1076-0962

Gordon-Reed, Annette, "Uncle Tom's Cabin" and the Art of Persuasion: How Harriet Beecher Stowe helped precipitate the Civil War, *The New Yorker*, June 6, 2011. Accessed October 30 at <https://www.newyorker.com/magazine/2011/06/13/the-persuader-annette-gordon-reed>.

Gore, Al, "Rachel Carson and *Silent Spring*," in *Courage for the Earth* (Peter Matthiessen ed.) New York: Houghton Mifflin, 2007.

Hay, Peter, *ECOFEMINISM, A Companion to Environmental Thought*, Scotland: Edinburgh University Press, 2002.

Keats, John, "La Belle Dame sans Merci: A Ballad," *Poetry Foundation*, 1819. Accessed October 28 at: https://www.poetryfoundation.org/poems/44475/la-belle-dame-sans-merci-a-ballad.

Koehn, Nancy, "From Calm Leadership, Lasting Change," *The New York Times*, October 27, 2012. https://www.nytimes.com/2012/10/28/business/rachel-carsons-lessons-50-years-after-silent-spring.html

Lear, Linda, *Rachel Carson, Witness for Nature*, New York: Houghton Mifflin Harcourt, 1997.

Lear, Linda, *The Life and Legacy of Rachel Carson*, 2020. Accessed September 16, 2020 <https://www.rachelcarson.org/>.

Lepore, Jill, "The Right Way to Remember Rachel Carson," *The New Yorker*, March 26, 2018. https://www.newyorker.com/magazine/2018/03/26/the-right-way-to-remember-rachel-carson

Lockwood, Alex, "The Affective Legacy of *Silent Spring*," *Environmental Humanities*, Vol. 1 (2012): 123–140, https://doi.org/10.1215/22011919-3610003

Lytle, Mark H., "Rachel Carson: Saint or Sinner?" *Human Ecology Review*, Volume 23, Number 2 (2017): 55–64, https://doi.org/10.22459/HER.23.02.2017.06

Lytle, Mark H., *The Gentle Subversive, Rachel Carson, Silent Spring and the Rise of the Environmental Movement*, New York: Oxford University Press, 2007.

Nightingale, Andrea, "The nature of gender: work, gender, and environment," *Environment and Planning D: Society and Space*, volume 24 (2006): 165–185, https://doi.org/10.1068/d01k

Quaratiello, Arlene, *Rachel Carson: a biography*, Westport USA: Greenwood Press, 2004.

Rachel Carson and the Romantics, radio broadcast, Radio National, February 9, 2013. Accessed 29 October 2020 <https://www.abc.net.au/radionational/programs/archived/poetica/2013-02-09/4467452>.

Renner, Rebecca, "Silent Spring is More than a Scientific Landmark: It's Literature, On the Underrated Poetry of Rachel Carson's Masterpiece," *Literary Hub*, April 20, 2018. Accessed 29 October 2020 <https://lithub.com/silent-spring-is-more-than-a-scientific-landmark-its-literature/>.

Reports with Eric Sevareid, Documentary, *CBS* (April 3, 1963) New York.

Souder, William, "100 Women of the Year, 1963: Rachel Carson," *Time Magazine*, March 5, 2020. Accessed 16 September, 2020 <https://time.com/5793611/rachel-carson-100-women-of-the-year/>.

Stein, Karen, "Introduction," *Rachel Carson: challenging authors*, Rotterdam: Sense Publishers, 2012.

Sturgeon, Noël, "Ecofeminist Movements," *Ecology: Key Concepts in Critical Theory*, edited by Carolyn Merchant, 237–249. Buffalo: Prometheus Books, 1994.

Twidle, Hedley, "Rachel Carson and the Perils of Simplicity: Reading Silent Spring from the Global South," *Ariel: a review of international English literature*, vol. 44 no. 4 (2014): 49–88, https://doi.org/10.1353/ari.2013.0028

Vance, Linda, "Ecofeminism and the politics of reality," *Ecofeminism*, edited by Greta Gaard, Philadelphia: Temple University Press, 1993.

Walker, Kenny, "Without Evidence, there is No Answer": Uncertainty and Scientific Ethos in the Silent Spring[s] of Rachel Carson, *Environmental Humanities*, 2 (2013): 101–116, https://doi.org/10.1215/22011919-3610369

Webster, Bayard, "Rachel Carson – a scientist with a poet's wonder," *The New York Times*, May 25, 1982. Accessed 1 November, 2020. <https://www.nytimes.com/1982/05/25/science/rachel-carson-a-scientist-with-a-poet-s-vision.html>.

Alive to the Slow Terror of the Megadams: The Literary Journalism of Jacques Leslie's *Deep Water*

Callie Long

Introduction

U.S. literary journalist Jacques Leslie invites an ecocritical rethinking of the vast repositories called megadams in *Deep Water: The Epic Struggle over Dams, Displaced People, and the Environment*. Taking his time—a distinguishing feature of literary journalism—to draw attention to the affective lives of the people whose stories he tells, Leslie focuses on three megadams on three continents. Through human stories, he shows how, in his own words, "cultures [are] shattered" when communities are dispersed

C. Long (✉)
Hamilton, ON, Canada

P. Calvi (ed.), *The Literary Journalist as a Naturalist*, Palgrave Studies in Literary Journalism,
https://doi.org/10.1007/978-3-031-56634-9_8

and displaced by the massive projects and "the battle between dam builders and opponents [...] rejoined."[1]

The vast edifices of Leslie's text—India's Sardar Sarovar, Lesotho's Katse Dam, and Australia's Hume—plumb watery depths filled with latent power. All tend to the overwhelming in a social, political and environmental sense. All are superstructures that seem to swallow everything up in their path in the service of progress, tokens of "[one] of our most ambitious efforts to harness the power of nature [that exposes us] to nature's vagaries."[2] Within this context, Leslie's deep dive into megadams rests on his pivotal idea: "[To] see dams whole, and in doing so to glimpse the fate of the earth."[3] Hazarding a guess that the water-filled giants will one day be thought of as no more than "relics of the twentieth century, like Stalinism and gasoline-powered cars, [...] symbols of the [...] delusion that humans are exempt from nature's dominion, of greed and indifference to suffering,"[4] Leslie's unsettling narrative may well be read as apocalyptic, given his final summation in *Deep Water* that "all dams will die."[5] Nevertheless, while narrativizing megadam projects' impact locally, regionally and globally, the author also extends to his readers a respite to consider how a text such as *Deep Water* can "mobilise affect and private feelings to make possible new public cultures in responding to crisis" through a particular brand of nonfiction literariness that engages with the politics of affect.[6] Such an invitation brings into view the corrosive and asymmetrical power of multinational corporate forces underpinning megadam projects that threaten the livelihoods and prospects of the global poor.

Foregrounding the "accretive, [...] attritional, [...] exponential [and] incremental" forms of "slow violence" enacted against "unimagined communities" by megadam projects, Leslie brings to life Rob Nixon's

[1] Jacques Leslie, *Deep Water: The Epic Struggle Over Dams, Displaced People, and the Environment,* (Picador: New York, 2005a), 347; 8.

[2] Rapoza, Kenneth, "China Floods Call into Question Sustainability of Massive Three Gorges Dam," *Forbes* (July 20, 2020), <https://www.forbes.com/sites/kenrapoza/2020/07/20/china-floods-calls-into-question-sustainability-of-massive-three-gorges-dam/?sh=5ede08b34d50>.

[3] Leslie, *Deep Water,* 9.

[4] Leslie, *Deep Water,* 347.

[5] Leslie, *Deep Water,* 345.

[6] Alex Lockwood, "The Affective Legacy of *Silent Spring,*" *Environmental Humanities* 1, 2012, 124.

construct of slow violence.[7] Therefore, reading *Deep Water* through Nixon's lens is to gain insight into the idea that such violence happens "gradually and out of sight, a violence of delayed destruction that is dispersed across time and space."[8] Leslie illustrates why this kind of violence often remains obscured, overlooked, ignored and forgotten precisely because the non-spectacular forms of violence perpetrated upon the environment and its dwellers run counter to what violence is generally expected to be, namely stunning, explosive and above all, sensationally visible. By inviting his audience to consider megadams within the context of such insidiously corrosive conditions, Leslie's narrative restores the excised bodies—Nixon's "unimagined [or] surplus people [who are] deemed superfluous to the labor market and to the idea of national development"—rendering them real and visible for all to see.[9] Leslie quotes the U.S. anthropologist Thayer Scudder as saying that "the worst thing you can do to people other than kill them"[10] is to force resettlement on them, which invariably forms part of developing and building megadams, as happened with the "Gwembe Tonga people who'd lived for centuries along the Zambezi River in what is now Zambia."[11] Considering such development as a form of slow violence and the people whose lives are impacted by megadams, the conditions that cause such cultural disintegration cannot be ignored or brushed aside as simply the unfortunate but necessary cost of progress.

The Literary Journalist as Naturalist

Leslie is part of a growing cohort of literary journalists who have pointed out that for too long, "nature" has been used as a punchbag. Specifically, the long-form, evidence-based (and science-informed) narrative journalism of writers such as John McPhee, Elizabeth Kolbert[12] and the revolutionary, environmental *and* foundational work of Rachel Carson has spurred contemporary journalistic responses to concerns about our environment. Within this context, Leslie's narrative examines the claims that

[7] Nixon, Rob, *Slow Violence,* (Harvard University Press: Cambridge, MA and London, England. 2011), 2.

[8] Nixon, *Slow Violence,* 2.

[9] Nixon, *Slow Violence,*150.

[10] Leslie, *Deep Water,*108.

[11] Leslie, *Deep Water,* 4.

[12] Elizabeth Kolbert. *The Sixth Extinction.* (Henry Holt & Co.: New York, New York, 2014).

the benefits of megadams—the expansion of irrigated land, improving drinking water supply, controlling floods and producing hydropower—again and again, outweigh their cost yet are "generally justified with projects that are impossible."[13] Leslie juxtaposes these perennially forecasted paybacks against the assumption that these water-filled behemoths "are eternal, as if they'll never topple over or be dismantled or fill with sediment or lose their financial rationale," pointing out that megadams are no more than "loaded weapons aimed down rivers, pointed at ourselves; [...] proof of the gambling nature of the societies that build them."[14] Leslie's text delivers its most severe condemnation of megadams within the context of such lofty but mostly unachievable aims. Leslie proposes that anthropologist "[Thayer] Scudder's distress" arises not only from watching villagers of long acquaintance suffer and die" but from a more "fundamental" level, eroding "the assumptions on which [the anthropologist's] career in dams is based."[15] He also notes that Scudder would admire "a dam in plan," lend his "support to its construction as a consultant," only to grow "disillusioned," having to document "its failures of implementation and its social and environmental damage" time after time.[16] Nonetheless, "in the five years following the [World Commission on Dams', 2000] report," some "thirty-five very large dams (crests over 150 meters) and three hundred dams with crests over 60 meters started construction."[17] As Rhodante Ahlers, Margreet Zwarteveen and Karen Bakker note, "Not only has the moratorium on dam construction been lifted, but the dams built are larger than ever before," with governments mobilizing "a 'right to development' discourse to promote" these gargantuan water repositories, all the while buttressed by a "global call for alternative, non-fossil energy sources."[18]

As Leslie sets out to document the stories of three megadams and those of the three commissioners related to each project, he also declares that he hopes to discern the "fate of the earth."[19] His goal, while ambitious, does

[13] Rhodante Ahlers, Margreet Zwarteveen, and Karen Bakker, "Large Dam Development: From Trojan Horse to Pandora's Box," in *The Oxford Handbook of Megaproject Management*, ed. Bent Flyvbjerg, (Oxford University Press: Oxford, 2017), 1.

[14] Leslie, *Deep Water*, 345–346.

[15] Leslie, *Deep Water*, 108.

[16] Leslie, *Deep Water*, 108.

[17] Ahlers et al., "Large Dam Development," 2.

[18] Ahlers et al., "Large Dam Development," 2.

[19] Leslie, *Deep Water*, 9.

not seem so farfetched, given our world buckling (and bucking) recently under the strain of a global viral pandemic and runaway climate change—all increasingly non-abstract and impacting the lives of communities and cultures worldwide. A perfect storm arises given how inflexible megadams are when we live in a time when "climate change catastrophizes the weather," as Namwali Serpell points out in *The New York Times Magazine*.[20] Changing weather patterns must then concretize any environmental abstraction that a literary journalist (as a nascent naturalist) may think to dally with.[21] It is, therefore, a timely reminder that unable to narrow such dams "enough to eke more force from less water during droughts [and unable to expand them] to accommodate floods," these engineering Goliaths, such as the Kariba Dam, for instance, force one of two solutions, especially when waters rage: "let the water flow over the top of the dam or […] open up a spillway for controlled release."[22] Should the dam fail, either solution could spell doom for the region downstream, including another megadam further downriver.

Reading *Deep Water,* I am reminded of the dams of my childhood—magical touchpoints that still dot the parched and dust-dry farming landscapes of my youthful memories. Growing up in a semi-arid region of South Africa with its red soil, these dams, often no more than large, relatively shallow, muddy pools, presented no threat I was aware of at the time. Instead, they served as a source of life for the white-held, apartheid-era farms they irrigated. Megadams were the stuff of fiction and the future. My only memory and idea of a dam so big that its wall contained elevators came from a line drawing of the Boulder Dam in a chapter in *America* from the *Wonderful Worlds of Walt Disney* series.[23] For a rural child in the 1970s, such a massive edifice was entirely beyond comprehension, yet the image fired up my imagination in any number of ways. If the word soothing had a synonymic image, I would have imagined it to be a vast stretch of fresh, open water, held at bay by a massive wall, yet still benign and

[20] Namwali Serpell, "Learning from the Kariba Dam," *The New York Times*: July 22, 2020, sec. *Magazine*. <https://www.nytimes.com/interactive/2020/07/22/magazine/zambia-kariba-dam.html?referringSource=articleShare>

[21] Serpell, "Learning from the Kariba Dam," (July 22, 2020). <https://www.nytimes.com/interactive/2020/07/22/magazine/zambia-kariba-dam.html>

[22] Serpell, "Learning from the Kariba Dam," (July 22, 2020). <https://www.nytimes.com/interactive/2020/07/22/magazine/zambia-kariba-dam.html>

[23] Walt Disney Productions. "The Grand Canyon," in *America,* (House of Grolier/Star House/Potters Bar/Herts: England, 1965), 227.

life-giving; nothing of the cynical-sounding utterance that "[a] rock-filled dam is just an ugly pile of rock that you put a concrete slab on to make it watertight.[24] Furthermore, nowhere in this juvenile mental landscape could I have imagined that a dam could be likened to a "very long, very heavy axe head with the blade burying itself in the valley floor" with all the violent symbolism that must accompany such a simile.[25]

If we are fortunate, we grow up relatively unscathed, learn new things, and expand our ways of thinking, including having long-held beliefs shattered. In my case, the stunted, skewed, and saccharin Disney-fied impressions of big and bold dams, to gain insight into how the concept of nature has been conscripted to pay service to racism, sexism, colonialism and essentialism, for instance.[26] By the time *Deep Water* was published in 2005, I had some inkling of the political machinations and power struggles that accompany the construction of these behemoths, having been introduced to *Encounters with the Archdruid: Narratives About a Conservationist and Three of His Natural Enemies* by John McPhee[27]—a literary, journalistic journey through three wildernesses: a coastal island, a Western mountain range, and the Colorado River in the Grand Canyon, as well as the pitting of the U.S. environmentalist, David Brower, against three of his ideological enemies. As a young print and radio reporter covering environmental stories for the mainstream media in the late 1980s and early 1990s, McPhee's 1971 text stayed with me in part because I had heard environmentalists and journalists sound the alarm about water outstripping oil as the earth's most sought after and precious commodity. Such a dire prediction has long been attributed to past World Bank staffer Ismail Serageldin, whose words that while "the wars of this century were fought over oil, the wars of the next century will be fought over water" are still cited as often as they are contested.[28] Although Leslie refers to Serageldin in *Deep Water*, he does not quote this prophecy. He does not have to. Leslie's prescient narrative, while alarming, is not alarmist in any sense of the word. The

[24] Leslie, *Deep Water*, 139.

[25] Leslie, *Deep Water*, 138.

[26] Stacey Alaimo, *Exposed: Environmental Politics & Pleasures in Posthuman Times.* (University of Minnesota Press: Minneapolis, MN, 2017), 11.

[27] John McPhee, *Encounters with the Archdruid: Narratives About a Conservationist and Three of His Natural Enemies.* (Farrar, Straus and Giroux: Canada and the United States of America, 1971).

[28] Ismail Serageldin, "Water Wars? A Talk with Ismail Serageldin." *World Policy Journal*, 2009/10, 25.

literary journalist reminds the reader that imagination and knowledge combined (in this case of the contexts within which megadams come into being) provide some measure of resistance against market-centred state corporatism that objectifies the environment wholesale, treating it as an entirely expendable commodity, creating fertile ground within which the systemic and insidious form of Nixon's slow violence proliferates.

Creating an Affective Threshold

While Leslie's original intention was to focus on the story of the World Commission on Dams, he writes that by shifting from "one commission history to three commissioner narratives," he "felt freed" to "turn a startling but opaque statistic—that at least 40 to 80 million people had been displaced by dams, usually with disastrous consequences—into stories with flesh and bones."[29] The result of this reframing is that Leslie, in *Deep Water,* selects three members of the independent World Commission on Dams to tell the story of megadams, following the framework adopted by the organizers of the commission that "sorted nominees [...] into three categories—pro-dam, mixed and antidam."[30] Leslie writes that he was compelled to adopt a similar approach: choosing "as subjects one commissioner from each group."[31] One of the stories is that of U.S. anthropologist Thayer Scudder, whose field of study is the impact of megadams on the people of Southern Africa. Scudder's narrative shows that even as megadams can signal hope, they all too often end up as a blight on the landscape, engendering despair for those who consider their locations home. Other protagonists in *Deep Water* include the Indian activist Medha Patkar, who opposes the Sardar Sarovar that threatens to displace thousands of people in Western India, and an Australian water manager, Don Blackmore, whose aim is to reverse the effects of drought in such a way as to allow Australia to follow in the footsteps of California's earlier so-called successes when it comes to tackling the scarcity of water.

By telling the "flesh and bones"[32] stories of his protagonists, Leslie also carves out a critical space for the megadams and the rivers that feed them

[29] Jacques Leslie, "Using Narrative to Tell Stories About Water." *Nieman Reports*, 3. <https://niemanreports.org/articles/using-narrative-to-tell-stories-about-water/>

[30] Leslie, *Deep Water*, 3.

[31] Leslie, *Deep Water*, 3.

[32] Leslie, *Deep Water*, 3.

and, in a sense, speaks to his audience as the edifices acquire their unique descriptive traits and personas: the monumental, "voluminous" Sardar Sarovar of India, the "graceful arch" of the Katse Dam of Lesotho and the Australian Hume that suggests "a Botoxed and face-lifted dowager," among others.[33] In Leslie's narrative, the megadams and their environs always loom large, disrupting a traditional "anthropocentric framework that renders the natural world always only in relation to the human."[34] In this way, Leslie takes the singular topic of the massive constructions, megadams, and threads his characters' lives through the debate that fuels much of the controversy surrounding the behemoths. While simultaneously focused on the human element of the story, but never at the expense of the dams and rivers that feed them, Leslie shows how irrevocably interconnected humans, non-humans, and the environment are, inviting his readers to consider whether such dams are indicators of progress, destroyers of lives or something more nuanced within the context of pressing contemporary climate concerns.

Leslie's literary journalism, which revolves around the fate of megadams within the context of the earth itself, is significant. However, in tackling the stories of the commissioners' work as a means to make sense of the dams themselves, his paradigmatic journalism takes the reader on an immediate and intimate, temporal and spatial journey that focuses on the massive dams, these "modern pyramids" of our time that form the very "core of every water debate."[35] Thus, megadams are the problem to which he applies his journalistic model of puzzle-solving to grapple with the "collisions of the burgeoning human population and the planet's unchanging supply of freshwater [that] has already started and will grow worse."[36] Combining scientific research with creative prose, the text thus engages with more than simply a reader's dread of a world precariously perched on the edge of an abyss.[37] Instead, Leslie's narrative generates an affective response—more than emotions or feelings—characterized as the space within which the body's spontaneous capacity plays out over a "qualitative

[33] Leslie, *Deep Water*, 14; 128; 229.

[34] Robert Alexander, "Literary Journalism as Ecocriticism," in *The Routledge Companion to American Literary Journalism*, eds. William E. Dow and Roberta S. Maguire. (Routledge: New York, 2020), 482.

[35] Leslie, "Using Narrative to Tell Stories About Water." *Nieman Reports*, 2. <https://niemanreports.org/articles/using-narrative-to-tell-stories-about-water/>

[36] Leslie, *Deep Water*, 8.

[37] Lockwood, "The Affective Legacy of *Silent Spring*," 124.

threshold."[38] Such are the "forces of encounter," as Gregory Seigworth and Melissa Gregg argue, given how affect builds up in our "capacities to act and be acted upon."[39] As Lockwood notes, Carson showed us how "capitalist atomisation of the natural world was (and is) avoidable, and that the anger felt about that destruction could (and can) be channelled into political action, away from the flow of dull, empty drifting that goes nowhere."[40] To think through Leslie's text as inviting affect, as opposed to emotions—nameable and stable as a concept—is to "inventively" embark on a journey mapped as a "threat [...] from the future [...] what might come next."[41] Therefore, to gain insight into the shocking displacement and wholesale erasure of communities, scattered "like the chicks of a bird"[42]—in the words of an elderly woman from the highlands of Lesotho—it is necessary to follow Leslie into a narrative space attuned to the affective "resonances that circulate about, between, and sometimes stick to bodies and worlds" as Seigworth and Gregg argue.[43] In this way, Leslie invites his readers to go against the flow, to examine, along with him, the subject of the megadam, and to fully experience the forces of encounter brought to life by his narrative.

Meticulously weaving the voices of the three commissioners and other voices that would under most circumstances never make the news into and through his narrative, what surfaces is the notion that there is "an author at work," as Norman Sims puts it.[44] In a similar vein to literary journalist John McPhee's triadic structure in *Encounters with the Archdruid*, Leslie threads the three commissioners' stories in such a way that his distinct voice comes to the fore, given his recognition that his attraction to this sub-genre of journalism is that it represents the "antithesis of long-distance reporting."[45] Inserting himself deftly into the story, Leslie recounts how

[38] Brian Massumi, *Parables for the Virtual: Affect, Movement, Sensation*. (Duke University Press: Durham and London, 2002), 84.

[39] Gregory Seigworth, and Melissa Gregg, "An Inventory of Shimmers," in *The Affect Theory Reader*, eds. Melissa Gregg and Gregory Seigworth, Duke University Press, 2010), 2; 1.

[40] Lockwood, "The Affective Legacy of *Silent Spring*, 138.

[41] Brian Massumi, "The Future Birth of the Affective Fact," in *The Affect Theory Reader*, eds. Melissa Gregg and Gregory Seigworth, (Duke University Press, 2010), 53.

[42] Leslie, *Deep Water*, 132.

[43] Seigworth and Gregg, "An Inventory of Shimmers," 2.

[44] Norman Sims, *The Literary Journalists*. (New York: Ballantine Books, 1984), 2.

[45] Leslie, "Using Narrative to Tell Stories About Water," 2.

walking on the wall of the Katse Reservoir[46] feels as though he is engaged in something illicit, "like strolling across the deck of an aircraft carrier, windswept, feeling something alive beneath my feet."[47] As much as the "violence resides beneath the deck" of an aircraft carrier, "evidence of the dam's violence is hidden underwater.[48] In moments such as this one, Leslie illustrates not only the scale of the dams concerning the humans who find a home within the shadows of these dams but that, high or low, the masses of water contained by megadams pose a threat to all species. By invoking the force of the water as violent, Leslie draws attention not so much to the "scale of the modern edifice[s]" as to show his readers "the scale of the very modern forced removals" and other deprivations the dams incur, much as Nixon does in describing the writer and activist Arundhati Roy's advocacy around megadams.[49]

Leslie is also an author with a specific point of view—a view from somewhere—as opposed to the more traditional thinking that would consider such a view as no more than a skewing of an objective stance or a "view from nowhere" as Canadian journalist and author Candice Callison describes objectivity in discussing the limits and possibilities of journalism.[50] Callison's point is that objectivity, a cornerstone of journalism, is often "interpreted [...] as a way of not acknowledging [one's] social location as a journalist."[51] Instead, Leslie's literary journalism powerfully draws on a combination of the everyday stories "of our neighbours" and news, usually "from far-off centers of power that hardly touched our lives," yet invites the reader into the lives of these individuals and communities with significant effect.[52] Filled with "feelings and dramatic moments," this is a narrative that encompasses not only "a special power because we know the stories are true," but an intimate and balanced approach that vivifies his subject matter as an author connected with his

[46] In *Deep Water,* Jacques Leslie writes that in the early 2000s, Kariba is Africa's tallest dam, and one of two constructed in the tiny landlocked southern African kingdom of Lesotho; the other is Mohale dam.

[47] Leslie, *Deep Water*, 138.

[48] Leslie, *Deep Water*, 138.

[49] Nixon, *Slow* Violence, 161.

[50] Candace Callison, "Objectivity is 'the View from Nowhere' and Potentially Harmful." Interview by Anthony Germain. *The Sunday Edition*, Canadian Broadcasting Corporation, (July 10, 2020), Audio, 27:31.

[51] Callison, "Objectivity is 'the View from Nowhere,'" (July 10, 2020).

[52] Sims, *The Literary Journalists,* 1–2.

subjects, and not afraid to let them speak.[53] The result is that Leslie does not stand separate from his narrative. The journalist is part of his story, engendering a profound sense of interconnectedness and relationality between the material, the historical, the symbolic, and critically, the "felt" lives of those whose stories he tells.[54]

Leslie takes his readers with him as he traverses the planet in search of a dam that would qualify as a good one—an idea both contained in the title of a chapter and referenced when he describes Scudders' anthropological work related to megadams and resettlement.[55] Foregrounding the stories of the people whose lives are entangled with the vast stretches of dammed water and applying the fundamental principles of literary journalism: fact and truth-telling presented as a writer would fiction, including dialogue and scene construction to frame the story, the journalist illustrates how "knowledge [can be gained] through personal experience."[56] Instead of exclusively relying on the rational and logical—the basis for scientific modes of inquiry—Leslie's immersive approach to journalism demonstrates profoundly how such a style can be generative when writing about contemporary political and economic corporatism. Carefully threading the needle between ecology and geopolitics and the journalist as literary writer and naturalist allows Leslie to address the intricate and complex conditions that make megadam stories multi-valenced, multidimensional and multifaceted. Leslie's ecocritical style of journalism elaborates "a range of theoretical challenges to anthropocentric thinking," as Robert Alexander points out, quoting Pippa Marland's recent reworking of the definition of ecocriticism, which points to the "destructive agency she [assigns to] 'the human.'"[57] One of Leslie's protagonists, Medha Patkar, takes on the Indian Sardar Sarovar, which the journalist describes as "a wall, a monument, a modern ziggurat" that will "shatter" the "tribal cultures" of the people, who will be forced to resettle on "infertile lowland plots," or "end up on city streets."[58] Describing Patkar as "wearing plaid in a polka-dot

[53] Sims, *The Literary Journalists,* 1–2.

[54] Mark Kramer, "Breakable Rules for Literary Journalists," in *Literary Journalism: A New Collection of the Best American Nonfiction*, eds. Norman Sims and Mark Kramer, (Ballentine, 1995), 22.

[55] Leslie, *Deep Water*, 105.

[56] L. Wilkins, "Humankind, Nature and the New Journalism: A Return to the Mythopoeic." *ECQUID NOVI*, 10:1–2, 1989, 179.

[57] Alexander, "Literary Journalism as Ecocriticism," 483.

[58] Leslie, *Deep Water*, 14.

universe," Leslie paints a picture of an activist who embodies integrity.[59] Constantly under attack from the Indian government, Patkar used her "own suffering [through fasts and jail time] to advance [her] cause"—that the Sardar Sarovar is not a good dam by any stretch of the imagination, nor is any "ill-considered dam [...] and for that matter, all rapacious development, which to her was just another phrase for globalization."[60] By foregrounding Patkar's larger-than-life convictions to underscore his respect for her beliefs and knowledge, Leslie's approach to writing about the complex debate around megadams translates into a narrative steeped in affect that plays out across what can only be described as an unhurried kind of journalism. Unlike more traditional forms of journalism, literary journalism thus offers the space to place "equal emphasis [...] on what is known [as well as] how that knowledge came to be acquired."[61] Arising from the "intellectual tradition of Romanticism," literary journalism and its use of "literary techniques" usually limited to fictional accounts, offers journalists the opportunity to show how "subjective perception" is not the archenemy "of the omniscient observer" and thus craft a narrative space that considers humans as indivisible from nature.[62]

Leslie places the factual story into a more profound cultural context throughout his narrative. In this way, he does what Wilkins describes as connecting the facts of the story and events with the cultural facts of symbols and myth.[63] Leslie writes, "[of] the roughly thirty thousand people in the Katse and Mohale basins, about a tenth would be resettled."[64] The rest of the communities who call the mountains home are left without their farms and "grazing land enriched by the alluvial soil near the rivers, forcing them farther up the barren slopes" where the small, round, perfectly thatched homes "seemed to lose their snap, as the struggle for survival intensified."[65] The same can be said for the communities Leslie focuses on in India and, to some extent, Australia—neglected, discounted and dismissed, their stories, too, demand an affective response. For the tribal communities, whose lives and livelihoods are at stake, India's Sardar Sarovar—one of the world's largest dams—represents what Leslie refers to

59 Leslie, *Deep Water*, 21.
60 Leslie, *Deep Water*, 22; 25.
61 Wilkens, "Humankind, Nature and the New Journalism," 179.
62 Wilkens, "Humankind, Nature and the New Journalism," 179.
63 Wilkens, "Humankind, Nature and the New Journalism," 181.
64 Leslie, *Deep Water*, 132.
65 Leslie, *Deep Water*, 132.

as a "wailing wall."[66] The massive dam is the archnemeses of Patkar, "the world's foremost antidam activist, an Indian firebrand whose protests against dams repeatedly involved courting her own death."[67] However, it is not only Medha's voice that is heard. Leslie also includes the often-unheard voices of villagers who are deeply affected by the dam, such as one man from Maharashtra in India, who recounts that "the first time his home was submerged, he moved to higher ground," only to have to move again when "that place, too, was inundated," and finally discovering "that in government records my house wasn't even affected by the dam."[68] In Australia, Leslie casts his focus on the Murray River, its "infinitely complex" rehabilitation, and its weirs and dams, including that of Hume Dam, which he describes as a "blunt instrument" and the dam of "last resort," Dartmouth Dam.[69] Australia, he writes, presents a paradox. The country is one of extremes, with its "harsh, unforgiving environment and a terrain so deficient in water and nutrients that its plants and animals developed radical, hyperefficient strategies to survive."[70] These scenes let Leslie's narrative flow like the waters he writes about, constantly invoking the imagination and always narratively chipping away at "the barrier of public indifference," as Carson noted.[71] In this way, Leslie does what literary journalists can and must do: explore the factual in tandem with our affective responses, instead of always being forced to choose between the two. As such, Leslie's skill in engendering affect through his prose is a powerful means to invite the reader to understand that the arrogation of ecological reserves asks of the human species to consider acting in different ways if we are to prevent disaster and persevere.

In a "world of science, where knowledge is hierarchically organized, fact [...] privileged over anecdote, and observation over narrative," Leslie's lyrical approach to his craft lifts the stories of all impacted by megadams, including humans, non-humans, *and* the environment to a level that is "alive to the ways knowledge is both enabled and constrained by these positions of power."[72] Indeed, the journalist's narrative "works to take

[66] Leslie, *Deep Water*, 14.

[67] Leslie, *Deep Water*, 6.

[68] Leslie, *Deep Water*, 102.

[69] Leslie, *Deep Water*, 229–30.

[70] Leslie, *Deep Water*, 249.

[71] Lockwood, "The Affective Legacy of *Silent Spring*," 124.

[72] Lindsay Morton, "Conceptualizing an Ecological Approach to Ethical Literary Journalism," in *The Routledge Companion to American Literary Journalism*, eds. William E. Dow and Roberta S. Maguire, (Routledge: New York, 2010), 361.

'subjectivity, cognitive agency, and geographical-material-historical-cultural location seriously into account.'"[73] Such is the power *and* the ethics of Leslie's ecocritical journalism as he untangles the complex and messy debate surrounding megadams.

Deep Water is an intensely immersive kind of literary journalism that plays out over space and time, uses the stylistic persuasive power of telling stories in sets of threes, rises to the challenge of tackling the complex history and debate about megadams, and offers a restorative, prolonged plunge into the very deep and troubled waters of the megadam debate. The outcome of Leslie's style of journalism is suffused with "profound implications for the types of reporting journalists attempt as well as the results of that reporting," as Lee Wilkins argues.[74] Leslie's journalism demonstrates this point with its emphasis on taking time to "find things out [and notice] stories that others [may] miss, [communicating] it all to the highest standard," entangling our notions of "scale and time," as Alexander notes in referring to Susan Greenberg's definition of "slow journalism."[75] Such a notion of meticulous accounting of the finest detail is also underscored by Nixon, who "explores the toxic mismatch between the frenzied pace of the contemporary news cycle and the creeping destructive force" brought on by increasing climate-related events and lethal fallout of wars, as Alexander notes.[76]

As geopolitics roil, Leslie's text thus gets around the perennial problem of the news machine's insatiable 24/7 appetite for the spectacular, "event focused, time-bound, and body bound" information, always asking for more, by doing that which Nixon suggests—complicating and troubling "conventional assumptions about violence as a highly visible act that is newsworthy."[77] We need only think of a massive dam bursting to understand that this spectacular context is the antithesis of Leslie's slow journalism, which converts "into [...] narrative the disasters that are slow moving and long in the making" and that results in what Nixon refers to as the "long dyings" of those impacted by ecological disasters, such as the ones

[73] Morton, "Conceptualizing an Ecological Approach to Ethical Literary Journalism," 362, quoting Lorraine Code.

[74] Wilkens, "Humankind, Nature and the New Journalism," 179.

[75] Alexander, "Literary Journalism as Ecocriticism," 484.

[76] Alexander, "Literary Journalism as Ecocriticism,"484.

[77] Nixon. *Slow Violence*, 3.

brought on by megadams.[78] Indeed, as Nixon urges, when the spectacular is all and "when public policy is shaped primarily around perceived immediate need," *Deep Water* illustrates the cumulative wholesale threats megadams pose to the ecosystems and the people who rely on them when viewed as the fodder that feeds the mainstream media.[79] Reading *Deep Water* in tandem with *Slow Violence* is to gain an elevated understanding of what is at stake—not only for those who dwell along the planned paths of such edifices, the environment, ***and*** all humans and non-humans whose home is this planet.

On its own merits, the takeaways from Leslie's literary journalism are stark and telling. Following what he refers to as "the imperatives of narrative nonfiction [that] carried me like a current to the book's last words."[80] Leslie's narrative, through its story-telling qualities, provides a deep sense of how to come to grips with the logic of the ambitions of megadams as progress, stacked up against what the journalist Serpell describes as representing no more than "human folly" in the *New York Times Magazine* (2020). For precisely this reason, reading the text through a lens of "slow violence" is to gain insight into what Nixon means by the notion of "staggered" and staggering ecological "long dyings" within the context of race, poverty, anger, activism, helplessness and courage.[81] Critically, Nixon considers his concept of slow violence and its elusive qualities within the context of representation, especially in creating "stories, images, and symbols" that rise to the task of communicating the insidiousness of the "delayed effects" of such forms of violence.[82]

Nixon, who foregrounds the storytellers or writer activists, such as Arundhati Roy, Ken Saro-Wiwa and Abdulrahman Munif, as their "narrative imaginings [...] may offer us a different kind of witnessing: of sights unseen."[83] Leslie does something similar by choosing to use the three commissioners' stories to frame his narrative about megadams in a way that is "alive to the inhabited impact of corrosive transnational forces [including the megadam industry, which] jeopardizes the livelihoods, prospects, and memory banks of the global poor" and make them

[78] Nixon. *Slow Violence*, 3.
[79] Nixon. *Slow Violence*, 2–3.
[80] Leslie, "Using Narrative to Tell Stories About Water," 5.
[81] Nixon. *Slow Violence*, 1–3.
[82] Nixon. *Slow Violence*, 3.
[83] Nixon. *Slow Violence*, 3; 15.

invisible.[84] These are the bodies of people who are perpetually and constantly racialized and erased, which Leslie, as a literary journalist, puts front and centre in his text for all to acknowledge.

Published in 2005, *Deep Water* can thus be read as a portent of our radically changing times in which the environment plays such a leading role yet is still used as a punching bag by humanity. Leslie describes the massive repositories of water as the "largest structures built by humans and the sites of so many different sorts of drama, where development's tentacles reach into remote valleys and upend the existing cultural and environmental regimes."[85] The journalist also reminds the reader that "[water] is one the great looming subjects of the twenty-first century," and megadams, the massive and costly edifices that generate electricity and provide water for irrigation at the cost of the environment, are indicators of social disaster.[86] In turn, some scholars liken the megadams of the last century to "Trojan Horses, [...] important embodiments of political and ideological spatial strategies," whereas the "large dams of the twenty-first century" can be thought of as "Pandora's Boxes due to a proliferation of private and quasi-private actors involved in their development."[87] What characterizes these projects are often the "conjoined processes of imaginative expulsion and forced removal": conditions that Leslie's narrative brings vividly to life as a rebuttal to the "spatial amnesia" that follows when "communities, under the banner of development, are physically unsettled and imaginatively removed, evacuated from place and time." In this way, people are "uncoupled from the idea of both a national future and a national memory" and thus become the Nixon-like "unimagined," untethered from existence.[88,89]

Three iconic megadams around the world—Kariba, of which Leslie writes so lyrically, the Three Gorges and the Grand Ethiopian Renaissance—have featured prominently in headlines relatively recently at a time when news cycles seem primarily concerned with the damaging social, political and economic fallout brought on by the far-reaching COVID-19 pandemic. Floods in China are putting the Three Gorges Dam's sustainability

[84] Nixon. *Slow Violence*, 5.
[85] Leslie, *Deep Water*, 8–9.
[86] Leslie, *Deep Water*, 8.
[87] Ahlers et al., "Large Dam Development," 1–2.
[88] Nixon. *Slow Violence*, 51.
[89] Nixon. *Slow Violence*, 2.

in question.[90] Serpell writes that "climate change and neglect" have forced Southern Africa's Kariba Dam on the border of Zambia and Zimbabwe "to the brink of calamity," imposing a "crisis prefigured in the dam's troubling colonial history."[91] Moreover, the Grand Ethiopian Renaissance Dam threatens fragile political relationships between Ethiopia, Egypt and Sudan—an already volatile part of the world—with on-again/off-again talks between the countries over water flows in the Blue Nile.[92] Nevertheless, despite all evidence to the contrary, dams are still being planned and built.

Leslie invited his readers in 2005 to imagine the dams of *Deep Water* and others "in five hundred or a thousand years after their useful life has ended [...] when every one of them suffers from an altered hydrological regime as a result of climate change."[93] If tourism still exists in this imagined future, Leslie writes that while the "people who lived around the dams will have been dispersed [...] and their cultures shattered," the tourists will continue to visit these relics.[94] The megadams of our time may still be around, but only as "reminders of an ancient time when humans believed they could vanquish nature and found themselves vanquished instead."[95] Leslie's text has all the hallmarks of the prophetic, nearly two decades after it was first published. His is a narrative steeped in affect, creating a space that engages with the "threat as a *felt quality*," a "nagging potential of [...] a still worse next again after that."[96] Except, as Leslie's narrative illustrates, the worst has already happened for millions of people who have suffered the brutal and devastating consequences of Nixon's slow violence.

[90] Rapoza, "China Floods," 2020.

[91] Serpell, "Learning from the Kariba Dam," (July 22, 2020). <https://www.nytimes.com/interactive/2020/07/22/magazine/zambia-kariba-dam.html>

[92] Reuters, "Egypt Agrees with Ethiopia, Sudan on the Priority of Reaching a Dam Binding Deal," (July 21, 2020). < https://www.reuters.com/article/us-ethiopia-dam-egypt-presidency/egypt-agrees-with-ethiopia-sudan-on-the-priority-of-reaching-a-dam-binding-deal-idUSKCN24M2QB>

[93] Leslie, *Deep Water*, 347.

[94] Leslie, *Deep Water*, 347.

[95] Leslie, *Deep Water*, 347.

[96] Brian Massumi, "The Future Birth of the Affective Fact," 61; 53.

References

Ahlers, Rhodante, Margreet Zwarteveen, and Karen Bakker. "Large Dam Development: From Trojan Horse to Pandora's Box." In *The Oxford Handbook of Megaproject Management*. Edited by Bent Flyvbjerg, Oxford: Oxford University Press, 2017.

Alaimo, Stacey. *Exposed: Environmental Politics & Pleasures in Posthuman Times*. Minneapolis, MN: University of Minnesota Press, 2017.

Alexander, Robert. "Literary Journalism as Ecocriticism" In *The Routledge Companion to American Literary Journalism*, 482–497. Edited by William E. Dow and Roberta S. Maguire. New York: Routledge, 2020.

Callison, Candace. "Objectivity is 'the View from Nowhere' and Potentially Harmful." Interview by Anthony Germain. *The Sunday Edition*, Canadian Broadcasting Corporation (CBC), July 10, 2020. Audio, 27:31.

Kramer, Mark. "Breakable Rules for Literary Journalists." In *Literary Journalism: A New Collection of the Best American Nonfiction*, 21–34. Edited by Norman Sims and Mark Kramer, Ballentine, 1995.

Leslie, Jacques. *Deep Water: The Epic Struggle Over Dams, Displaced People, and the Environment*. New York: Picador, 2005a.

Leslie, Jacques. "Using Narrative to Tell Stories About Water." *Nieman Reports*, 2005b.

Lockwood, Alex. "The Affective Legacy of Silent Spring." *Environmental Humanities* 1, (2012), 123–140.

Massumi, Brian. *Parables for the Virtual: Affect, Movement, Sensation*. Durham and London: Duke University Press, 2002.

Massumi, Brian. "The Future Birth of the Affective Fact," in *The Affect Theory Reader*. Edited by Melissa Gregg and Gregory Seigworth, 52–70. Duke University Press, 2010.

McPhee, John. *Encounters with the Archdruid: Narratives About a Conservationist and Three of His Natural Enemies*. Canada and the United States of America: Farrar, Straus and Giroux, 1971.

Morton, Lindsay. "Conceptualizing an Ecological Approach to Ethical Literary Journalism." In *The Routledge Companion to American Literary Journalism*, edited by William E. Dow and Roberta S. Maguire, 360–372. New York: Routledge, 2010.

Nixon, Rob. *Slow Violence and the Environmentalism of the Poor*. Cambridge, MA and London, England: Harvard University Press. 2011.

Rapoza, Kenneth. "China Floods Call into Question Sustainability of Massive Three Gorges Dam." Forbes, June 20, 2020.

Reuters. 2020. "Egypt Agrees with Ethiopia, Sudan on the Priority of Reaching a Dam Binding Deal." sec. *Environment*. July 21, 2020.

Seigworth, Gregory, and Melissa Gregg. "An Inventory of Shimmers," in *The Affect Theory Reader*, edited by Melissa Gregg and Gregory Seigworth, 1–25. Duke University Press, 2010.

Serageldin, Ismail. "Water Wars? A Talk with Ismail Serageldin." *World Policy Journal*, 26(4), 2009/10.

Sims, Norman. *The Literary Journalists*. New York: Ballantine Books, 1984.

Serpell, Namwali. "Learning from the Kariba Dam." *The New York Times*: July 22, 2020, sec. *Magazine*.

Walt Disney Productions. "The Grand Canyon," in *America*, England, House of Grolier/Star House/Potters Bar/Herts, 1965.

Wilkins, L. "Humankind, Nature and the New Journalism: A Return to the Mythopoeic." *ECQUID NOVI*, 10 (no. 1–2), 1989. https://doi.org/10.1080/02560054.1989.9653016

Revisiting Places and People: How Immersion and Cohesion Are Created in Danish Digital Long-Form Journalism That Deals with Climate Change

Steffen Moestrup

In 2004, the Danish newspaper *Jyllands-Posten* made a series of long-form investigative reports on environmental changes in places such as Tuvalu, Delhi and Copenhagen.[1] In 2019, the newspaper repeated the reporting by revisiting the same places, spanning five different continents.[2] The new series of long-form digital productions which was named "The Climate Tour 2019" was produced by the same journalists as the 2004 series, but the 2019 series made use of new digital storytelling tools. In this chapter,

[1] Parts of the 2004 edition can still be accessed here: https://jyllands-posten.dk/international/ECE11296711/genoplev-klimaekspeditionen-kloden-og-klimaet-fra-2004/

[2] The 2019 version can be accessed here: https://jyllands-posten.dk/international/ECE11296705/tag-med-paa-klimarejsen-2019/

S. Moestrup (✉)
Danish School of Media and Journalism, Aarhus, Denmark
e-mail: sdm@dmjx.dk

P. Calvi (ed.), *The Literary Journalist as a Naturalist*, Palgrave Studies in Literary Journalism,
https://doi.org/10.1007/978-3-031-56634-9_9

I will analyze both the 2004 series and the 2019 series, focusing on the latter since the aim is to demonstrate the literary strategy of *revisiting* using digital elements, which allow for a more cohesive and immersive experience.

Theoretical Framework and Methodology: Modes and Materialities

The theoretical framework of this chapter is rooted in a media aesthetic approach. Building on the work of Liv Hausken,[3] I subscribe to an understanding of the media aesthetic approach as explorative and anti-essentialist. It offers a conceptualization of media that focuses on the uses of a medium and finds media materialities based on use rather than expectations of general, universal characteristics.[4] This is underlined by an anti-essentialist understanding of media as opposed to an essentialist understanding which would strive "toward defining one, single essence of a particular medium and a keystone, and, with them, the promise of a unified theory."[5] The media aesthetic approach is particularly interested in the medium's importance for how something appears and makes sense in context. Media materiality is not determined in advance but is rather examined in relation to the focus of the analysis.[6] This means that the main idea of a media aesthetic approach is not to investigate what a medium *is* but rather what it *does* in practice.[7] The materiality of a particular medium should thus be defined as the characteristics of that particular medium with respect to its particular context and use.

As an interdisciplinary research area, the media aesthetic approach leaves room for other disciplines to build a stronger analytical departure point. In this chapter, I combine the media aesthetic approach with elements from multimodal research due to the multimedia focus of the analysis. Multimodal research supports the explorative nature of the media

[3] Hausken, Liv. 2009. *Medieestetikk. Studier i estetisk medieanalyse.* Oslo: Spartacus - Scandinavian Academic Press. Hausken, Liv. 2013. *Thinking Media Aesthetics. Media studies, film studies and the arts.* Frankfurt: Peter Land Academic Publishing.

[4] Hausken, "Medieestetikk," 9.

[5] Hausken, "Thinking Media Aesthetics," 36.

[6] Hausken, 36.

[7] Hausken, "Medieestetikk," 10.

aesthetic approach and focuses on the interplay between different modes of communication.[8] A mode is defined by Kress as "a socially shaped and culturally given semiotic resource for making meaning. *Image, writing, layout, music, gesture, speech, moving image, soundtrack and 3D objects* are examples of modes used in representation and communication."[9] The idea is to examine the mode in itself and consider the context of the mode and its surrounding modes, because "different modes offer different potentials for making meaning. These differing potentials have a fundamental effect on the choice(s) of mode in specific instances of communication."[10] Again, we should notice how specificity is a fundamental idea in multimodal research, in line with the anti-essentialist notion of the media aesthetic approach. To address the idea of specificity, the analyst should determine the focus of the analysis in advance. In the present analysis, I am particularly interested in how different modes of communication using a range of media materialities express *revisiting* as a narrative strategy and how the modes work in combination to create a cohesive, immersive multimedia product.

The analytical focus is substantiated by recent scholarship on digital long-form journalism, which supports the fundamental characteristics of immersion and cohesiveness. Jacobson et al. claim that the genre of digital long-form journalism is a "driving force behind a new period of literary journalism"[11] and point to the interplay between literary and multimedia techniques to create stories that are immersive, dramatic and captivating. Dowling and Vogan use the term "cognitive container"[12] to characterize the way digital long-form journalism tries to capture the reader's attention in order to improve immersion in the story. Instead of opening up the story and allowing readers to freely explore on their own by, for instance, clicking on links to external websites, the authors create a cognitive

[8] See, for example, Kress, Gunther. 2010. *Multimodality – a social semiotic approach to contemporary communication*. London: Routledge and Knox, John. 2009. *Multimodal Discourse on Online Newspaper Home Pages: A Social-semiotic Perspective*. PhD thesis. Sydney: University of Sydney.

[9] Kress, "Multimodality," 79.

[10] Kress, 79

[11] Jacobson, Susan, Jacqueline Marino, and Robert E Gutsche jr. 2015. "The digital animation of literary journalism." *Journalism*, 17(4), 528. https://doi.org/10.1177/1464884914568079.

[12] Dowling, David, and Travis Vogan. 2015. "Can We 'Snowfall' This? Digital Longform and the Race for the Tablet Market." *Digital Journalism* 3 (2): 211. https://doi.org/10.1080/21670811.2014.930250.

container, an "enclosed environment"[13] to focus "[on] the space of the story world and its embedded multimedia elements without being scattered onto the open web via hyperlinks."[14] This cohesive, enclosed and distraction-free environment attempts to unify the different modes of communication with the purpose of creating a more immersive kind of digital long-form journalism.

The experience of immersion is also the central concept in van Krieken's analysis of the digital long-form urtext *Snow Fall*, in which narrative techniques such as scene reconstruction, event structures and viewpoint techniques are addressed to build an argument, pointing to multimedia usage "intensifying the experience of immersion."[15] Hiippala's case study of 12 long-form digital articles published in 2012–2013 concludes that the use of transitions and organizing content in linear structures "attempts to captivate [audiences] by creating a distraction-free environment."[16]

By building the present analysis on two cases spanning 15 years of digital development, this paper aims to investigate how and to what extent the cognitive container strategy manifests itself in digital storytelling, and how it is different from storytelling strategies developed in the earlier days of the convergent digital media landscape.

Two Cases and a Methodological Challenge

Before introducing the two journalistic projects that we will analyze below, we should first offer an introduction to the Danish media system and *Jyllands-Posten*, the newspaper which published them both.

The Danish media system can be placed within what Hallin and Mancini call the Democratic Corporatist Model,[17] or the Nordic Welfare Model.[18] Both models describe media systems defined by a politically independent

[13] Dowling, David. 2019. *Immersive longform storytelling: media, technology, audience.* New York: Routledge. 129.

[14] Dowling, 148.

[15] Van Krieken, Kobie. 2017. "Multimedia Storytelling in Journalism: Exploring Narrative Techniques in Snow Fall." *Information*, 2018, 9, 13. 1.

[16] Hiippala, Tuomo. 2017. "The Multimodality of Digital Longform Journalism." *Digital Journalism* 5 (4): 420–442. https://doi.org/10.1080/21670811.2016.1169197. 2

[17] Hallin, Daniel. C, and Paolo Mancini. 2004. *Comparing Media Systems.* Cambridge: Cambridge University Press.

[18] Syvertsen, Trine, Gunn Enli, Ole J. Mjøs, and Hallvard Moe. 2014. *The Media Welfare State: Nordic Media in the Digital Era.* Ann Arbor: University of Michigan Press.

press, despite several media being partly state-funded. Access to information about current affairs is considered a public good, and there is a rather normative focus on the media's democratic role[19] and its ability to inform citizens.[20] As in other countries, the press in Denmark is struggling with decreasing circulation, but it is still considered an important pillar of democracy.[21] The national daily newspaper *Jyllands-Posten* belongs to Denmark's largest legacy print organization, JP Politikens Hus, which publishes three national newspapers (*Jyllands-Posten* with 347,000 weekly readers, *Ekstra Bladet* with 267,000 weekly readers and *Politiken* with 501,000 weekly readers) and runs several publishing houses and a large online bookstore.

Both the 2004 and the 2019 versions of the Climate Tour were launched by *Jyllands-Posten* as prestigious media projects with high value and were followed by costly advertisements in both print and television media. Recent scholarship has argued that the highly competitive news market has led to legacy news media trying new approaches to reach audiences, including the production of in-depth researched high-quality storytelling.[22] The publication of the two climate tours can be read as an example of this trend. The case was selected for three main reasons. Firstly, the case spans a 15-year-long period, which makes it suitable for investigating how digital journalism has evolved and changed over a larger time span. Secondly, the case makes distinct use of one narrative strategy, namely, that of revisiting, which seems like a fruitful choice considering the topic of climate change. Bødker and Morris,[23] drawing on Adams,[24] have emphasized how the different temporalities of climate change (larger

[19] Blach-Ørsten, Mark, and Ida Willig. 2013. "Print og Netaviser i Danmark, USA og Frankrig." *Journalistica*, nr. 1., 2013. 198.

[20] Kammer, Aske. 2017. "Market Structure and Innovation Policies in Denmark." In *Innovation Policies in the European News Media Industry – a comparative study*, edited by Han Van Kranenburg. Springer. 37.

[21] Kammer, Aske. 2019. "Denmark." In *The SAGE International Encyclopedia of Mass Media and Society*, edited by Debra L. Merskin. London: SAGE. 2.

[22] Van Krieken, "Multimedia storytelling," 1. Also see Shim, Hoon. 2014. "Narrative journalism in the contemporary newsroom: The rise of new paradigm in news format." *Narrative Inquiry*, 24 (1) and Neveu, Erik. 2014. "Revisiting narrative journalism as one of the futures of journalism." *Journalism Studies*, 15, 533–542.

[23] Bødker, Henrik, and Hanna E. Moris. 2021. *Climate Change and Journalism. Negotiating Rifts of Time*. London: Routledge.

[24] Adams, Barbara. 1998. *Timescapes of Modernity. The Environment and Invisible Hazards*. London: Routledge.

time span, temporal complexity, distanced/abstract) and journalism (shorter time span, deadline-driven, immediacy, specific/concrete) often create an incompatible relationship. The strategy of revisiting can be seen as a narrative attempt to counter this incompatibility. The narrative strategy of revisiting is naturally only carried out in the 2019 version, which is the main reason why the analytical focus is on the 2019 version. Thirdly, the Climate Tour production can be read as an example of media potentiality when it comes to raising awareness and engaging users in societal issues such as climate change.

An important note on the methodological difficulty of assessing the 2004 version: in Denmark, access to archived versions of web sites is given by the Royal Danish Library. However, the oldest archived websites date back to June 2005, which is more than a year after the online publication of the 2004 Climate Tour. The methodological difficulties and challenges when studying prior versions of websites has been discussed by Brügger[25] and Schneider and Foot,[26] for example, and it is beyond the scope of this chapter to develop the discussion further. However, one important notion from the discussion is that the idea of going back to an original version of a website in the past is more or less impossible since the archived version of the website is likely to be different than the original website. The solution in the present analysis was to make use of republished articles from the 2004 version, which the newspaper republished when the 2019 version was launched. From communication with the newspaper, it was made clear that the newspaper has made an effort to create the republished versions of articles from the 2004 Climate Tour with the same content as the originals. This is by no means an optimal solution, but it is the best available.

[25] Brügger, Niels. 2008. "The Archived Website and Website Philology A New Type of Historical Document?" *Nordicom Review*, 29(2), January 2008, https://doi.org/10.1515/nor-2017-0183

[26] Schneider, Steven, and Kirsten Foot. 2004. "The Web as an Object of Study." *New Media & Society* 6: 114–122.

Analysis of 2004 Version: Text-Driven, Adventurous, and Personal

Jyllands-Posten pitches the 2004 production as a climate expedition. They use a concept brand called JP Explorer which had been used in previous years to label explorative travel reporting, such as the article series "500 days around the world" (1998–1999), and the Galathea 3 expedition in 2006–2007, where researchers, students and journalists sailed around the globe carrying out scientific tests along the way, which the journalists then reported. The explorer brand is regularly used in journalistic projects that have an adventurous departure point. The Climate Tour 2004 project is presented as the "biggest investment since the JP Explorer trip around the world." The opening article begins with the following lines:

> *On a daily basis and from around the globe, record breaking temperatures, extreme rain showers, fierce storms and the consequences for millions of people are reported. But it is also beyond doubt that the planet's climate stirs up debates and strong emotions.*

The opening lines establish the project as a fact-finding mission based on visiting people that are directly affected by climate change. The opening uses dramatic language, which is quite common in climate reporting,[27] and it underlines that the climate is a topic involving many emotions and much discussion. Hence, it becomes clear that the Climate Tour 2004 mission is to get closer to the truth in this morass of opinions and emotions. The articles that introduce the Climate Tour 2004 mission present the different locations that the expedition will visit and the stories they are going to tell, including the main points of each article. Compared to the page that introduced the Climate Tour in 2019 which will be analyzed below, the 2004 reporting is much more text-heavy and makes less use of suspense and withheld information. Instead, the 2004 page sums up the points from the different articles and makes the mission clear: journalists want to speak to people living in areas affected by climate change. The Climate Tour 2004 consists of a journalistic website with several articles, an educational site targeting school children, and a television show made in collaboration with the national Danish TV station, DR. Due to the

[27] Höijer, Birgitta. 2010. "Emotional Anchoring and Objectification in the Media Reporting on Climate Change." *Public Understanding of Science* 19, no. 6 (November 2010): 717–31. https://doi.org/10.1177/0963662509348863.

scope of this chapter and the idea of comparing digital storytelling across time, I will only focus on the digital journalistic site.

The main page is subdivided into two main parts that move chronologically forward the further down the page one scrolls. The articles from the series are on the left-hand side. On the right-hand side, we find a series of elements called "logbog" which is a travel log with more personal reflections about the journey. The travel log is visually presented on the main page with a similar layout to an old treasure map, underlining the adventurous spirit of the project. The logbook—which is also an element used in the 2019 version—works as a more personal, diary-like component that allows reporters to reflect on the expedition in a more personal manner. There are eight logbook entries and eight articles so we can speak of an evenly balanced mix between the journalistic pieces and the logbook pieces. The logbook consists of small anecdotes from the expedition that often describe obstacles and how the team overcame them. For instance, in the entry "Wrong turn in the desert" (published April 27, 2004), the reporters describe how they missed an important sign on the road between Windhoek, Namibia and Maun, Botswana and instead of driving northeast ending up driving south, resulting in a detour of 400 kilometers. In the entry "Playing pool with the chief," we hear the story of how the Explorer team ended up having a game of pool with the local chief in Ji Parana, Brazil. These small, observations of life on the road add to an impression of familiarity between the reporters and the media users. We get to know them a little better, and they can present everyday life experiences within a journalistic report that deals with a very serious topic.

The media materiality of the main page layout supports a linear consumption of the media product. The user is free to click on any article but the articles are ranked in chronological order, starting with the departure article and ending with the home-coming article, which guides the user to click on the articles in the same order as the journey took place. The layout of the articles, which can be clicked on the left-hand side of the overview screen, are in the same style, signaling the same mode of communication. A photo that has the same width as the subheading text is on the top of an article's individual page. All of the photos are clickable and when clicked, enlarge to a full screen version. There are usually between one and three photos in the articles. The main impression when scanning the articles is that text is the main component. This is further emphasized by the size of the images that are mostly the same rather small width of the text block on the website. Only if the user clicks on the images do they become enlarged to a full screen view. This weakens the feeling of immersion and cohesiveness while giving the impression of reading a traditional newspaper, without taking advantage of digital opportunities.

Analysis of the 2019 Version: Humanizing Climate Change by Using Revisiting as Literary Strategy

As in the 2004 version, the 2019 version also uses the JP Explorer brand. It is placed on the top of the landing page alongside a potpourri of blue-toned photos from the 2019 climate expedition displaying some of the stories in the multimedia production. Whereas the 2004 version used text to sum up some of the stops on the tour, the 2019 version uses photos to signal some of the locations that will be visited along the way. Unlike the 2004 text, the 2019 photos do not contain any information about the content of the stories, but only work as visual appetizers.

Four small icons at the bottom of the photos signal four different types of user engagement: Facebook (where the user can share the story), Twitter (where the user makes an automated tweet about the story), LinkedIn (where an automated post is created) and an email-icon which generates an email containing automated text where the user invites an email contact to click on a link that leads to the production. All of the icons point to sharing functions, which work as a commercial tool that is likely to generate more clicks when the material is shared on various social media platforms. The multiple sharing options are in line with recent scholarship on sharing news on social media that points to the function of a sharing button as a way for the user to commit himself to the story and become part of the branding of the media text. The users signal to friends and SoMe connections that they approve of the story and can also use sharing in their personal SoMe-branding.[28]

Below the four sharing icons, we see the text "Start the Journey" and a moving arrow pointing downwards, which leads the user in the correct direction. Generally, the production uses a number of navigational

[28] Lee, Chei, Sian, and Long Ma. 2012. News sharing in social media: The effect of gratifications and prior experience. *Computers in Human Behavior*, 28 (2). 331–339. Retrieved from https://doi.org/10.1016/j.chb.2011.10.002. Also see boyd, danah, Scott Golder, and Gilad Lotan. 2010. "Tweet, tweet, retweet: Conversational aspects of retweeting on Twitter." *HICSS '10 Proceedings of the 2010 43rd Hawaii International Conference on System Sciences* (pp. 1–10). Honolulu, HI. Retrieved from https://doi.org/10.1109/HICSS.2010.412

guidance tools in order to lead the user forward in the story and make sure, or rather make more likely, that the user will keep reading. The text called "Start the journey" addresses the user in a personal way, signaling that he or she is invited on the trip and just needs to decide when it is time to immerse himself or herself in the trip.

Once the user starts scrolling downwards, the images change into one photo showing a Land Rover moving toward the reader's point of view in a muddy landscape. Inside the Land Rover, we see a female driver (perhaps a stringer) and one of the Jyllands-Posten reporters. This is the first time we encounter the reporter visually. On several occasions in the 2019 version, we see the reporters and photographer on the photos; this works both as visual proof of the fact that the reporters are really on site and not just doing desk journalism, and it builds proximity between the sender of the media product and the receivers.

Embarking on an adventure, the introductory text that is centered on the photo of the Land Rover reads:

> *15 years ago JP Explorer travelled around the world and visited people on locations that were already affected by climate change.*

The reader scrolls ahead, and a graphic map appears, showcasing the 2019 route. Above the map, we see the text "Now we do it again" and a bit further down, it states: "What has happened to the climate and the people? JP Explorer tries to find answers."

Whereas the 2004 introductory text was based on the idea of getting to the hard facts and visiting people directly affected by climate change, the 2019 introduction situates the new tour as a follow-up to the 2004 version. Here we witness the explicit narrative strategy of revisiting. The 2019 version is presented as a continuation of the 2004 version and as revisiting the places and people that where originally explored 15 years ago. Instead of working from scratch, the revisiting strategy is used to highlight the strength of building upon existing material. This is done by repeatedly using the revisiting strategy throughout the production and by, for instance, using the idea of microlevel comparison, which is only possible because the reporters visited the same places and people 15 years previously. One example of microlevel comparison can be seen in the two photos (Figs. 1 and 2):

Fig. 1 Here the user has used the slider function to scroll downwards on the photo and made the 2019 photo of Talo Mila Laiva take up most of the space, while the 2004 photo is underneath

Fig. 2 In this image, the user has used the slider function to scroll upwards and made the 2004 photo of the same person at the same location appear while the 2019 photo is underneath

The two photos work as the backbone of an article portraying the Tuvalu inhabitant Talo Mila Laiva. The reporters visited her on the 2004 expedition and in the 2019 article. The report reads as follows:

> *15 years ago, the JP Explorer team met 11-year-old Talo Mila Laiva as she was climbing up a palm tree on the seashore. Since then, the palm tree has been taken by the waves and the coast has moved two meters closer to her house which is now situated only three meters from the edge of the roiling Pacific Ocean.*

The text elaborates on the change of which we have visual proof in the photograph. Visual proof thus becomes a primary media materiality in the photo. Next to each other, with a vertical scrolling transition between them, the 2004 photo and the 2019 one become entwined to an extent in which they create a cohesive and immersive multimedia element. The two photos are taken from the same point of view and use the same perspective, which creates an accessible way of comparing as well as linking the two of them together, into almost one single image, underlining the feeling of cohesion. It is by scrolling that the user gets one image to disappear and another to appear. The user can scroll between the photos, which contributes to an immersive experience.

The use of revisiting as a digital narrative strategy also takes place at levels other than micro. The use of the slider in the following instance illustrates this (Fig. 3):

A slider is a web element that allows the user to screen between two images stacked in the same frame. When moving the slider, one image appears and the other one disappears. In the 2019 version, the slider is used between two images that are created using data from various sources such as Google Earth and satellite photos from Landsat and Copernicus. One of the images shows the amount of rainforest in an area in 1984 while the other one shows the same area in 2016. The contrast between the two and, in particular, the substantial deforestation, is clearly visualized thanks to the slider. The media materiality of the slider is a specific and illustrative clarification of climate change in a delimited area.

However, the Climate Tour 2019 mostly contains comparisons at a micro level, by addressing the life of one individual and changes in her/his immediate surroundings. By doing so, the climate change discussion becomes less abstract and gains a human face. It takes on a more universal,

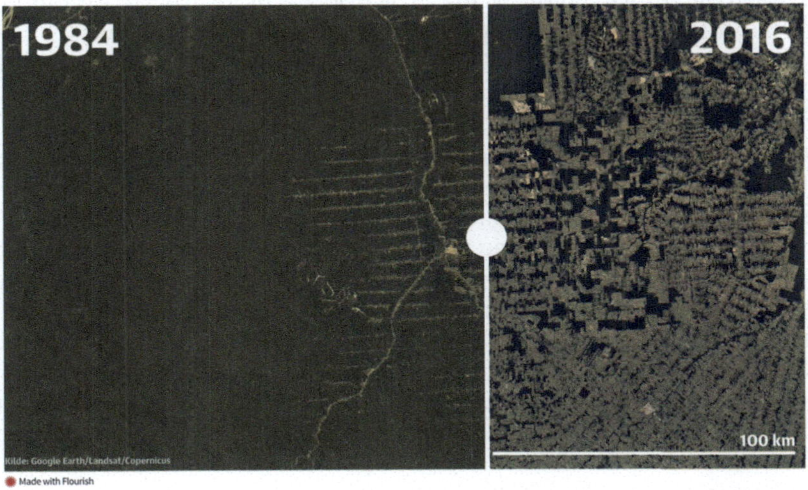

Fig. 3 By moving the slider, the user can see deforestation of the same area in 1984 and in 2016

compassionate tone, making it easier for a wider public to relate to. The human aspect is also underlined by the way the reporters textually use the notion of reunion in several articles. For instance, when they meet with Kwinnie Vila on Tuvalu, the caption under one of her photos reads:

> *Kwinnie Vila is looking out of a window with dough on her hands, exactly as she did when the JP Explorer team met her 15 years ago. Only now, her house is white and her hair a bit more gray.*

The media materiality of the image caption is descriptive but in a way that emphasizes comparison. We witness that the reporters not only report on climate change but also the changes that people universally undergo as time goes by. Hair has turned gray, for instance. This is a way of humanizing the sources and building on the compassionate notion of reunion: meeting each other again and building on an existing relationship. Humanizing the sources is a way to create a more emotional connection between the reader and the source. The reader is more likely to care about the lives of the sources when we get to know them in familiar ways.[29]

The notion of 'getting to know' is also used when it comes to the reporters and the photographer. As in the 2004 version, in the 2019 version there's also an item labeled "Log Book." In the logbook, we once again read more personal, diary-like entries from the reporters and the photographer bringing them and their journey closer to the reader. On several occasions, we hear about their obstacles along the way. For instance, in a short video, they examine their car from which large pieces of plastic have fallen. Another blog entry deals with the photographer being verbally assaulted and chased in the streets of New Delhi, India. The most important difference between the 2004 logbook and the 2019 version is that the latter is more visual than the former. The logbook is clearly demarcated from the other elements in the series, underlining that this is a special part of the production and should not be interpreted in similar ways to more traditional reporting. This separation is likely chosen to maintain an objective and factual approach. The producers' goal seems to deliver unbiased reporting, and not be accused of activism.[30]

In general, the 2019 articles all display the same structure, layout, and recurring elements which support a sense of cohesion. Usually, the 2019 articles either have a top photo that spreads across the entire width of the screen (the 2004 articles only had images only as wide the text) or a looping video, as visual access to the stories. Then the body of the text follows, subdivided into smaller sections by photos, graphics, maps and looping videos. Most of the digital elements can be enlarged by pressing on the item with the mouse. At the end of each article we find a clickable graphic showing the route of the expedition which underlines the immersive notion of traveling with the reporters. After clicking on the graphics, the

[29] Varma, Anita. 2020. "Evoking Empathy or Enacting Solidarity with Marginalized Communities? A Case Study of Journalistic Humanizing Techniques in the *San Francisco Homeless Project*" *Journalism Studies*, https://doi.org/10.1080/1461670X.2020.1789495. Also see: Wahl-Jorgensen, Karin, and Mervi Pantti. 2021. "Introduction: The emotional turn in journalism." *Journalism*, 22(5), pp. 1147–1154.

[30] The notion of doing objective, non-activist climate reporting was also underlined by the newspaper's editor-in-chief, Jacob Nybroe, when he was interviewed about the project. For instance, he argued that the newspaper had decided not to carbon-offset the flights taken on the expedition because he felt this could be interpreted as an activist stance in the climate change discussion. The interview can be read here: https://jyllands-posten.dk/premium/indblik/Indland/ECE11234658/farvel-lille-danmark-goddag-til-verden-og-et-stort-klimaregnskab/

user is redirected to an overview page that showcases the different articles in the series.

Two digital elements should be further addressed: the use of looping videos and zoom-in effects on the photos.

Looping videos are used extensively throughout the production. Once the user scrolls down to the section of the site that contains the video, it automatically starts to play and continues to play in a loop until the user scrolls further on. The media materiality of the looping videos provides presence and dynamics. They allow the user to step further into the environment, and experience the life of the sources. The experience gives us an immersive sense of being present alongside the sources in whatever action they are performing. The looping video also brings dynamism to the story because the looping videos usually appear next to static items, often written text. As pointed out by Hiippala[31] and Planer and Godulla,[32] among others, the looping video is a reoccurring element in many contemporary examples of digital long-form journalism.

Another digital element bringing dynamism is the zoom-in effect in some of the photos. The zoom-in starts when the user scrolls near a photo that is predetermined to have the effect, a rather slow visualization of zooming inwards on the image. The effect is often used in photos featuring people, and adds a media materiality to the image that could be read as intimacy and compassion. Zoom-in allows the viewer to get close to the details in an image, and because this detail is often part of a human being, the effect adds to a feeling of intimacy and compassion with the sources. It is as if the viewer could stand in the same room as the photo subjects, and moving their heads and they could position their gaze a little closer to the portrayed subjects. It could be argued that there is an element of presence at play with the zoom-in effect. It gives the impression that the viewer is actually getting a little closer to the location where the photos were taken, creating a somewhat illusionary impression of being present and immersed in the reality of the story.

Both tools—looping video and zoom-in effect on photos—are played automatically, which causes a more immersive and cohesive outcome than if the two elements were only used by user activation. If, for instance, the

[31] Hiippala, "The Multimodality of Digital Longform Journalism."

[32] Planer, Rosanna, and Alexander Godulla. 2020. "Longform Journalism in the USA and Germany: Patterns in Award-Winning Digital Storytelling Productions." *Journalism Practice* 15, 566–582. https://doi.org/10.1080/17512786.2020.1742771.

user could zoom in and out of a photo by clicking with the mouse button or using fingers on a tablet, the narrative flow might be disrupted. This type of automation, allows for the producers to take some control away from the users, and instead create a more cohesive and immersive narrative.

Conclusion: Creating Immersion and Cohesion by Using the Old with the New

Compared to the 2004 version, the 2019 version is a clearer example of what Dowling refers to as a multimedia product with different digital elements: "blending those elements into a coherent whole in its own self-contained package (…) shielding the reader from the distractions of the open web."[33] The 2019 version presents a more cohesive and immersive experience compared to the 2004 version. The distraction-free experience is created by the interplay of several digital elements, including looping videos, zoom-in effects on photos, the use of a slider and the use of scrolling transitions between photos from 2004 and photos from 2019. The Climate Tour 2019 main page can be interpreted as a distraction-free media product. It allows the reader to stay on the same site and consume the story by simultaneously engaging with different digital elements cohesively.

The use of *revisiting* adds to the impression of immersion and cohesiveness, and results in creating an ongoing narrative strategy that connects the 2004 and the 2019 versions. Naturally, the strategy could not be used in the 2004 since this was the first climate expedition, but by situating itself as a follow-up and continuously drawing on the idea of revisiting and reunion, the 2019 version delivers further integration between the individual stories of the multimedia project, an integration that is more elusive in the 2004 version. By *revisiting*, the 2019 version creates a humanizing take on climate change. The temporal complexities of climate change can make the topic "less newsworthy and mediagenic,"[34] but, in the analyzed piece, the 15-year period has been accompanied by specific changes to the climate that allow for a specificity in the revisiting strategy. The idea of revisiting places and people is an emotional and human activity which

[33] Dowling, "Immersive long-form storytelling," 31.

[34] Adams, Barbara. 2021. "Foreword: timescapes of climate change." in *Climate Change and Journalism. Negotiating Rifts of Time*. Edited by Henrik Bødker and Hanna E. Morris. London: Routledge. xiii

helps to strengthen the sense of immersion. In other words, it is by integrating bits and pieces from the 2004 version and the original material, that the 2019 version succeeds in creating a cohesive and immersive piece of storytelling. And it does so by drawing on documenting both the actual effects of climate changes and by utilizing the technological and digital developments that took place in the 15-year interval between the two versions of the story.

Digital technology has created more ways to tell stories online, and recent scholarship supports the idea of a connection between cohesiveness and immersion and the ability to raise awareness among users of the multimedia product. Östman[35] documented news media's important role in raising awareness and engaging citizens in environmental issues. Planer and Godulla[36] have stated that the digital long-form genre has substantial potential for raising awareness of issues like climate change. The comparative analysis of "The Climate Tour" in its 2004 and 2019 versions shows how the use and interplay of various digital elements form a cohesive, immersive experience that exemplifies how digital long-form storytelling can raise climate awareness.

References

Adams, Barbara. 2021. "Foreword: timescapes of climate change." in *Climate Change and Journalism. Negotiating Rifts of Time*. Edited by Henrik Bødker and Hanna E. Morris. London: Routledge.

Adams, Barbara. 1998. *Timescapes of Modernity. The Environment and Invisible Hazards*. London: Routledge.

Blach-Ørsten, Mark, and Ida Willig. 2013. "Print og Netaviser i Danmark, USA og Frankrig." *Journalistica*, nr. 1., 2013.

Boyd, Danah, Scott Golder, and Gilad Lotan. 2010. "Tweet, tweet, retweet: Conversational aspects of retweeting on Twitter." *HICSS '10 Proceedings of the 2010 43rd Hawaii International Conference on System Sciences* (pp. 1–10). Honolulu, HI. Retrieved from https://doi.org/10.1109/HICSS.2010.412

Brügger, Niels. 2008. "The Archived Website and Website Philology A New Type of Historical Document?" *Nordicom Review*, 29(2), January 2008, https://doi.org/10.1515/nor-2017-0183

[35] Östmann, Johan. 2014. "The Influence of Media Use on Environmental Engagement: A Political Socialization Approach." *Environmental Communication A Journal of Nature and Culture* 8 (1): 92–109.

[36] Planer and Godulla, "Longform journalism."

Bødker, Henrik, and Hanna E. Moris. 2021. *Climate Change and Journalism. Negotiating Rifts of Time.* London: Routledge.

Dowling, David, and Travis Vogan. 2015. "Can We 'Snowfall' This? Digital Longform and the Race for the Tablet Market." *Digital Journalism* 3 (2): 209–224. https://doi.org/10.1080/21670811.2014.930250.

Dowling, David. 2019. *Immersive longform storytelling: media, technology, audience.* New York: Routledge.

Hausken, Liv. 2009. *Medieestetikk. Studier i estetisk medieanalyse.* Oslo: Spartacus - Scandinavian Academic Press.

Hausken, Liv. 2013. *Thinking Media Aesthetics. Media studies, film studies and the arts.* Frankfurt: Peter Land Academic Publishing.

Hallin, Daniel. C, and Paolo Mancini. 2004. *Comparing Media Systems.* Cambridge: Cambridge University Press.

Hiippala, Tuomo. 2017. "The Multimodality of Digital Longform Journalism." *Digital Journalism* 5 (4): 420–442. https://doi.org/10.1080/21670811.2016.1169197.

Höijer, Birgitta. 2010. "Emotional Anchoring and Objectification in the Media Reporting on Climate Change." *Public Understanding of Science* 19, no. 6 (November 2010): 717–31. https://doi.org/10.1177/0963662509348863.

Jacobson, Susan, Jacqueline Marino, and Robert E Gutsche jr. 2015. "The digital animation of literary journalism." *Journalism*, 17(4), 527–546. https://doi.org/10.1177/1464884914568079.

Kammer, Aske. 2017. "Market Structure and Innovation Policies in Denmark." In *Innovation Policies in the European News Media Industry – a comparative study*, edited by Han Van Kranenburg. Springer.

Kammer, Aske. 2019. "Denmark." In *The SAGE International Encyclopedia of Mass Media and Society*, edited by Debra L. Merskin. London: SAGE.

Knox, John. 2009. *Multimodal Discourse on Online Newspaper Home Pages: A Social-semiotic Perspective.* PhD thesis. Sydney: University of Sydney.

Kress, Gunther. 2010. *Multimodality – a social semiotic approach to contemporary communication.* London: Routledge.

Lee, Chei, Sian, and Long Ma. 2012. News sharing in social media: The effect of gratifications and prior experience. *Computers in Human Behavior*, 28 (2), 331–339. Retrieved from. https://doi.org/10.1016/j.chb.2011.10.002.

Neveu, Erik. 2014. "Revisiting narrative journalism as one of the futures of journalism." *Journalism Studies*, 15, 533–542.

Östmann, Johan. 2014. "The Influence of Media Use on Environmental Engagement: A Political Socialization Approach." *Environmental Communication A Journal of Nature and Culture* 8 (1): 92–109

Planer, Rosanna, and Alexander Godulla. 2020. "Longform Journalism in the USA and Germany: Patterns in Award-Winning Digital Storytelling

Productions." *Journalism Practice* 15, 566–582. https://doi.org/10.1080/17512786.2020.1742771.

Schneider, Steven, and Kirsten Foot. 2004. "The Web as an Object of Study." *New Media & Society* 6: 114–122.

Shim, Hoon. 2014. "Narrative journalism in the contemporary newsroom: The rise of new paradigm in news format." *Narrative Inquiry*, 24 (1).

Syvertsen, Trine, Gunn Enli, Ole J. Mjøs, and Hallvard Moe. 2014. *The Media Welfare State: Nordic Media in the Digital Era*. Ann Arbor: University of Michigan Press.

Van Krieken, Kobie. 2018. "Multimedia Storytelling in Journalism: Exploring Narrative Techniques in Snow Fall." *Information*, 2018, 9, 13.

Varma, Anita. 2020. "Evoking Empathy or Enacting Solidarity with Marginalized Communities? A Case Study of Journalistic Humanizing Techniques in the *San Francisco Homeless Project*" *Journalism Studies*, https://doi.org/10.1080/1461670X.2020.1789495.

Wahl-Jorgensen, Karin, and Mervi Pantti. 2021. "Introduction: The emotional turn in journalism." *Journalism*, 22(5), pp. 1147–1154.

Environmental Activism and Resistance in Latin America: Literary Journalism's Portrayal of the Struggle of Environmental Leaders

Dolors Palau-Sampio

Introduction

The murder of the Brazilian environmental activist Chico Mendes in 1988, a long time defender of Amazonia against deforestation, awakened environmental consciousness.[1] However, his assassination did not lead to the protection of environmentalists, but rather to an increasing impunity.

Funded by the Conselleria d'Innovació, Universitats, Ciència I Societat Digital (Generalitat Valenciana. Ref. IAICO/2021/125).

[1] Andrew Revkin, "The burning season: the murder of Chico Mendes and the fight for the Amazon rain forest." (Washington, DC: Island Press, 2004), 18–19.

D. Palau-Sampio (✉)
Universitat de València, València, Spain
e-mail: Dolors.Palau@uv.es

P. Calvi (ed.), *The Literary Journalist as a Naturalist*, Palgrave Studies in Literary Journalism,
https://doi.org/10.1007/978-3-031-56634-9_10

The 2019 Global Witness report underlines that, despite increasing environmental awareness, 2019 was "the highest year ever [as long as records exist] for killings [of environmental activists]," in addition to rising levels of non-lethal threats and criminalization.[2] According to the report, a total of 212 land and environmental defenders were killed around the world.

The situation was particularly bloody in Latin America, the deadliest region for local communities fighting extractive industries, deforestation, or large infrastructure projects. Colombia headed a ranking that included ten Latin American countries among the 20 most dangerous for environmental defenders.[3] The report also revealed that "a significant number of murdered land and environmental defenders belong to indigenous communities."[4]

Using a qualitative methodology, this chapter focuses on how environmental activism is portrayed by narrative journalism. The objective is to analyze the resources invested in the reporting process and how narrative journalism can provide insightful knowledge of the complex circumstances surrounding environmental conflicts, which often involve "influential business and economic interests, political power battles, criminal activities," and also "sensitive issues concerning Indigenous rights to land and natural resources."[5] Long-form profiles dealing with activism and resistance in seven Latin American countries with the highest murder rate have been analyzed.

Media, Environmental Struggles, and Resistance

Rachel Carson's book *Silent Spring* ([1962] 2002) is considered to be a foundational work of modern environmental journalism.[6] It was significant not only because of Carson's exposure of the use of pesticides and chemical insecticides, but also due to her great skill to engage readers with

[2] Global Witness, *Defending tomorrow* (online, 2020), 6.

[3] Global Witness, 9.

[4] Global Witness, 7.

[5] Eric Freedman, "In the crosshairs: The perils of environmental journalism," *Journal of Human Rights* 19, no. 3(2020): 275.

[6] David B. Sachsman and Johan M. Valenti, *Routledge Handbook of Environmental Journalism* (Abingdon: Routledge, 2020).

a combination of facts and feelings.[7] Carson's pioneering approach emphasized the important role played by the media in the rise of the modern environmental movement[8] and influenced later generations of reporters and environmental disseminators, notably in the USA.

In recent years, together with global awareness on climate change issues, two other main factors have created a fertile breeding ground to foster the development of narrative environmental journalism in Latin America. On the one hand, the increasing calls for environmental justice, mainly by local and indigenous communities which have been suffering from the extraction of their natural resources for centuries. On the other hand, the emergence of a generation of digital native media[9] aimed at quality and investigative journalism, portraying themselves as committed to independence.[10]

Human Rights Defenders and Violence

The United Nations (UN) uses the term "environmental human rights defenders" to refer to "individuals and groups who, in their personal or professional capacity and in a peaceful manner, strive to protect and promote human rights relating to the environment, including water, air, land, flora and fauna."[11] Their role has gained in strength because of the scarcity of natural resources,[12] while communities affected by large extractive projects have become more aware and have started to demand more involvement in a democratic decision-making process to decide the future of their territories.

[7] Alex Lockwood, "The affective legacy of Silent Spring," *Environmental Humanities* 1, no. 1(2012): 123–140.

[8] Anders Hansen, "Communication, media and environment: Towards reconnecting research on the production, content and social implications of environmental communication," *International Communication Gazette* 73, no. 1–2 (2011).

[9] Janine Warner, Mijal Iastrebner, David LaFontaine, James Breiner and Adriana Peña Johannson, *Inflection point: impact, threats, and sustainability: a study of Latin American Digital Media Entrepreneurs* (online, 2017).

[10] Summer Harlow and Ramón Salaverría, "Regenerating Journalism: Exploring the 'alternativeness' and 'digital-ness' of online-native media in Latin America," *Digital Journalism* 4, no. 8 (2016).

[11] Michel Forst, *They spoke truth to power and were murdered in cold blood* (online, 2016), 8.

[12] Armin Scheidel, Daniella del Bene, Juan Liu, Grettel Navas, Sara Mingorría, Federico Demaria, Sofía Ávila, Brototi Roy, Irmak Ertör, Lea Temper, and Joan Martínez-Alier, "Environmental conflicts and defenders: A global overview" (online, 2020).

The policies of economic liberalization undertaken by some Latin American countries since the 1980s, the relocation of mining capitals to the Global South in the 1990s,[13] and the growing demand by the global economic structures[14] all led to a "neoextractivism" stage with the advent of the new millennium.[15] Mining conflicts affecting indigenous and rural communities more than doubled in less than a decade in Latin America, reaching 274 in 2019, according to the Observatory of Mining Conflicts in Latin America.[16]

The demands for access to and control over natural resources and territory and the claim of new forms of citizen participation are linked to the emergence of the "environmental justice movement" in the 1980s.[17] The concept of "environmental justice" is rooted in the theory of social justice, aimed at a socially just distribution of goods in a society.[18] However, far from just looking for a distributive social justice, the environmental justice movement also focuses on the idea of recognition and participation in the [decision-making process about extraction],[19] including economic (distribution), cultural (recognition), and political (participation), in order to overcome the "democratic deficit."[20]

Environmental defenders are a diverse pool of grassroots activists increasingly facing different forms of human rights violations at the hands of states and corporations. These violations range from the criminalization of their protest to violence against themselves and their communities, and even assassination. These violations are more notable when indigenous groups are involved in the activism.[21] The UN 2016 special report

[13] Gavin Bridge, "Mapping the Bonanza: Geographies of Mining Investment in an Era of Neoliberal Reform," *The Professional Geographer* 56 (2004).

[14] Roldan Muradian, Mariana Walter and Joan Martinez-Alier, "Hegemonic transitions and global shifts in social metabolism: Implications for resource-rich countries. Introduction to the special section," Global environmental change 22, no. 3 (2012).

[15] Maristella Svampa, "Commodities Consensus: Neoextractivism and Enclosure of the Commons in Latin America," *South Atlantic Quarterly* 114, no. 1 (2015).

[16] Svampa; OCMAL, *Conflictos mineros en América Latina* (online, 2019).

[17] Robert D. Bullard, "Race and environmental justice in the United States," *Yale Journal of International Law* 18(1993).

[18] John Rawls, *A theory of justice* (Cambridge, MA: Harvard University Press, 1971).

[19] David Schlosberg, "Reconceiving Environmental Justice: Global Movements and Political Theories," *Environmental Politics* 13, no. 3 (2004).

[20] Nancy Fraser, *Scales of justice: Reimagining political space in a globalizing world* (New York: Columbia University Press, 2009), 13.

[21] Scheidel, "Environmental conflicts and defenders," 9.

highlights four root causes underlying violations against those who "spoke truth to power and were murdered in cold blood."[22] First, the imbalance of power in local communities "extremely vulnerable to exploitation and abuse," communities who are "marginalized and excluded from decision-making," environmental impact assessments and consultations[23] by corporations and state authorities. Second, the stigmatization of defenders, who are considered "anti-development" when they oppose the prioritization of economic growth and midterm profits over the welfare of their communities.[24] Third, the corruption created by the lack of transparency in projects that require vast financial investments and the absence of independent investigations which leads to continued impunity.[25] Finally, the vulnerable conditions of the environmental defenders, who have achieved their role by "accident" or "necessity."[26]

Impunity is a "multidimensional phenomenon" which involves security, justice and human rights systems.[27] "Green impunity,"[28] which is related to crimes against the environment, is deeply worrying in Latin America. Research highlights the fact that its roots are ingrained in the culture of lawlessness and organized crime in the region, as corruption is not only a consequence of criminality, but is also the cause and "enabler" of such unlawful behavior.[29] Among the recommendations for reducing impunity, the Global Impunity Index 2020 suggests strengthening institutions of justice, professionalizing security institutions, increasing the capacity to address and gain access to justice in cases involving environmental crimes (as well as implementing protective mechanisms for environmental activists and defenders), supporting freedom for investigative journalism, and boosting international cooperation and supervision.[30] The issue of transnational environmental crime has gained momentum in

[22] Forst, *They spoke truth*, 21.

[23] Forst, 21.

[24] Forst, 24.

[25] Forst, 25.

[26] Forst, 26.

[27] Juan Le Clercq Ortega and Gerardo Rodríguez Sánchez Lara, *Escalas de impunidad en el mundo. Índice Global de Impunidad 2020* (online, 2020), 25.

[28] Celeste Cedillo and Juan A. Le Clercq, "Green impunity: measuring ecojustice, institutional capacities and policy design as an approach to environmental security," in *Handbook of Security and the Environment* (Cheltenham, UK: Edward Elgar Publishing, 2021).

[29] Sonia Alda Mejías, "El combate a la corrupción para combatir el crimen organizado," *Real Instituto Elcano*, 5, no.16 (2016), 26.

[30] Le Clercq and Rodríguez, *Escalas de impunidad*, 28–30.

Table 1 Authors, media, and issues

Author	*Title*	*Defender*	*Environmental issue/ country*
Carlos Dada	*Berta Cáceres passed through here*	Berta Cáceres	Hydroelectric dam *Honduras*
Joseph Zárate	*The Man who chose the forest, and died for it*	Edwin Chota	Illegal logging *Perú*
Juan Miguel Álvarez	*Choco: Chava didn't come here to fall in love*	Elizabeth Moreno Barco, Chava	Illegal mining industry and coca crops *Colombia*
Alejandra Gutiérrez Valdizán	*Fishermen lost patience*	Carlos Maaz	Mining industry *Guatemala*
Thelma Gómez Durán and Patricia Mayorga	*Without territory we are nothing*	Julián Carrillo Martínez	Mining industry Illegal logging *Mexico*
Lisseth Boon and Lorena Meléndez	*Attacks against indigenous communities aim to militarize our territory*	Lisa Lynn Henrito Percy	Mining industry *Venezuela*
Mayuri Castro	*The secret the mine swallowed*	Jose Tendetza	Mining industry *Ecuador*

recent years as a way of highlighting the intertwined dimensions of harm, as well as the need to develop a criminological framework through which eco-global criminality can be interpreted.[31]

Extractive industries, hydroelectric dams and illegal logging are the sectors in which the largest number of violations among environmental human rights defenders are observed.[32] This chapter analyzes seven profiles, published between 2015 and 2020 which deal with the above issues in Latin American countries, to understand how narrative journalism can capture the complexity of environmental conflicts and the role of the defenders (Table 1).

[31] Robert Douglas White, *Transnational environmental crime: Toward an eco-global criminology* (London: Willan, 2018).

[32] Forst, *They spoke truth*, 2016.

Narrative Journalism's Interpretation of the Environmental Fight

Narrative journalism provides exceptional conditions for tackling a complex portrait of environmental issues[33] and narratives aimed at social change. Resorting to both journalistic methods and literary resources,[34] Latin American narrative journalists are addressing the challenges of sustainable development, exposing environmental offenses and denouncing violence against activists, from threats to murder.

The journalistic profile focuses on a main character and mirrors their personality and life path. Through the profile, the journalist can reflect on the fight of environmental defenders, their achievements, relevance, symbolism, dramatic nature or unconventional lives.[35] The journalistic profile has its origins in the biographic tradition started by Plutarch to narrate the *Parallel lives* of the noble Greeks and Romans, and entered into the journalistic field in the second half of the nineteenth century. It is linked to the professionalization of the press in the UK and the USA.[36] However, its definitive consolidation arrived with the emergence of such magazines as *The New Yorker* (1925)—which coined the term *profile* to give a name to a section—and *Esquire* (1933).

De Rosendo defines the profile as a journalistic genre aimed at accurately narrating who a public character is by means of the articulation of life actions and personality (2010). The genre has also been presented as the art of "drawing a personality with words"[37] not just to stay on the surface, but to go deeper. "It seeks to illuminate a hidden place of the character, seeks to reveal the unknown," as highlighted by one of the best-known authors of this genre, the journalist Jon Lee Anderson.[38] In this sense, Julio Villanueva Chang underlines the idea of "deciphering" a

[33] Robert Alexander, "Literary Journalism and Ecocriticism", in *The Routledge Companion to American Literary Journalism* (New York: Routledge, 2019).

[34] Norman Sims, *The literary journalists* (New York: Ballantine, 1984); David Abrahamson, "A narrative of collegial discovery on some conceptual essentials," *Literary Journalism Studies* 2, no. 2 (2010).

[35] Helen Benedict, *Portraits in print: a collection of profiles and the stories behind them* (New York: Columbia University Press, 1992), 4.

[36] Sue Joseph and Richard L. Keeble, *Profile Pieces: Journalism and the 'Human Interest' Bias* (Abingdon and New York: Routledge, 2016).

[37] Danilo Moreno Hernández, "El arte de dibujar, con palabras, a una persona" (online, 2005).

[38] Moreno Hernández, "El arte de dibujar."

person, digging into their life and personality. This implies an in-depth multidimensional type of work, involving interviews with the subject, and also with sources close to them, as well as their in-depth description, the description of their home, workplace and general environment.[39] This is "either a significant insight into the subject or a point of larger, symbolic value."[40] A literary profile is a combination of narrative aspects and intensive reporting, "precise in the gathering of its facts and the detail," as Gay Talese emphasized.[41] The Nobel Prize winner Mario Vargas Llosa argued that the biographical profile shows that journalism can be an artistic work "without renouncing its main core obligation, which is to inform."[42]

Portraying Environmental Activism

Narrative journalism represents a symbiotic territory in which journalism and literature coexist and overlap, combining the portrayal of real characters through the expressive means of literature.[43] The reporting methods and production routines are associated with slow journalism's philosophy[44] allowing for narrative journalism to go beyond the event,[45] while showing a clear commitment with context, in order to "trigger a better understanding of social processes and social change."[46]

Rights to Territory and Identity

Landscape and ancient traditions are key elements in presenting lifestyles and natural habitats that need to be protected. Carlos Dada invokes them to introduce the conditions of the Zapotal river, jeopardized by mining and hydroelectric projects. Berta Cáceres, the Honduran environmental

[39] Benedict, *Portraits in print.*

[40] Benedict, 3.

[41] Mick Brown, "'I wanted to elevate journalism': Gay Talese, the writer who nailed Frank Sinatra" (*The Telegraph*, 2015).

[42] Mario Vargas Llosa, "El periodismo y la creación" (*El País*, 2013), 41.

[43] Albert Chillón, *La palabra facticia: literatura, periodismo y comunicación* (Bellaterra/ Castelló /València: Universitat Autònoma de Barcelona/Servei de Publicacions de la Universitat Jaume I/Publicacions de la Universitat de València, 2014).

[44] Megan Le Masurier, "What is slow journalism," *Journalism Practice* 9, no. 2 (2016).

[45] Katherine Fink and Michael Schudson, "The rise of contextual journalism, 1950s–2000s," *Journalism* 15, no. 1 (2014).

[46] Erik Neveu, "Revisiting the 'story vs. information' model," *Journalism Studies* 18, no. 10 (2017).

activist, who was a symbol of the indigenous Lenca resistance, fought against those projects until her assassination in 2016.

> A river died here. Only its carcass remains. A dry bed. A dusty trench with enormous boulders and rocks of all sizes, where fish once swam...
>
> Down below, we walk on what was once a river with Magdaleno Aguilar, a sturdy 61-year-old Lenca who fished here his whole life; who bathed here his whole life. Even though the law requires the hydroelectric plants to keep part of the river flowing to preserve the ecosystem, there's no water here anymore.[47]

Juan Miguel Álvarez chooses the San Juan River in Colombia to show its centrality in the life of the community: "As there are no roads, everyone depends on the ability to move on the water. Children are first taught to jump into rivers and swim." Álvarez points at the extraordinary threats the river faces while describing a boat trip with the defender Elizabeth Moreno Barco, familiarly known as "Chava" or "the iron woman" due to her leadership skills:

> She [Chava] had already gone through the 250 navigable kilometers and clearly distinguished the threats to life in the river: State abandonment, armed conflict, glyphosate sprayed on coca crops that ends up damaging their subsistence and poisoning the forest mass, indiscriminate logging for timber trade, illegal mining that destroys the soils and contaminates the water with residual chemicals, and the non-existent upstream management of domestic garbage.[48]

Description and enumeration are useful resources to offer an overview of the main environmental problems created by the external invasion and destruction of a sustainable ecosystem. The causes that defenders support, such as the case of Peruvian Edwin Chota—killed in 2015—are recounted through such means:

> When the loggers come in, the animals flee from the sound of the chainsaws. Community members must walk into the forest for days to hunt for food...As the tractors drag logs through the forest, they leave in their wake

[47] Carlos Dada, "Por aquí pasó Berta Cáceres," *El Faro* (2016).

[48] Juan Miguel Álvarez, "'Chocó: Chava no vino a enamorarse'," *El Malpensante* (2000).

unserviceable land that cannot be used to plant crops. Loggers have even brought diseases that the indigenous groups had never encountered before.[49]

When describing the special links to territory, narrative journalists usually combine documentary and personal sources to offer a complete picture of the defenders' mission. The profile of the indigenous *rarámuri* Julián Carrillo Martínez draws from data on the forest area dimension and how it was reduced as a result of logging and illicit drug crops. It also borrows the voice of an anonymous member of the community also threatened to death:

> —Before they began to cut down trees there was lots of water; there was a lot of snow; there was a life for us: a good harvest, it rained a lot, the streams did not run dry. If the wood disappears, the springs of the land will dry up. It will be the end of the wild animals. Everything will be finished. Where wood exists, everything lives.[50]

José Tendetza had worked in the mining industry for four years before becoming the leader of the Shuar community, "an ardent opponent of mining in his land," located in the Condor Mountain Range of southern Ecuador.[51] "Like all the elders in his family," he preferred to earn a living through farming, hunting, and fishing on the river. Castro affirms that since 2006, Tendetza had traveled "wherever anti-mining assemblies were held" in order to denounce the increase in poisoning of water sources, changes of river courses, and deforestation of entire areas.[52] The day he disappeared, on November 29, 2014, Tendetza was supposed to attend a meeting with the Mining Regulation and Control Agency. He never showed up. Nor did he attend the International Tribunal for the Rights of Nature in Lima, where he was expected to report on the large-scale mining project known as The Mirador. Castro writes that the project had had tremendous impact. "[It was] one of the greatest pollutants and violators of nature's right." Exploitation was carried out "with excavations of between 300 and 500 meters deep" in order to extract minerals, where

[49] Joseph Zárate, "El hombre que eligió el bosque y lo asesinaron," *Etiqueta Negra* (2015).
[50] Thelma Gómez Durán and Patricia Mayorga, "Sin territorio no somos nada," *Mongabay* Latam (2019).
[51] Mayuri Castro, "El secreto que la mina se tragó," *GK* (2021).
[52] Castro, "El secreto."

"for every ton of copper obtained, two tons of debris are removed, such as stones and sand, which end up in tailing ponds."[53]

Life Missions

Narrative journalism provides the strands to weave solid storytelling based on the motivations that led some people to dedicate—and even lose—their lives to environmental causes.[54] The opportunity to delve into the biographies of environmental leaders and gain access to different sources allows for the main traits of Berta Cáceres to be drawn.

Cáceres is presented a woman who advocated on behalf of her indigenous identity—"People…remember her as particularly insistent on revering nature, a belief inherited from the Lenca world view; and on recuperating indigenous dignity and pride"—but also was keen on using the tactics she had learned from revolutionary movements in Latin America, "by participating in regional networks, appealing to international solidarity, and exchanging experiences."[55]

Six of the seven profiles of the environmental defenders analyzed in this study show that the leaders became so as a way to continue the fight for their ancestral territory, sometimes by necessity and often without the necessary skills to assume such leadership, as expressed in the UN report.[56] This was the case of the Guatemalan Carlos Maaz, a fisherman assassinated when fighting the pollution caused by a nickel mine in a lagoon traditionally associated with the Maya *q'eqchi'* community.[57] Another such case was that of Julián Carrillo, killed in 2018 in the Mexican state of Chihuahua for opposing the illegal logging of the forest inhabited by the *rarámuri* indigenous community. Carrillo used to say: "We learned to defend the land by watching and by hearing…our parents and the elderly people."[58]

Charisma and leadership play a major role in the profiles of the three women analyzed in this study, even if their fight against the patriarchy has not been easy.[59] Venezuelan Lisa Lynn Henrito Percy, the only one with

[53] Castro, "El secreto."

[54] Neveu, "Revisiting the 'story'."

[55] Dada, "Por aquí pasó."

[56] Forst, *They spoke truth*, 26.

[57] Alejandra Gutiérrez Valdizán, "Los pescadores perdieron la paciencia," *Agencia Ocote* (2019).

[58] Gutiérrez Valdizán, "Los pescadores."

[59] Forst, 27.

higher education, fought for the integrity of the Pemón community and territory without fearing the military's retaliation, as shown in the following scene:[60]

> A short woman with long, dark hair faces a major of the Venezuelan Army who is guarded by three men in olive-green uniforms. Neither the man's insignias nor his height seem to intimidate her. "If you want to do in this municipality what you are doing in the rest of the country, you are very much mistaken, because there are people here and we are not going to allow it," she tells him, defiantly.[61]

Álvarez highlights Chava's leadership through the testimonies gathered during his stay in the Chocó community: "[The co-workers and people who know her] highlight her charisma, her capacity to listen to all voices without losing patience; her courage to face political discussions, even with armed leaders; and her willingness to work for the well-being of the communities she represents."[62] But Álvarez also resorts to his own observations in situ:

> [F]or her family, Chava is the matriarch. She is the guide and the source. She is a single mother of four, a grandmother of two, a daughter looking after her mother, the third of five siblings, the most admired and beloved aunt.[63]

Details complete the portrait of this "Iron Woman": "They call her Chava or Chavita. Although the diminutive is pure affection because she is enormous: 1.75 meters, wide back, long arms and legs, hands like mallets."[64]

One of the main leadership skills pointed out in the profiles are dialectical abilities. Even though Castro did not have the opportunity to meet the Ecuadorian Jose Tendetza in life, she uses one of the last video recordings of the environmental defender, in December of 2013, to show the conviction he addressed his audience: "In the middle of an assembly, he tells the

[60] Tom Wolfe, *El nuevo periodismo* (Barcelona: Anagrama, 1976), 50.

[61] Lisseth Boon and Lorena Meléndez, "Ataques contra comunidades indígenas tienen como objetivo militarizar nuestro territorio," *Runrun* (2020).

[62] Álvarez, "'Chocó'."

[63] Álvarez, "'Chocó'."

[64] Álvarez, "'Chocó'."

audience 'It is time to say [to Ecuacorriente] enough is enough'."[65] The company, financed by Chinese capital, is in charge of the mega-mine that Tendetza opposed. A project which, according to political authorities and investors, "is Ecuador's new golden ticket to prosperity." The Shuar defender is remembered as a "very brave [person] and able to persuade and raise awareness about the territory's defense and the opposition to mining in those who listened to him."[66]

Among the motivations that led environmental defenders to adopt this life mission, Edwin Chota illustrates how the inspiration often transcends the origins. "The man who died for his Asháninka community had not always been Asháninka,"[67] Zárate, who interviewed many sources to reconstruct how his life changed, writes. Another shows the detail in Zárate's profile:

> How could it be possible that the man...who loved to dance to the Bee Gees and John Travolta, the man who never walked out the door without his shirt ironed and his shoes shined, could now wear a tunic, feathered crown and sandals, and live in a straw hut in the middle of the Amazon?[68]

Stigmatization, Prosecution, and Perpetrators' Impunity

By means of a series of narratives showing the stigmatization when fighting illegal practices and corruption, the journalists reconstruct the context of impunity in which the aggressions toward environmental defenders operate. Dada is particularly talented when underlining how powerful interests act to create division and conflict in the community: "From outside the forest, it appears to be a conflict between corporations and poor communities; but once you enter the trees it looks more like the poor fighting the poor."[69] Dada reports how some families accuse Berta Cáceres' environmental organization COPINH (Council of Popular and Indigenous Organizations of Honduras) of impeding community development by opposing the Agua Zalca hydroelectric dam. In a region forgotten by the government, environmental defenders denounce "a plan on the part of powerful groups to take strategic control of Honduras's natural resources."

[65] Castro, "El secreto."
[66] Castro, "El secreto."
[67] Zárate, "El hombre que eligió."
[68] Zárate, "El hombre que eligió."
[69] Dada, "Por aquí pasó."

After years of killings, including that of Cáceres, "differences between people are far too wide to expect them to quickly return to being a single community."[70]

Gutiérrez Valdizán presents an environment of fear and violence that seems inspired by a horror novelist. He stresses that "a novel could be written in a Stephen King tone about El Estor [...] a territory of the tropics that seems built on gunpowder," with "a timeline of deaths –assassinations– recounted by criminal complaints filed or due to demonstrations that ended in brutal evictions."[71] Among those deaths, there is Carlos Maaz's killing, a case few people seem to know something about. "The witnesses point to the Police, to the State, but in Guatemala, that is no reason for scandal," Gutiérrez Valdizán writes.[72]

A trail of deaths and multiple threats lead most defenders to assume their vulnerability. Lisa Henrito ponders as her "unavoidable fate" that "the battle she has fiercely waged since the '90s will end in her death or incarceration": "I always tell my people: [...] if I'm imprisoned, don't try to free me. You need to understand that what [the military] want is the land, not me."[73]

Jose Tendetza had feelings similar to those of Lisa Henrito. "He was aware of his [impending] death," Castro, who relies on the anecdote narrated by one of the environmental leader's brothers, states.[74] Even though the result of Tendetza's autopsy was asphyxiation by strangulation, along with the fact that two Ecuacorriente field workers were arrested, Tendetza's murderers remain unpunished. "His death would mark the beginning of the slow decline of the resistance to mining in the area," Castro explains, while "the mining to which he was so fervently opposed marches on in the region."[75] The journalist quotes one of the experts who investigated Tendetza's murder in order to emphasize that the lack of recognition of the Shuar's rights over their ancestral land, together with the fact that many inhabitants of the area work in the mining sector, are kindling for social conflict.

The vulnerability and defenselessness of environmental leaders are lived with fear not only by themselves, but also by their families. Zárate

[70] Dada.
[71] Gutiérrez Valdizán, "Los pescadores."
[72] Gutiérrez Valdizán.
[73] Boon and Meléndez, "Ataques contra comunidades."
[74] Castro, "El secreto."
[75] Castro.

reproduces a graphic sentence to underline Edwin Chota's words to answer his father's fears and pleadings for him not to return to the territory of the indigenous community: " 'I can't,' he said. 'They'd have to kill me to stop me'."[76] And the prophecy was fulfilled. Zárate starts Chota's profile narrating his end by means of a little but revealing anecdote, which combines psychological and physical traits: his "wide, exaggerated and contagious smile, with a prominent gap where a front tooth was missing," was the clue to identify his skull.[77]

The helpless condition of environmental leaders carrying out their mission in a scenario of impunity is transferred into their closest relatives. Such was the case of Isidro Carrillo, brother of the also murdered environmental defender Julián Carrillo. First prosecuted and imprisoned, Isidro was acquitted after Amnesty International's intervention. However, he had to leave the community because he continued to receive threats and was finally killed when returning to visit sick relatives. Even the surviving relatives are forced to flee their homes. The day after Julián's murder, his wife and family members had to leave their house, belongings, and animals.[78] Displacement and uprooting are often the only way to survive.

Conclusions

Narrative journalism provides a fertile soil to deal with environmental issues, as it allows an in-depth approach, combining ethnographic and contextual information, gathered on the ground or from documentary sources, observing and interviewing defenders and community members. The long-form profiles of these activists reflect not only the roots of their fight, but also how they turn it into a life mission. By delving into the defenders' personal stories, journalists are able to show how their biographies are closely interwoven with their struggle against the harassment of mining companies and illegal logging corporations. The defenders' intimate relationship with the territories to be preserved is conveyed by means of descriptive language that provides vivid access to the scenes. However, beyond the indisputable aesthetic beauty, the descriptions also emphasize the connection of the communities with their territories, which are affected by corporate projects. In other words, the aim is to highlight these

[76] Zárate, "El hombre que eligió."

[77] Zárate.

[78] Gómez Durán and Mayorga, "Sin territorio."

people's fundamental right to a territory so they can maintain their ancestral lifestyle, culture, and traditions.

Among the techniques of narrative journalism, scenes have the power to reproduce the defenders' actions and highlight how determined they are to overcome the multiple challenges they face. The profiles selected deal with some of the most serious environmental conflicts in Latin America.[79] Through a combination of investigative reporting and narrative journalism techniques, they are an excellent resource to reflect on the environmental risks associated with mining and illegal logging, but also on the precarious conditions faced by the defenders.[80] The defenders' vulnerability is highlighted by accounts of several episodes involving threats, prosecution, and human rights violations with tragic outcomes, including their assassination or displacement.

The long-form profiles analyzed offer extraordinary insights into structural problems by relying on community leaders and local voices. The wealth inherent to the environment and mineral deposits has turned Latin America into a target for large international corporations, while precarious socioeconomic conditions prevent opposition to such projects. Fueled by the promise of private investment that acts as a balance for the absence of public funding, a considerable amount of damage is hidden. On-site reporting illustrates the social conflict surrounding these projects, and particularly the "green impunity" linked to environmental crime.[81] Corruption and organized crime in the region,[82] as well as the inability of institutions of justice and security to combat crime and protect environmental defenders, highlight the challenges faced by those who fight against global eco-criminality.

References

Abrahamson, David. "A narrative of collegial discovery on some conceptual essentials." *Literary Journalism Studies* 2, no. 2 (2010): 85–95.

Alda Mejías, Sonia. "El combate a la corrupción para combatir el crimen organizado" [Fighting corruption to fight organized crime]. *Real Instituto Elcano*, working paper 6, no. 16 (2016). Accessed 9 August 2020. https://bit.ly/3Plh9RX

[79] OCMAL, *Conflictos mineros en América Latina.*

[80] Forst, *They spoke truth.*

[81] Cedillo and Le Clercq, "Green impunity."

[82] Alda Mejías, "El combate a la corrupción."

Alda Mejías, Sonia. "Buen gobierno y cultura de la legalidad, componentes esenciales de las políticas de seguridad contra el crimen organizado" [Good governance and culture of legality, essential components of security policies against organized crime] In *El crimen organizado en América Latina: manifestaciones, facilitadores y reacciones* [Organized crime in Latin America: demonstrations, facilitators and reaction], 123–153. Edited by Carolina Sampó and Valeska Troncoso. Madrid: Universidad Nacional de Educación a Distancia, 2017.

Alexander, Robert. "Literary Journalism and Ecocriticism.", In The *Routledge Companion to American Literary Journalism*, 482–497. Edited by William E. Dow and Roberta S. Maguire. New York: Routledge, 2019.

Álvarez, Juan Miguel. "Chocó: Chava no vino a enamorarse" [Choco: Chava didn't come here to fall in love], *El Malpensante* 2020. Accessed 5 August 2020. https://tierraderesistentes.com/en/index.php/2020/03/27/choco-chava-no-vino-a-enamorarse/

Benedict, Helen. *Portraits in print: a collection of profiles and the stories behind them*. New York: Columbia University Press, 1992.

Boon, Lisseth and Lorena Meléndez. "Ataques contra comunidades indígenas tienen como objetivo militarizar nuestro territorio" [Attacks against indigenous communities aim to militarize our territory], *Runrun* 2020. Accessed 5 August 2020. https://tierraderesistentes.com/en/index.php/2020/03/13/lisa-lynn-henrito-percy-guardiana-del-pueblo-pemon/

Bridge, Gavin. "Mapping the Bonanza: Geographies of Mining Investment in an Era of Neoliberal Reform." *The Professional Geographer* 56 (2004): 406–421.

Brown, Mick. "'I wanted to elevate journalism': Gay Talese, the writer who nailed Frank Sinatra." *The Telegraph*, 18 November 2015. http://www.telegraph.co.uk/books/authors/gay-talese-interview

Bullard, Robert D. "Race and environmental justice in the United States." *Yale Journal of International Law* 18 (1993): 319–335.

Carson, Rachel. *Silent spring*. Boston and New York: Houghton Mifflin Harcourt, 2002.

Castro, Mayuri. (2019) 'El secreto que la mina se tragó' [The Secret the Mine Swallowed], *GK*. https://tierraderesistentes.com/es/2019/10/30/el-secreto-que-la-mina-se-trago/, accessed 21 August 2021.

Cedillo, Celeste, and Juan A. Le Clercq. 2021. "Green impunity: measuring eco-justice, institutional capacities and policy design as an approach to environmental security." In *Handbook of Security and the Environment*, edited by Ashok Swain, Joakim Öjendal and Anders Jägerskog, 115–131. Cheltenham, UK: Edward Elgar Publishing.

Chillón, Albert. *La palabra facticia: literatura, periodismo y comunicación* [The factitious word: literature, journalism and communication]. Bellaterra/Castelló /València: Universitat Autònoma de Barcelona/Servei de Publicacions de la Universitat Jaume I/Publicacions de la Universitat de València, 2014.

Dada, Carlos (2016) 'Por aquí pasó Berta Cáceres' [Berta Cáceres passed through here], *El Faro*. https://elfaro.net/es/201610/centroamerica/19401/Berta-Cáceres-passed-through-here.htm, accessed 5 August 2020.

De Rosendo, Belén. *El perfil periodístico* [The journalistic profile]. Madrid: Tecnos, 2010.

Fink, Katherine and Michael Schudson. "The rise of contextual journalism, 1950s–2000s." *Journalism* 15, no. 1 (2014): 3–20.

Forst, Michel. 2016. *They spoke truth to power and were murdered in cold blood*. https://bit.ly/3RAzGvB, accessed 5 August 2020.

Fraser, Nancy. *Scales of justice: Reimagining political space in a globalizing world*. New York: Columbia University Press, 2009.

Freedman, Eric. "In the crosshairs: The perils of environmental journalism." *Journal of Human Rights* 19, no. 3 (2020): 275–290.

Global Witness. *Defending tomorrow* (Report), 2020. Accessed 20 July 2020. https://bit.ly/3yS2ZCo

Gómez Durán, Thelma and Patricia Mayorga. "Sin territorio no somos nada" [Without territory we are nothing], *Mongabay Latam* 2019. Accessed 5 August 2020. https://tierraderesistentes.com/en/index.php/2020/03/25/sin-territorio-no-somos-nada/

Gutiérrez Valdizán, Alejandra. "Los pescadores perdieron la paciencia" [Fishermen lost patience], *Agencia Ocote* 2019. Accessed 5 August 2020. https://tierraderesistentes.com/en/index.php/2020/03/24/los-pescadores-perdieron-la-paciencia/

Hansen, Anders. "Communication, media and environment: Towards reconnecting research on the production, content and social implications of environmental communication." *International Communication Gazette* 73, no. 1–2 (2011): 7–25.

Harlow, Summer and Ramón Salaverría. "Regenerating Journalism: Exploring the 'alternativeness' and 'digital-ness' of online-native media in Latin America." *Digital Journalism* 4, no. 8 (2016): 1001–1019.

Joseph, Sue and Richard L. Keeble (editors.) *Profile Pieces: Journalism and the 'Human Interest' Bias*. Abingdon and New York: Routledge, 2016.

Kramer, Mark. "Breakable rules for literary journalists." In *Literary journalism: A new collection of the best American nonfiction*, 21–34. Edited by Norman Sims and Mark Kramer. New York: Ballantine Books, 1995.

Le Clercq Ortega, Juan A. and Gerardo Rodríguez Sánchez Lara. *Escalas de impunidad en el mundo. Índice Global de Impunidad 2020* [Scales of impunity in the world. Global Impunity Index 2020], 2020. Accessed 20 August 2021. https://bit.ly/3c3feTr

Le Masurier, Megan. "What is slow journalism." *Journalism Practice* 9, no 2 (2016): 138–152.

Lockwood, Alex. "The affective legacy of Silent Spring." *Environmental Humanities* 1, no. 1 (2012): 123–140.

Moreno Hernández, Danilo. *El arte de dibujar, con palabras, a una persona* [The art of drawing a person with words], 2005. Accessed 7 August, 2020. https://bit.ly/3nOGmbN

Muradian, Roldan, Mariana Walter and Joan Martinez-Alier. "Hegemonic transitions and global shifts in social metabolism: Implications for resource-rich countries. Introduction to the special section." *Global environmental change* 22, no. 3 (2012): 559–567.

Neveu, Erik. "Revisiting the 'story vs. information' model." *Journalism Studies* 18, no. 10 (2017): 1293–1306.

OCMAL, Observatory of Mining Conflicts in Latin America. *Conflictos mineros en América Latina* [*Map of mining conflicts in Latin America*], 2019. Accessed 8 August 2020. https://bit.ly/3uEs80N

Rawls, John. *A theory of justice*. Cambridge, MA: Harvard University Press, 1971.

Sachsman, David B. and Johan M. Valenti. *Routledge Handbook of Environmental Journalism*, 1–15. Abingdon: Routledge, 2020.

Revkin, Andrew. *The burning season: the murder of Chico Mendes and the fight for the Amazon rain forest*. Washington, D.C: Island Press, 2004.

Scheidel, Armin, Daniella del Bene, Juan Liu, Grettel Navas, Sara Mingorría, Federico Demaria, Sofía Ávila, Brototi Roy, Irmak Ertör, Lea Temper, and Joan Martínez-Alier. "Environmental conflicts and defenders: A global overview." *Global Environmental Change*, 63 (2020), https://doi.org/10.1016/j.gloenvcha.2020.102104

Schlosberg, David. "Reconceiving Environmental Justice: Global Movements and Political Theories." *Environmental Politics* 13, no. 3 (2004): 517–540.

Sims, Norman. *The literary journalists*. New York: Ballantine, 1984.

Svampa, Maristella "Commodities Consensus: Neoextractivism and Enclosure of the Commons in Latin America." *South Atlantic Quarterly* 114, no. 1 (2015): 65–82.

Vargas Llosa, Mario. "El periodismo y la creación." *El País*, 19 Mai 2013.

Villanueva Chang, Julio. *La silueta del perfil* [Profile silhouette], 2014. Accessed 8 July 2020. https://bit.ly/3IpZfv5

Warner, Janine, Mijal Iastrebner, David LaFontaine, James Breiner and Adriana Peña Johannson. *Inflection point: impact, threats, and sustainability: a study of Latin American Digital Media Entrepreneurs*, 2017. http://data.sembramedia.org/download-the-study/

White, Robert Douglas. *Transnational environmental crime: Toward an eco-global criminology*. London: Willan, 2018.

Wolfe, Tom. *El nuevo periodismo* [The new journalism]. Barcelona: Anagrama, 1976.

Zarate, Joseph, "El hombre que eligió el bosque y lo asesinaron" [The man who chose the forest, and died for it], *Etiqueta Negra*, 2015. Accessed 5 August 2020. https://www.internationalboulevard.com/the-man-who-chose-the-forest-and-died-for-it

An Ecologist's Personal Investigation: Sandra Steingraber on Danger, Ecology, and Writing

Mileta Roe

On an ultrasound table in a Boston hospital, ecologist and writer Sandra Steingraber awaited an amniocentesis. In the very same place, years earlier, she had been scanned for signs of recurring cancer. This time, thirty-eight years old and pregnant, she reminded herself that visible growth would be a good result. Moments later, she held aloft a pair of vials glistening with amniotic fluid and watched on the monitor while the baby executed a surprising flip turn. Months later, while contemplating how to recount that scene for her book on pregnancy and environmental toxins, she studied a passage from *Silent Spring*[1] on late arctic storms: in graceful prose Rachel Carson constructed a wide-angle view of the landscape, an entire ecosystem, only to refocus closer and closer until bringing the reader inside the frozen eggs of an owl's nest to reveal one reason—fewer

[1] Rachel Carson, *Silent Spring* (Boston: Houghton Mifflin, 1962).

M. Roe (✉)
Bard College at Simon's Rock, Great Barrington, MA, USA
e-mail: mroe@simons-rock.edu

P. Calvi (ed.), *The Literary Journalist as a Naturalist*, Palgrave Studies in Literary Journalism,
https://doi.org/10.1007/978-3-031-56634-9_11

owls—why lemmings populations increase after such late-spring storms. In her own passage, Steingraber employed a similar strategy but reversed the shift in focus, beginning inside the enclosed womb and gliding ever outward to consider the origins of the amber liquid:

> Before it is baby pee, amniotic fluid is water. I drink water, and it becomes blood plasma, which suffuses through the amniotic sac and surrounds the baby—who also drinks it.
>
> And what is it before that? Before it is drinking water, amniotic fluid is the creeks and rivers that fill reservoirs. It is the underground water that fills wells. And before it is creeks and rivers and ground water, amniotic fluid is rain. When I hold in my hands a tube of my own amniotic fluid, I am holding a tube full of raindrops.... Whatever is in the world's water is here in my hands.[2]

Here and in her other writings on environmental topics, Steingraber explains complex science with the perennial goal to show readers the 'exquisite communion between the outside environment of the world and the inside of us'.[3] As a contributor to *Orion Magazine* and as the author of several books, Steingraber pulls out the 'telling details, translating data into narrative and making a case'[4]—her own description of Rachel Carson's rhetorical prowess in Steingraber's introduction to the Library of America's 2018 edition of *Silent Spring*. Steingraber's personal relationship to her topics, her vivid writing, and her recent activism related to fracking all work to revise traditional notions of scientific and environmental reporting. With a mind attuned to both science and poetry, a far-reaching ecological awareness, and a fearlessness in confronting truths and using herself as an example, Steingraber stands out amidst a field of science writers who have followed in Rachel Carson's footsteps.

[2] Sandra Steingraber, *Having Faith: An Ecologist's Journey to Motherhood* (Cambridge, Massachusetts: Perseus, 2001), 66–67.

[3] Sandra Steingraber, Keynote Address presented at International Women's Day Conference, Bard College at Simon's Rock, Great Barrington, Massachusetts, March 5, 2011.

[4] Sandra Steingraber, introduction to *Rachel Carson: Silent Spring & Other Writings on the Environment*, ed. Sandra Steingraber (New York: Literary Classics of the United States, 2018), xxii.

Writing Science

Born five decades after Rachel Carson, Sandra Steingraber struggled with a similar dilemma as an undergraduate: whether she should pursue science or creative writing. Ultimately both disciplines prevailed, as Steingraber went on to earn a doctorate in biology and a master's degree in creative writing (and publish a volume of poetry) before finding her way to science writing. Yet it was only after years of university teaching and research that, inspired by *Silent Spring*, Steingraber 'left the laboratory and became a science writer'[5] to create a bridge between science and the public's knowledge of environmental dangers. Since then, she has earned multiple honors for her research and writing, including the Heinz Award for work on environmental health, the Environmental Health Champion Award from Physicians for Social Responsibility, and the Rachel Carson Leadership Award from Chatham College, Carson's alma mater. In addition, she has been the subject of two feature documentary films by Chanda Chevannes, and Steingraber's books have been acclaimed across the globe.

Her first book, *Living Downstream: An Ecologist's Personal Investigation of Cancer and the Environment*,[6] details her family's history of cancer, her own diagnosis of bladder cancer at age twenty, and the very real connection between toxic chemicals and bladder cancer rates in her hometown of Rockford, Illinois. The larger report—investigating exposures and risks and potential correlations to the newly available cancer registries—is, of course, bigger than Steingraber or her family. Without being maudlin or overwrought—one critic wrote that she is 'frighteningly forthright'[7]—she draws the reader in to process her own diagnosis and treatment in lyrical scenes that intersect the grim but beautifully described science, which is thoroughly annotated. The elegance of her writing is counterbalanced by scrupulous documentation: included in the 400-page book are eighty-eight pages of source notes. And along with the science, she weaves in social criticism, emphasizing the implicit human right to clean water and a childhood without the fear of contamination.

[5] Sandra Steingraber, "Silent Spring: A Father-Daughter Dance" in *Rachel Carson: Courage for the Earth*, ed. Peter Matthiessen (New York: Houghton Mifflin, 2007), 56.

[6] Sandra Steingraber, *Living Downstream: An Ecologist's Personal Investigation of Cancer and the Environment* (Cambridge, Massachusetts: Da Capo, 1997 [2010]).

[7] Robert K. Musil, *Rachel Carson and Her Sisters: Extraordinary Women Who Have Shaped America's Environment* (New Brunswick, New Jersey: Rutgers University Press, 2014), 166.

Perhaps an ecologist is invariably hyperaware of her surroundings, perpetually analyzing the external and internal experiences of what it is to be human in this world. In Steingraber's writing world, topics appear to emerge naturally from her own life, her health, her family, her community, and so the autobiographical stance of her prose feels natural too. She is in the writing and the science because she is living it. In subsequent books, she has taken on topics of similar personal resonance. In *Having Faith: An Ecologist's Journey to Motherhood*[8] she looks with wonder and warning at the ecology of pregnancy and embryology via her own experience and some very present dangers; in *Raising Elijah: Protecting Children in an Age of Environmental Crisis*[9] she examines the hazards of environmental toxins on the young—again, primarily through the lens of motherhood and her own family, informed by research, scientific data, and analysis. Each of her books presents a version of this unusual combination of autobiographical narrative with rigorous science (for a non-specialist audience)—a 'multivalent discourse',[10] in the words of one reviewer of *Living Downstream*.

The World's Ugliest Gerund

Beyond her books and other writing, Steingraber has been a contributor to *Orion Magazine*—a respected American journal dedicated to environmental and cultural concerns. For *Orion* she wrote a group of eight pieces between 2010–2013 that now look like her initiation, reckoning, and call to action against fracking—the blasting of natural gas from underground shale. To my reading, these fairly brief pieces feel like profound moments of meditation, in which the author makes small observations that ripple out, join others, and sluice back toward their origins, having touched the wider waves. Beginning with 'Shale Game: Playing with unknowns underground',[11] Steingraber rhapsodizes about moving her family from

[8] Steingraber, Sandra, *Having Faith: An Ecologist's Journey to Motherhood* (Cambridge, Massachusetts: Perseus, 2001).

[9] Steingraber, Sandra, *Raising Elijah: Protecting Children in an Age of Environmental Crisis* (Cambridge, Massachusetts: Da Capo, 2011).

[10] Michael A. Bryson, "It's Worth the Risk: Science and Autobiography in Sandra Steingraber's *Living Downstream*," *Women's Studies Quarterly* 29, no. 1&2 (2001): 178.

[11] Sandra Steingraber, "Shale Game: Playing with unknowns underground," *Orion Magazine* 29, no. 3 (2010): 12–13.

the woods of western New York state to a nearby village and thus a return to civilization:

> ... it felt like a faith healing. I could walk again. A sidewalk stretched from my door out to a craggy maple tree and then connected with another sidewalk that headed down the block toward Main Street. Here was a track, upon which the wheels of a double stroller could roll, that linked me to coffee, library books, postage stamps, hardware displays, bank tellers, and a bus line. Hallelujah.[12]

Those hundred-year-old sidewalks, she learns, are made of shale, a sedimentary rock originating in marine mud and related, via the same ancient sea, to a deep deposit of methane, also called natural gas. Lying a mile or more underground, beneath that very region of New York state, the Marcellus Shale extends 600 miles to Pennsylvania and New Jersey and is the largest methane deposit in the United States. A few paragraphs into the article, we know about her new locale and its emotional import to the author, a bit about the village and her geologist neighbor, and more than a little about how a village sidewalk connects its citizens' feet to the depths of geology, a 'vanished ocean', and fracking—'the world's ugliest gerund'.[13]

At this point Steingraber tells us four stories about fracking, using oral markers: 'the geological story goes like this', then the engineering story, the environmental story, and the human story.[14] She covers a lot. Somehow, in these four mini narratives, she weaves in pollination systems, ozone, Marcellus, New York—the namesake village of the shale deposit—roadbuilding, endocrine disruptors, 10–30 tons of chemicals, drilling leases, a trillion-dollar payout, and the fact that fracking is not subject to any environmental regulations at the federal level. In the alarming words of an energy company representative whom she quotes, 'The shale army has arrived. Resistance is futile'.[15] She has brought her readers to the edge, and we prepare for the worst, despite a final paragraph describing a couple of public meetings at the village firehouse and library during which some lively opposition to fracking is voiced. We breathe again, until reading her closing to the piece:

[12] Steingraber, 12.
[13] Steingraber, 12.
[14] Steingraber, 12.
[15] Steingraber, 13.

> On the way home, walking on an unbroken sidewalk made of shale above an as-of-yet unshattered bedrock made of shale, my son said, 'We shouldn't wreck this place down, right, Mom?' And his words drew a battle line across my heart.[16]

It is a moving ending, a masterful circling around to the sidewalk image and the sense of belonging with which she began, as well as some not-so-subtle social criticism delivered via a child's innocent observation. But the article demonstrates Steingraber's way of linking, for herself and her readers, two distinct moments and feelings: the moving back to civilization and feeling of social and cultural connectedness at the beginning, and, at the end, the anguish and resignation in recognizing a wrong that must be combatted. This aspect of her writing is notably skillful, even novelistic. Here and elsewhere, these autobiographical moments, in which she recounts what was happening *and* what she was thinking, manage to unite her experience with a much wider sense of human experience, without explicitly doing so. By revealing her own thoughts and relatedness, she is showing us what ours could be, and that sets in motion our own trail of thoughts and associations and feelings.

The next of these *Orion* pieces is, however, much closer to a manifesto. Entitled 'The Whole Fracking Enchilada: Violating the bedrock, the atmosphere, and everything in between',[17] the text and its tone register that Steingraber is scared. She acknowledges that what she is writing is a departure from her usual inquiries into toxic chemicals—the focus of most of her prior work. And in case we were not sure of her current position, she states full on, 'Fracking is linked to every part of the environmental crisis—from radiation exposure to habitat loss—and contravenes every principle of environmental thinking'.[18] As she moves through the details, the gathering storm is palpable. Fracking, she writes, is already happening in Canada and the United States; Poland and Turkey are also at risk. In the United States, apart from drilling and earthquakes, the flowback—millions of gallons of fracking fluid made poisonous by unknown chemicals, heavy metals, and benzene—will be stored or dumped or left festering in pits. And the previously clean water that was used for fracking will never

[16] Steingraber, 13.

[17] Sandra Steingraber, "The Whole Fracking Enchilada: Violating the bedrock, the atmosphere, and everything in between," *Orion Magazine* 29, no. 5 (2010): 14–15.

[18] Steingraber, 14.

come back to the water table—it's too toxic—and so must 'disappear'. Her repeated, final refrain: 'There is no away'.[19]

Three years later, Steingraber published an *Orion* piece that established a new role for her as a public biologist in the fracking world. 'Coffee in Jail: When standing up for what's right lands you in a correctional facility'[20] is pretty much exactly that—a tale of her fifteen days in the Chemung County Jail for protesting the transportation and storage of volatile gases from fracking beneath Seneca Lake. The article is a tour de force, a four-part immersion report of her jail experience that is sometimes funny, sometimes serious, laced through with honestly competing emotions: on the one hand, her disbelief that she is actually there, and on the other, an assuredness that what she did was the right thing to do. In the following excerpt from the third section, entitled Civil Disobedience, she revisits the thought process preceding her decision:

> CIVIL DISOBEDIENCE is like planned parenthood. The second word in the phrase doesn't exactly align with the first one.
>
> You search your conscience. You reach a decision. You make preparations. You talk about your decision and your preparations in meaningful ways to all concerned. And then comes the time for action.
>
> You leave the condom in the drawer.
>
> Or you refuse the deputy's third order to get off the driveway. (The one owned by the nation's largest natural-gas storage and transport company. The one the truck with the massive drill head strapped to its flatbed wants to drive on. Right now.)
>
> Either way, where you end up bears little resemblance to the circumstances that kicked it off.[21]

The analogy is true, clever, and ominous. We can imagine the author in cellblock 5D having those very thoughts, and we and she know that she has felt differently, made a choice, crossed a line. On her personal website's list of contents at the time of writing this article, right below the category 'Fracking', was 'Civil Disobedience'—as though one inevitably leads to the other.

[19] Steingraber, 15.

[20] Sandra Steingraber, "Coffee in Jail: When standing up for what's right lands you in a correctional facility," *Orion Magazine* 32, no. 5 (2013): 15–16.

[21] Steingraber, 16.

Recalling Carson

In the fall of 2012, Steingraber published an unusual, telling feature for *Orion*, '*The Fracking of Rachel Carson: Silent Spring*'s *Lost Legacy, Told in Fifty Parts*'.[22] Printed with a well-known photo of Rachel Carson and nine accompanying photographs of fracking by Nina Berman, the fifty sections ostensibly honor the fiftieth anniversary of the publication of *Silent Spring*. Through these short texts, Steingraber creatively splices together clips of Carson's life and writings with her own reflections and dispatches on more recent environmental hazards. The numbered sections encourage readers to count along and ponder what has transpired since 1962, what enduring effect five decades of public, toxicological knowledge have wrought. The short passages also effectively comingle Carson and Steingraber's voices as the text zigs back and forth between snippets of Carson's career and context to excerpts from her speeches to Steingraber detailing the recent progress of fracking in Carson's beloved Pennsylvania. And in between, Steingraber allows herself to isolate facts or brief comments (in their own numbered sections) that read variously as headlines or poetic asides with added impact or irony. Then, as the weight of fracking's harm piles up with no intervention in sight and after recounting Carson's explanations for 'our collective reluctance to give up on poisonous technologies', Steingraber exclaims in the entirety of number forty, 'Oh, Rachel'.[23] Is it a sigh, an apology, or an invocation? Possibly all three. In sum, the fifty individual pieces create a mosaic relating past and present via verbal pivots and line-space pauses that require readers to do the same: rotate the pieces, glimpse the larger picture, understand how Pennsylvania then and now—from shale deposits to hawk refuges, benzene in the air, cancer rates, and loss of forest habitat—has suffered the effects of fracking, despite Carson's historical call for us to do better. Fittingly, number fifty reprises a well-known quotation from the last section of *Silent Spring*, giving Carson the final word. But the titular 'Fracking of Rachel Carson' still hovers, as if to ask throughout the article: Have we mined Carson's wisdom or blasted her legacy? The resulting composite is both tribute and lament, by parts gratitude, grief, and rage.

[22] Sandra Steingraber, "The Fracking of Rachel Carson: *Silent Spring*'s Lost Legacy, Told in Fifty Parts," *Orion Magazine* 31, no. 5 (2012): 14–23.
[23] Steingraber, 21.

The ties between the two writers' lives and their work have deeper resonance. If you have read *Living Downstream*, you know that the book begins with Steingraber sifting through Rachel Carson's papers at the Yale library, and that Carson's work and biography—especially her brief illness and later death from breast cancer—are interspersed throughout the book. Carson kept her cancer hidden, fearing that her critics might use her illness to discredit her claims about pesticides and cancer or attack her objectivity as a scientist. Carson was probably right to worry, given that corporate and government authorities are still unlikely to believe individuals' experience of environmental harm. As Kristen Abatsis McHenry wrote in a recent article after having interviewed women living adjacent to fracking sites in Pennsylvania: 'People's testimony of environmental harm and illness are documented extensively, yet it seems that these "stories" are not viewed as evidence by government agencies, energy corporations, and the scientific community'.[24] And all these years after *Silent Spring*, science journalist Cornelia Dean reports that scientists still contend with the 'culture of science that inveighs against researchers' participation in public debates'.[25]

Elsewhere in *Living Downstream*, Steingraber recounts being 'fascinated by the evident reciprocity between environmental activism and *Silent Spring* … that Carson was as influenced by activism and advocacy as the contemporary environmental movement was influenced—some would say inaugurated—by the publication of *Silent Spring*'.[26] But unlike Carson, Steingraber stepped into the fray with her feet as well as her writing and personal testimony. Steingraber, who helped to found New Yorkers Against Fracking and New York Health Professionals Against Fracking while writing those early pieces for *Orion*, backed up her writings by attending protests, getting arrested, giving interviews, testifying in public fora as a scientist and citizen. In another *Orion* piece she vents her frustration in attempting to enlist colleagues to speak out:

> I'm working on pulling together Biologists Against Fracking but am having a tough time of it. Even the scientists whose own research shows harm from drilling and fracking operations are hesitant to speak out. Some fear loss of

[24] Kristen Abatsis McHenry, "Fracking Women: A Feminist Critical Analysis of Hydraulic Fracturing in Pennsylvania," *International Journal of Feminist Approaches to Bioethics* 10, no. 2 (2017): 94.

[25] Cornelia Dean, *Making Sense of Science* (Cambridge, Massachusetts: Harvard University Press, 2017), 3.

[26] Steingraber, *Living Downstream*, 33.

grant money. Some believe their objectivity will be questioned—as though objectivity and neutrality were the same thing. Some are just so, so busy. And some believe fracking is inevitable. The result is political silence, a stunning position for physicians under oath and other medical professionals studying such things as cancer-causing chemicals.[27]

Facing ad hominem attacks from multiple directions, Steingraber's public written and oral testimony—as a biologist who knows the scientific literature and as a citizen who has been personally affected by the issues she elaborates on—is all the more valiant. And yet she has more than demonstrated through her own work and advocacy that such synergies as the activism that led to and was ignited by *Silent Spring* are still possible. Thanks to the successful grassroots efforts of the organizations mentioned above and allied groups, New York state has since banned fracking.

The Personal as Scientific and Political

Another synergy—between feminism and nonfiction writing—has contributed to the personal voice in Steingraber's and other writers' prose. Historians and critics, including Robert K. Musil in *Rachel Carson and Her Sisters: Extraordinary Women Who Have Shaped America's Environment*, have traced a lineage since the nineteenth century of American women who created organizations and wrote about nature in ways that 'paved the way for the conservation effort',[28] environmentalism, and various social movements. Starting with *Living Downstream*, Steingraber's bold decision to write 'autobiographically in the same book where [she is] objectively and dispassionately laying out the evidence for the link between cancer and the environment' was possible, she stated in an interview, because the feminist movement from the late 1960s onward promoted the 'idea that women's experience is not only interesting but also authoritative.…[providing] a narrative strategy to illustrate the data points'.[29] American writer-naturalist Terry Tempest Williams, best known for *Refuge: An Unnatural History of Family and Place*,[30] also figures

[27] Sandra Steingraber, "The Silence of Science," *Orion Magazine* 32, no. 1 (2013): 16.

[28] Musil, *Rachel Carson and Her Sisters*, 51.

[29] "Enviro Close-Up No. 626: an Interview with Sandra Steingraber," by Karl Grossman, *EnviroVideo.com*. March 21, 2018. https://www.youtube.com/watch?v=Xz83VjQsO6o.

[30] Terry Tempest Williams, *Refuge: An Unnatural History of Family and Place* (New York: Vintage Books, 1991 2001]).

prominently among contemporary American women trained in science who have written for the public sphere. Williams, too, has confronted themes of cancer, women, family, conservation, and the environment. Her narrative threads reveal her own experience, changing awareness, and a cyclical and spiritual frame of reference. In *Refuge*, Williams often relays a series of intense and personal observations followed by brilliant and relational summations that link her own or all human experience with the vast biotic and physical world.

For both writers the act of bearing witness is an operational principle. Steingraber has commented, 'I simply tell my own story, lay out the scientific evidence, and then step back.... I believe in speaking truth to power. I believe in personal testimony and the power of human utterance'.[31] Steingraber's personal ethos, therefore, adheres to a heuristic method; narrative is the delivery. Unlike many first-person nonfiction works that touch on the environment, Steingraber's personal stories never veer toward environmental or cultural memoir, because they are always in service of something bigger and more crucial, grounded in fact and analysis. Williams is more overt in offering sensory impressions and, at times, appealing directly to readers' sense of truth and moral actions, as in this excerpt from an article in *Orion* for which she and a photographer observed the effects of the disastrous 2010 oil spill in the Gulf of Mexico: 'This is the horror of what we are witnessing—the magnitude of our addiction and its lethal consequences for those who have no part or say in the decision'.[32] The sentence throws a gut punch, deftly shifting from visible horror to the unseen, fundamental horror of 'our addiction' to fossil fuels, by which Williams deliberately corrals readers into shared complicity and relatedness.

Ecology as Perception and Practice

As a science writer, Steingraber reports on real life, on what is happening to her and around her. She does not go looking for stories or try to find the emblematic, suffering family that will, in turn, become the public face of an epic disaster that might earn her a new book contract. Her method

[31] "Speaking Truth to Power: An Interview with Sandra Steingraber," by Simmons S. Buntin, *Terrain.org*. September 22, 2007. https://www.terrain.org/2007/interviews/sandra-steingraber/.

[32] Terry Tempest Williams, "The Gulf Between Us," *Orion Magazine* 29, no. 6 (2010), 48.

of writing, itself, appears to be ecological, providing narrative elements that model how 'ecosystems are dynamic'.[33] Just as our lives are interrelated with everything else on Earth, Steingraber structures her writing and her explanations through her own experience and her capacity to think beyond the present moment, mapping how our choices today have everything to do with a billion years ago and with whatever happens in the future. She collapses time, integrates past and present stories, and shows how lives are connected. In *Living Downstream*, when she is poring over cancer registries and trying to trace patterns in the incidence of disease in different populations and environments, she comments: 'I am writing in a moving stream of data here. Science writers always do'.[34] And in that sense, she is suggesting both that data is always being generated and that science is a process, a series of moments—a changing, moving stream (or story) that continues on.

In formulating responses to current environmental crises and anthropogenically induced climate change, the work of many academics and theorists run parallel to the relational thinking that Steingraber and other ecologists and science writers are practicing. Working within environmental humanities, for example, ethnographer Deborah Bird Rose has written of the dangers of human exceptionalism, that 'we are not a unique outlier but rather are part of various continua' of complex interactions.[35] And anthropologist Marianne Elisabeth Lien, studying human–animal ethnography, has detailed the 'multispecies entanglements'[36] that occur in communities that are constantly being made and remade. Philosopher Lorraine Code, informed by feminist epistemology, warns of the power dynamics at play in ecosystems and calls for ecological thinking and practices 'to produce "habitats" where people can live well together, locally and globally, and respectfully within the physical/natural world'.[37] Feminist theorist Donna Haraway's oft-quoted desire 'to stay with the trouble' via

[33] Steingraber, "The Fracking of Rachel Carson," 21.

[34] Steingraber, *Living Downstream*, 42.

[35] Deborah Bird Rose, "Shimmer: When All You Love is Being Trashed," in *Arts of Living on a Damaged Planet: Ghosts and Monsters of the Anthropocene*, eds. Anna Lowenhaupt Tsing et al. (Minneapolis: University of Minnesota Press, 2017), 55.

[36] Marianne Elisabeth Lien, "Unruly Appetites: Salmon Domestication 'All the Way Down'," in *Arts of Living on a Damaged Planet*, 108.

[37] Lorraine Code, "Flourishing," *Ethics and the Environment* 4, no. 1 (1999), 55.

'generative joy, terror, and collective thinking'[38] reflects the hold that such writings and readings can provide us during these precarious times.

Conclusion

Despite readers and reviewers' comparison of Sandra Steingraber to Rachel Carson—and Carson's canonization in the realm of literary journalism—Steingraber's writings have yet to receive the full attention they deserve. And there is certainly more work to be done on scientific writing as literary journalism, especially on writers—like Steingraber—whose scientific expertise and experience run deep. And while there has been an explosion of literary books and articles on environmental topics, few are by scientists who write for the general public or with literary tendencies. As Stephen Jay Gould explained, when writing, the majority of scientists 'tend to dismiss stylistic matters as essentially irrelevant to their work'.[39] But for Steingraber, language and story are as essential as what she is reporting. As a scientist, she remains committed to research and analyzing data. Her writing—at times direct and unflinching, at times more creative—conveys both gravity and imagination. Her techniques range from dialogues, scenes, wit, repetition, cadences that punctuate thoughts, to poetic phrases and asides, as though she is thinking to herself or speaking aloud to the reader. The informality is disarming; the facts are bracing. The more poetic passages encourage us to attend, as do the lines of great poets, 'by making us focus on particular groups of words in one mental moment'.[40] Although her use of personal narrative resembles the 'participant-observer voice'[41] in American advocacy journalism, her analysis is more pointed, her purview is much wider than issue-driven social reform, and she does not typically focus on relaying observations of others. As a writer, her role is compound: scientist-participant-writer or practitioner-reporter? The boundaries (and categories) are blurred. And rather than a transgression of so-called scientific or journalistic objectivity, casting herself as subject (or a data

[38] Donna J. Haraway, *Staying with the Trouble* (Durham and London: Duke University Press, 2016), 31.

[39] Stephen Jay Gould, *The Hedgehog, the Fox, and the Magister's Pox: Mending the Gap Between Science and the Humanities* (New York: Harmony Books, 2003), 163.

[40] Brian Boyd, *Why Lyrics Last: Evolution, Cognition, and Shakespeare's Sonnets* (Cambridge and London: Harvard University Press, 2012), 16.

[41] Nancy L. Roberts, "Meridel Le Sueur, Dorothy Day, and the Literary Journalism of Advocacy During the Great Depression," *Literary Journalism Studies* 7, no. 1 (2015), 47.

point) in her early works is meant to be illustrative and authentic—to recontextualize the uncertainty of potential harm, 'because behind every data point there is a life and we have to ask: How are we to protect life in the face of uncertainty'?[42] In other autobiographical moments, she does not perform but rather relates through a sequence of experiences, facts, thoughts, and emotions a way to comprehend connections and consequences. By acknowledging various stakeholders and perspectives, she contextualizes her topics and upholds journalism's potential to disrupt dominant discourses. Along with Carson and Williams, she would concur that public dangers require testimony and even activism, compelling action 'according to ecological good sense and ethical principles'.[43]

Steingraber's relational reporting stretches beyond the purely empirical and employs compelling personal narratives to communicate challenging science and keep readers involved. By including elements from her own life, thought process, and activism, she records a way to confront uncertainty, a way to dodge despair. By example, she is encouraging us to understand that if we, collectively, are to adapt and thus survive the climate crisis, we must revise many of the stories we tell ourselves. Those stories must reflect what is known about the dynamic and relational webs that transcend cultures and even time, linking us and our actions to all that is, was, and will be around us. There is no away.

References

Boyd, Brian. *Why Lyrics Last: Evolution, Cognition, and Shakespeare's Sonnets.* Cambridge and London: Harvard University Press, 2012.

Bryson, Michael A. "It's Worth the Risk: Science and Autobiography in Sandra Steingraber's 'Living Downstream'." *Women's Studies Quarterly* 29, no. 1&2 (2001): 170–182.

Code, Lorraine. "Flourishing." *Ethics and the Environment* 4, no. 1 (1999): 63–72.

Dean, Cornelia. *Making Sense of Science.* Cambridge, Massachusetts: Harvard University Press, 2017.

Gould, Stephen Jay. *The Hedgehog, the Fox, and the Magister's Pox: Mending the Gap Between Science and the Humanities.* New York: Harmony Books, 2003.

Haraway, Donna J. *Staying with the Trouble: Making Kin in the Chthuluecene.* Durham and London: Duke University Press, 2016.

[42] Steingraber, Keynote Address. See note 3 above.

[43] Bryson, "It's Worth the Risk," 171.

Lien, Marianne E. "Unruly Appetites: Salmon Domestication 'All the Way Down'." In *Arts of Living on a Damaged Planet: Ghosts and Monsters of the Anthropocene*, 107–124. Edited by Anna Lowenhaupt Tsing, Heather Anne Swanson, Elaine Gan, and Nils Bubandt. Minneapolis: University of Minnesota Press, 2017.

McHenry, Kristen Abatsis. "Fracking Women: A Feminist Critical Analysis of Hydraulic Fracturing in Pennsylvania." *International Journal of Feminist Approaches to Bioethics* 10, no. 2 (2017): 79–103.

Musil, Robert K. *Rachel Carson and Her Sisters: Extraordinary Women Who Have Shaped America's Environment.* New Brunswick, New Jersey: Rutgers University Press, 2014.

Roberts, Nancy L. "Meridel Le Sueur, Dorothy Day, and the Literary Journalism of Advocacy During the Great Depression." *Literary Journalism Studies* 7, no. 1 (2015): 44–57.

Rose, Deborah Bird. "Shimmer: When All You Love Is Being Trashed." In *Arts of Living on a Damaged Planet: Ghosts and Monsters of the Anthropocene*, 51–63. Edited by Anna Lowenhaupt Tsing, Heather Anne Swanson, Elaine Gan, and Nils Bubandt. Minneapolis, University of Minnesota Press, 2017.

Steingraber, Sandra. *Living Downstream: An Ecologist's Personal Investigation of Cancer and the Environment.* Cambridge, Massachusetts: Da Capo, 1997. 2nd edition, 2010a.

Steingraber, Sandra. *Having Faith: An Ecologist's Journey to Motherhood.* Cambridge, Massachusetts: Perseus, 2001.

Steingraber, Sandra. "Silent Spring: A Father-Daughter Dance." In *Rachel Carson: Courage for the Earth*, 49–62. Edited by Peter Matthiessen. New York: Houghton Mifflin, 2007a.

Steingraber, Sandra. "Speaking Truth to Power: An Interview with Sandra Steingraber." by Simmons S. Buntin. *Terrain.org.* September 22, 2007b. https://www.terrain.org/2007/interviews/sandra-steingraber/.

Steingraber, Sandra. "Shale Game: Playing with unknowns underground." *Orion Magazine* 29, no. 3 (2010b): 12–13.

Steingraber, Sandra. "The Whole Fracking Enchilada: Violating the bedrock, the atmosphere, and everything in between." *Orion Magazine* 29, no. 5 (2010c): 14–15.

Steingraber, Sandra. Keynote Address, presented at International Women's Day Conference, March 5, 2011a. Bard College at Simon's Rock, Great Barrington, Massachusetts. Digital Video Archive.

Steingraber, Sandra. *Raising Elijah: Protecting Our Children in an Age of Ecological Crisis.* Cambridge, Massachusetts: Da Capo, 2011b.

Steingraber, Sandra. "The Fracking of Rachel Carson: *Silent Spring*'s Lost Legacy, Told in Fifty Parts." *Orion Magazine* 31, no. 5 (2012): 14–23.

Steingraber, Sandra. "The Silence of Science." *Orion Magazine* 32, no. 1 (2013a): 15–16.

Steingraber, Sandra. "Coffee in Jail: When standing up for what's right lands you in a correctional facility." *Orion Magazine* 32, no. 5 (2013b): 15–16.

Steingraber, Sandra. Introduction to *Rachel Carson: Silent Spring & Other Writings on the Environment*, xix–xliii. Edited by Sandra Steingraber. New York: Literary Classics of the United States, 2018a.

Steingraber, Sandra. "Enviro Close-Up No. 626: An Interview with Sandra Steingraber." by Karl Grossman. EnviroVideo.com. March 21, 2018b. https://www.youtube.com/watch?v=Xz83VjQsO6o.https://www.youtube.com/watch?v=Xz83VjQsO6o

Williams, Terry Tempest. *Refuge: An Unnatural History of Family and Place.* New York: Vintage Books, 1991, 2nd edition 2001.

Williams, Terry Tempest. "The Gulf Between Us." *Orion Magazine* 29, no. 6 (2010): 35–53.

How Marjory Stoneman Douglas Saved Environmental Reporting (and maybe the Everglades)

Bret Schulte

The environment is not a beat that will cut you a quick path to journalistic glory. It's not the same as being a war correspondent scrambling through incoming fire to get to the truth of the conflict, or of human nature. Nor does it come with the prestige of the political reporter, jogging toward Air Force One with security credentials clacking. Environment reporting occasionally involves travel, sometimes to remote and exotic locales, but most often the only exploration a reporter gets is of data, or of scientific inquiry, or of regulatory policy. And this is how it often turns out:

> The National Academy of Sciences, the Intergovernmental Panel on Climate Change, and even, grudgingly, the Bush administration now believe Earth is warming. The roots of this emerging consensus go back to 1998, when climatologist Michael Mann used tree ring, ice core, and coral reef data to

B. Schulte (✉)
University of Arkansas, Fayetteville, AR, USA
e-mail: bjschult@uark.edu

P. Calvi (ed.), *The Literary Journalist as a Naturalist*, Palgrave Studies in Literary Journalism,
https://doi.org/10.1007/978-3-031-56634-9_12

> show relatively stable temperatures over the past millenniums, with a sharp spike in the 20th century.
>
> "All the new data are in the same direction, showing that warming is continuing," says Ralph Cicerone, an atmospheric scientist and president of the National Academy of the Sciences.
>
> Average surface temperatures have climbed about 1.4 degrees Fahrenheit since the early 20th century, coinciding with spiking atmospheric levels of carbon dioxide, which have ballooned 35 percent over the same period. Levels of methane, a far more potent heat-trapping gas, have jumped 152 percent since the preindustrial age. Last year was the hottest on record, and model projections show temperatures jumping anywhere from 2.7 to 10.7 degrees Fahrenheit over the next 100 years.[1]

This story ran in April 2006 in *U.S. News & World Report*, a magazine that purported to have about two million readers at the time and aimed for a middle-brow audience that was not particularly expert in governmental policy nor science. Still, this reportage provides no narrative nor description of scientist Michael Mann. It uses an expert source—faceless, a bureaucrat from the National Academy of Sciences—to basically repeat what the author has already asserted. The language is hedging. Then it continues to stone the reader to death with numbers about temperature changes. I know the journalist pored over the research. And I know the journalist believed that good journalism meant being able to tell a good story. I know this because I wrote it. What's particularly embarrassing is that by this time I had earned a Master's of Fine Arts in narrative nonfiction—and still this is what I produced. I wasn't particularly proud of the story, but I wasn't ashamed of it either. I got the facts straight. I covered a lot of ground. I talked to people who were recognized experts. My editor approved it; the wonks at the environmental nonprofits commended me for getting it right.

I did this kind of writing for almost five years at *U.S. News & World Report*, covering the environment and energy when I wasn't thrown onto something else the editors thought was more important, which could be anything. And it wasn't just me, and this writing wasn't just symptomatic of the period. Please try to engross yourself with the reporting below by the Associated Press:

[1] Bret Schulte, "Temperature Rising," *U.S. News & World Report*, April 10, 2006.

> JUNEAU - Critics and proponents are debating the merits of changes to federal environmental policy, including the effects on management of Alaska's Tongass National Forest.
>
> The largest U.S. national forest will probably be impacted by the Trump administration's recent revisions to the National Environmental Policy Act, Alaska's Energy Desk reported Wednesday. Supporters have said the changes streamline a regulatory process that hampers development on federal lands.
>
> Natalie Dawson, executive director of Audubon Alaska, said the updates engender less public engagement, which was a founding principle of the environmental act's framework.[2]

I certainly didn't think my journalism was part of a fundamental problem facing environmental journalists, which is that you are endowed with a responsibility to make a topic that is bogged down in science and public policy lively and relevant and readable so that audiences *want* to read it, and only then might someone be moved enough to act.

I've always wanted my writing to mesmerize, but I thought the science needed to take precedence. My editors did too. And there were so many facts, and only so much space. I wasn't alone. Plenty of science writing is focused on research instead of impact, a consequence of the journalist worried about getting the science right, or of getting lost in the tangled weeds of policy disputes rather than launching from the safety of the newsroom to the terrible terrain of toxic superfund sites, melting tundra, forests shaved to stubble, or the violent maw of mining operations. You face the task of tracking down locals who have suffered and summoning the courage to knock on their doors to find they are suspicious of you and have no interest in being the human face of an environmental catastrophe.

Furthermore, your editor doesn't want to pay for the expedition. A phone call to that reliable researcher at a desk is faster. And cheaper. Making phone calls to collect quotes from a disembodied voice over the phone might be the only thing possible in a lot of storytelling. But even so, even if you only have the phone, you can be crafty enough to reconstruct the course of events that led to a tragedy on a particular landscape and the creatures that lived there. A phone makes narrative possible, if imperfectly. And so does the abundance of amateur video and audio now sent streaming across social media by activists and concerned locals on the front lines. Narrative is the key to a reader's investment, no matter the

[2] Associated Press, "Federal Policy Change Sparks Debate over Alaska's Tongass," July 30, 2020.

topic. And there are few topics today freighted with such urgency as the condition of our planet. Bad science writing, the type of writing that focuses on numbers and policies rather than the living world itself, might be excused if brilliant writers had not already solved this problem. What they knew intuitively is what research has shown: our minds are wired to remember narrative because narrative is how humans have naturally communicated for eons, as brilliantly described by Lisa Cron in *Wired for Story: The Writer's Guide to Using Brain Science to Hook Readers from the Very First Sentence*.[3]

What Cron explored brilliantly, Marjory Stoneman Douglas, the mother of environmental writing, knew intuitively. Douglas's most famous work, *The Everglades: River of Grass*, published in 1947,[4] is regarded today—when it is regarded at all—as the treatise that defined her as a proto-environmentalist and feminist who worked for decades to rein in the frothing developers in Florida in the early twentieth century. But the book has another legacy, often forgotten: It demonstrated with stunning success the ability of narrative nonfiction to animate the history and significance of the natural world, to explain scientific concepts, and most importantly, to win over and energize audiences. Rachel Carson gets the lion's share of recognition for showing the power of environmental journalism to affect change. But Douglas pulled off the same feat with more style about fifteen years prior, for a much tougher cause than all the world's songbirds and flowers—saving a swamp. Douglas quite literally changed the future of the Everglades, and Florida. With the climate crisis today endangering every corner of the globe, it's worth examining how she did it.

Born Storytellers

Long before Cron's work, prominent journalists and editors, including Jack Hart of the *Oregonian*, saw narrative style journalism, or feature writing, as an antidote to the standard inverted pyramid style, which is viewed as boring and a corruption of traditional storytelling.[5] And according to Roy Peter Clark, who has authored several books about news feature

[3] Lisa Cron, *Wired for Story: The Writer's Guide to Using Brain Science to Hook Readers from the Very First Sentence* (Ten Speed Press: New York, NY, 2012).

[4] Marjorie Stoneman Douglas, *The Everglades: River of Grass* (Pineapple Press: Sarasota, 1947).

[5] John Hart, "Stories in the News," *Writer's Digest,* September 1995.

writing, the genre balanced several valences developed throughout the history of print news including: an element of human interest, narrative strategies, and that the story is inspired by the news.[6]

An impact study titled "The Power to Grow Readership"[7] by the Readership Institute at Northwestern University study found that feature-style writing increases reader satisfaction. That study supported research by the American Society of Newspaper Editors in cooperation with the Poynter Institute,[8] which found narrative stories generally outperformed other types of storytelling in communicating facts and getting readers interested enough in a story to follow it past the jump and onto the next page. This study was particularly useful because it laid out a strategy for papers with audiences they have long struggled to attract: young readers. The cohort favored the style over traditional newspaper writing, and narrative helped them better retain the information.

Research into how newspaper writing style affects reader enjoyment and comprehension is well established—and the outcomes are damning for the workhorse of the newspaper industry: the hard-news inverted pyramid style, which organizes stories by the most-important facts to least-important facts. The ASNE report points back to a 1979 study by media scholar Georgia Greene, which concluded that typical news-writing style aids in skimming but leads to clunky, disorganized writing that tries the patience of readers.[9] Typical news writing avoids storytelling; it specializes in story endings, Hart points out, meaning it puts the end of the story first in a typical summary lede.[10] This also means readers are given a conclusion without introduction to the material, its context, or an explanation of why it matters—all of which are cast deep in the text. The assumption of knowledge on the part of the reader by newspapers has resulted in alienated audiences who are left mystified by the very source whose aim is to clarify and illuminate. Frank Denton, author of the ASNE report, wrote: "When a reader picks up a newspaper and constantly collides with alien terms without which the stories have little meaning, it becomes even easier to turn on the television." Citing that many people have little literacy on

[6] Roy Peter Clark is an industry expert and consultant employed by media think tanks and newsrooms. For Poynter, he published the article "Reviving the Feature Story" in 2004.

[7] Northwestern University Readership Institute, "The Power to Grow Readership," 2001.

[8] American Society of Newspaper Editors. *Ways with Words: A Research Report of the Literacy Committee* (ASNE: Reston, 1993): 19.

[9] ASNE, 4.

[10] ASNE, 31.

such critical topics as how Congress works or what the Dow Jones Industrial Average means, he argues that "[It] is easy to understand why such a reader would find the newspaper incomprehensible, as if it were written in another language."[11]

Lost in the Woods

Good environmental writers understand that the story of the natural world is best told through the human interaction with it. They understand that the human intersection with nature provides history by generating drama. The story of nature is the story of people trying to understand it, trying to tame it, trying to kill it, trying to live with it, sometimes trying to save it.

This is not a new discovery, and while some journalists, such as John McPhee, David Quammen, and Elizabeth Kolbert,[12] masterfully deploy narrative in their reports, the idea of making environmental reportage a human story remains oddly novel, as if it's indulgent. In fact, the model for writing compelling environmental narrative was established long ago but got buried on bookshelves by more sensational genres: war (John Hersey's *Hiroshima*), murder (Truman Capote's *In Cold Blood*), and sex, drugs, and rock 'n' roll (Tom Wolfe's *The Electric Kool-Aid Acid Test*).

Douglas also understood the need for heart-pumping drama when she pioneered the use of human narrative to tell an environmental story. *River of Grass* is thick with graphic American prose that cast the making of the New World not as wide-eyed discovery and manifest destiny but as a centuries-long, blood-soaked war between natives and invaders. For example, when the Spaniards stormed a Florida beach in the early sixteenth century:

> Then there came that curious twang of a released bowstring, and in the next instant the thlock of an arrow reaching its mark, they stared at a man spinning and clutching at his throat, with an arrow sticking through it, stared as he fell and flopped like a fish, then they were all shouting together as the air thickened with hissing arrow shafts that found throats or joints or splintered

[11] ASNE, 5.

[12] Nearly any work by these journalists shows the power of human narrative in environmental reporting, though Quammen's article "Strawberries Under Ice" manages not just narrative but theme, misdirection, and metaphor typically found only in the masterpieces of fiction.

> and slashed like glass through chain mail to the vulnerable flesh. Spears came in fast and sure, spiking a man clean through before his blood could burst out, red as the flags, on the trampled sand.[13]

It's a masterful scene—a battle that goes from a single arrow to an all-out assault to annihilation in two sentences, alive with the verve of the Lost Generation but without the ennui. Douglas was a fiction writer as well as a journalist, who understood that good writing is genre agnostic: people, scenes, confliction, resolution. It's true for fiction, and it's true for stories of the natural world—at least, if you want people to read them.

As she recounts in her memoir, *Voice of the River*,[14] Douglas knew next to nothing about the Everglades when she started the project. In fact, the editor who approached Douglas originally wanted a book about the Miami River, an idea she rejected by calling the river "only about an inch long."[15] Still, she knew enough about the Everglades to know it should be preserved. And she knew what she didn't know, which meant, as a journalist, that she needed to do her research, and from her research she pioneered the way people—including scientists—thought about the Everglades. In her first visit to a state hydrologist, she learned the water was not stagnant but that it flowed, however slowly, and where that water crept among the saw grass you had, more or less, the Everglades.

As Douglas narrates in *Voice of the River*, she asked if the Everglades could be thought of as a river, and the hydrologist said that it technically fit the definition. It was moving primarily in one direction—at a rate so slow it was nearly imperceptible—but that's what made it interesting. "Do you think I could get away with calling it a River of Grass?" Douglas asked.[16] The answer was yes. Those three words were Douglas's solution to catching readers' interest. That solution, compelling and visual, is what we now call literary journalism, a metaphor grounded in research. Douglas set aside the novel she had been writing for six months and committed herself to four years of work researching and writing about an ecosystem that she had redefined, elevating it from a fetid swamp to a fantastical waterway.

[13] Douglas, *Everglades, 83.*

[14] Marjory Stoneman Douglas, *Voice of the River, an Autobiography with John Rothschild* (Pineapple Press: Sarasota, 1987).

[15] *Voice of the River*, 190.

[16] *Voice of the River*, 191.

The Untrained Environmentalist

Decades of experience as a fiction writer and a journalist had prepared Douglas to make the most difficult argument of her life: How and why to save a place widely regarded as a danger to human health and an obstacle to Florida's prosperity. There were other difficulties: she had hardly even been to the Everglades and knew next to nothing about ecology. So in telling the story of the Everglades, she connected the fauna and flora to the human history of the place. Her case to save the Everglades combined the science of the men who spent decades studying it, her own historical research of the people who had survived in and depended on the Everglades for centuries, and her talent as a writer. The result was a gorgeous narrative that brought human drama to a shunned landscape.

Perhaps Douglas's biggest asset was that she didn't consider herself an environmentalist when she wrote *River of Grass*. That advocacy work didn't start until around two decades later. When she stumbled across the topic of the Everglades in 1947, Douglas had moved on from journalism to fiction because she could make more money selling short stories to magazines like *The Saturday Evening Post*. She favored big plots with strong female protagonists, often in a natural setting.[17] The stories contained some of the strong social messages she wrote about her in her previous career as a crusading columnist for social justice at the *Miami Herald*.

> Long before she was the "voice of the river," she was an advocate of the dispossessed – of women, children, and the poor. As such she grew to understand that humans and nature were part of a larger whole, and that the good health of one depended on the other. She grew to care about the Everglades in part because she cared about people. It was that link, the connection of humans to the environment that formed the basis of Douglas's environmentalism.[18]

Though not a religious person, Douglas's views were formed by her father's Quaker upbringing that taught a moral responsibility for

[17] Jack Davis, "Conservation Is now a Dead Word: Marjory Stoneman Douglas and the Transformation of American Environmentalism," *Environmental History*, Vol. 8, No. 1, (2003) 53–76.

[18] Jack Davis, "Green Awakening: Social Activism and the Evolution of Marjory Stoneman Douglas's Environmental Consciousness," *The Florida Historical Quarterly* Vol. 80, No. 1, (2001): 44.

stewardship of the earth based on the theological concept that the earth is a creation of the divine.[19] Furthermore, that divine plan is designed so that human beings rely upon the earth and must be its stewards. As an advocate of social causes, rather than environmental ones, Douglas's natural approach to the story was to make a case for human investment in the Everglades by framing the environment as a natural source of concern for humankind.[20] And the story she told was not just of the ecology of the place, but the people who lived there. That tethering of people and place drew new and sudden interest in the Everglades, the book worked as a Florida speed bump to the century's unfettered development of natural landscapes, development fueled by progressive ideas that humankind could improve society by taming and sanitizing the land.

Douglas's reframing of the Everglades from swamp to natural wonder made the best use of all her talents and interests: a fiction writer's literary flair and development of scene and character; a reporter's obsession with details, motivations, and consequences; and finally, a cause that called for a crusader.

Here's a sample of the literary journalism from *River of Grass* that made Douglas an indispensable advocate for the Everglades by loading characters, history, color, humor, scene-setting, personification, and an understanding of drama into the story of a place that deserved saving:

> Richard Bushrod Turner, who had faced the Indians fighting with bows and arrows in the battle of Royal Palm Hammock, moved up in his boat to clear the land under that tremendous mangrove, ancient, solitary, utterly impressive. He built a shack and a rain-water cistern, and started some vegetables and a cane patch. He trapped raccoons and lived alone with the owls and the wildcats and the Indians slipping down the river in their canoes from the Big Cypress. They got along all right now.
>
> "Captain Hall went up and down to Key West with his schooner, picking up mails and skins and delivering supplies. Cuban fishing smacks put in at Caxambas for water, as they always had. Captain Bill Collier, north on Marco Island, began to put up a store and wharf after his father's boat was wrecked in the storm. There had been a preacher aboard besides the two of them when the storm came up, his father had given the preacher an ax and fell to his knees in prayer. The boat capsized, and Captain Bill's father

[19] Davis, "Green Awakening," 45.
[20] Douglas, *Everglades*.

drowned. Only he and the preacher were saved. Captain Bill claimed to have been an atheist ever after. He'd stop anything he was doing to argue about atheism with anybody, he was so religious about it."[21]

Notice the lack of data.

Human Nature

Stoneman Douglas's defense of the Everglades is, in fact, largely the tale of human discovery of the place, and the subsequent sagas of fear and misunderstanding but also awe and appreciation. Her chapters are arranged historically, not scientifically. After establishing the ecosystem of the land in a compact and wildly described opening chapter that is so visual it reads like a graphic novel, she moves into the exchange of land and people. The remainder of the book—the next fourteen chapters, in fact—is a human saga of expansionism, slavery, the clash of cultures, and the fury of colonialization as it was woven into the tangle of south Florida.[22] That interplay reveals the essential co-dependence of man and earth, even in a landscape as forbidding as the Everglades. Stoneman Douglas shares the story of Juan Ortiz, a Spaniard captured and eventually accepted into a tribe of Florida natives around 1528.

> He learned to eat with relish the Indian food, fish and shellfish roasted in the fires, and the sharp palmetto berries and hearts of palm, and deer steaks broiled and savory, and the good meat of queer little animals the Spanish had thought were dogs. He went without all clothing but a breechclout, and turned brown in the constant sun, and learned Indian words and the skills of the fish spear and net, and worked with the women at their tasks of pottery making and skin scraping and woodcutting, and constant carrying of water.[23]

The effect of telling the story of the Everglades through the people of Florida was a national sensation. In fact that technique—an idea that sounds so basic today—put *River of Grass* in the New York Herald Tribune's "What America Is Reading" list. The reviewer recommended

[21] Douglas, *Everglades*, 209.
[22] Douglas, *Everglades*.
[23] Douglas, *Everglades*, 91.

the book "to all readers concerned with... the great relations of man to nature."[24]

"I wrote from emotion," Douglas would later reveal in her memoir, "but not with that inner absorption that characterizes the novelists. That same love and emotion, I think, which was held somewhat in abeyance in my earlier periods of journalistic voyeurism, was ready to be channeled to the Everglades. It was almost as if the Everglades had waited for me. It was a cool subject, to which I could apply my passion."[25]

The passion courses through the book in writing that renders a dramatic tableau of life that is saturated in color. The fight for life in the Everglades is vivid and sometimes violent. Calling on the power of the novelist, Douglas often anthropomorphized the 3500 square-mile run of saw grass, or the churning society of plant and animals therein. At other moments, the Everglades is lyrical—a land of lush wonder and gothic growth.

> The saw grass in its essential harshness supports little else. It repelled man but on the rock the crowding forms made life abundant so that between the two the chronicle is balanced ... Southwest it was all custard apple, a subtropic, rough-barked inconspicuous tree with small pointed leaves and soft fruits. It grew fiercely, crowded on roots that became gnarled trunks twisted and arched into bracing roots in the drag of the water.[26]

What Flowed from *River of Grass*

Critics have noted that Douglas devoted more space in *River of Grass* to the human stories of the Everglades than to the ecology of the place or the ways that human interference has damaged the ecosystem. But perhaps this emphasis demonstrated that people could and should coexist with a terrain that to many seemed terrifying, impeded the manifest destiny of Anglo–American civilization, and threatened human health.

> Throughout her narrative, she encouraged readers to see the reciprocal exchanges between culture and nature as the basis for the integrity of place. In south Florida, Douglas argued, using the language of ecology, "the old subtle balance ... had been destroyed." Some students of her magnum opus

[24] Davis, "Conservation," 61.
[25] Douglas, *Voice*, 225.
[26] Douglas, *Everglades*, 28.

> have equated Douglas's open contempt for human "greed," inertia," and foolishness" as a tone that anticipated by fifteen years Rachel Carson's memorable statements in *Silent Spring* that the "control of nature" is a phrase conceived in arrogance…[27]

Our finest literary journalists who have made the environment their topic have, in one way or another, followed Douglas's lead.

"Atchafalaya," John McPhee's 1987 masterpiece, provides a meticulously researched history of America's most important waterway, the Mississippi River, as a man vs. nature narrative told in sequence with the writer's own discovery of the river's tangles, turns, and intersections with local communities from a boat with the men of the Army Corps of Engineers, who attempt to tame the water's will.[28] The result is something akin to a fable, a tale of human futility and environmental destruction wrought by man's insistence that he cleave from the earth, dominate the earth, and should therefore be able to reshape the earth. But the earth, or in this case the river, continues to move and it is humankind that acquiesces.

In 1996's *The Song of the Dodo*, David Quammen documents the devastation of fragmented and fraying ecosystems by finding himself translators, bushwhacking guides, and bemused locals to push him deep into thick and dangerous lands.[29] The story of Bali's jungle doesn't take us straight in the jungle. We wind through byzantine trails dotted with desperate heaps of shacks. There are bribes to be paid by waylaid travelers, there are cultural miscues. Human drama is the story of the environment and it's the story of reporting on the environment. Quammen reports on ecological devastation the way Anthony Bourdain eats food. The prize is the people. Each episode is a quest. In Tasmania, Quammen grows obsessed with the thylacine, a carnivorous marsupial that resembles a tiger-striped dog and has been assumed extinct since around World War II. Still, scattered sightings of thylacine have been reported for decades, some by credible sources such as park rangers, throwing Quammen into a quest for a sighting of his own. Sure, we learn about the history and the likely extinction of the animal, but we also learn about Marrawah, a pile of a few dozen people that passes as a village. It means nothing to the story of the thylacine except that it's a landing spot for an obsessed journalist,

[27] Davis, "Conservation," 62.

[28] John McPhee, "Atchafalaya," *The New Yorker,* February 23, 1987.

[29] David Quammen, *The Song of the Dodo* (Scribner: New York, 1996).

and an opportunity to show another species that can't be extracted from the story of ecological Armageddon: humankind.

Quammen finds a pub.

> "The jukebox is whomping out 'Wild Thing' by, if I'm not mistaken, The Troggs. A rangy young thug, presently in command of the pool table though drunk as snot, plays air guitar on his cue and pantomimes the vocal. Someone hollers, Take yer bloody shot, Mike. From comments tossed, from eyes rolled, I'm given to understand that Mike is an infamously tiresome character, well known if not well loved locally. He does not take his shot. He's onstage, he's transported: Waald Thang, and ah wanna know fuh shuh. He's busy. Hot licks on the cue..."[30]

Thus Quammen, with painful vividness and scientific powers of observation shows the self-absorption, the callow ignorance of the animal that eliminated the thylacine from the natural world. *The Song of the Dodo* won the John Burroughs Medal for Natural History Writing.

In 2015, Elizabeth Kolbert won the Pulitzer Prize for her masterpiece of literary, environmental journalism, *The Sixth Extinction*—an honor that eluded even *Silent Spring*. Her book is a cry of alarm gathered from each corner of the earth—Panama, Italy, Scotland, the Great Barrier Reef, Brazil, documenting the cataclysm facing flora and fauna from human interference.[31] Each destination provides both travelogue and tragedy, populated by scientists and activists desperately working against the unstoppable forces that have resulted in wild takeovers of invasive species, the decline of the diversity of fauna and flora because of international commerce, tourism, and climate change. Notably, Kolbert manages to provide a thorough rendering of the history of the concept of extinction—the decades-long debate over evolution and the origins of life that consumed scientists for much of the eighteenth and nineteenth centuries. Even while doling out a history lesson for much of the first hundred pages, Kolbert takes us somewhere—sets a scene, with full-blooded people engaged in academic pursuit, in this case, it's the Museum of Natural History in Paris. While talking to the inheritors of evolutionary science, she brings their forebears to life as she shows them engaged in a sometimes vicious debate over the nature of life and the controversial concept of extinction.

[30] Quammen, *Dodo,* 298–299.

[31] Elizabeth Kolbert, *The Sixth Extinction: An Unnatural History* (Picador, New York, 2004).

Survey Says…

What Douglas postulated in 1947, and what McPhee, Quammen, and Kolbert have emphatically proven since then, is that the path to writing about science and the natural world is through people. Still, environmental reporting largely languishes in the opaque world of theory, jargon, and policy. Difficult concepts are explained with difficulty because there are so few examples of why those concepts are relevant, of who is impacted. In 2015, the American Press Institute surveyed news consumers on areas they are most passionate about. The tops were sports and politics. Science lay at the bottom, alongside celebrities—despite the fact that most readers worldwide consider science to be of utmost importance, particularly climate science.[32] An international poll conducted by the Reuters Institute and the University of Oxford of forty media markets that span the globe showed 69 percent of respondents consider climate change an extremely or very serious problem.[33]

While it can't be said that Douglas alone helped preserve the Everglades through the power of storytelling, her work proved vital in changing public attitudes. When Congress finally appropriated the money to conserve the Everglades, the Florida senator Claude Pepper nodded to Douglas's work by referring to the ecosystem as the "River of Grass" in his speech at the ceremony (Davis 2003).[34]

Today, Marjory Stoneman Douglas is called the Defender of the Everglades on the website of the National Park Service. The institution credits her for recognizing that the swamp is a complex interaction of ecosystems. And if you want to visit the Everglades today, you'll find most tours don't market the Everglades as the Everglades. They're selling people on something else. They like to call it the River of Grass.

References

American Society of Newspaper Editors, *Ways with Words: A Research Report of the Literacy Committee*. Reston: ASNE, 1993.

Associated Press, "Federal Policy Change Sparks Debate over Alaska's Tongass," (July 30, 2020).

[32] Tom Rosenstiel, et al. "Twitter and the News: How People Use the Social Network to Learn about the World," (American Press Institute, 2015). https://www.americanpressinstitute.org/publications/reports/survey-research/how-people-use-twitter-news/

[33] Nic Newman, *Reuters Institute Digital News Report*, (Reuters Institute, 2020.)

[34] Davis, "Green Awakening," 50.

Beasley, Berrin, "Journalists' Attitudes Toward Narrative Writing," *Newspaper Research Journal* Vol. 19, No. 1 (1998) 78–89.

Clark, Roy Peter, "Reviving the Feature Story," 2004. Accessed Sept. 10, 2021 at https://www.poynter.org/archive/2004/reviving-the-feature-story/

Cron, Lisa, *Wired for Story: The Writer's Guide to Using Brain Science to Hook Readers from the Very First Sentence.* New York: Ten Speed Press, 2012.

Davis, Jack, "Green Awakening: Social Activism and the Evolution of Marjory Stoneman Douglas's Environmental Consciousness," *The Florida Historical Quarterly* Vol. 80 No. 1 (2001) 43–77.

Davis, Jack, "Conservation Is now a Dead Word: Marjory Stoneman Douglas and the Transformation of American Environmentalism," *Environmental History* Vol. 8 No. 1 (2003) 53–76.

Hart, John, "Stories in the News," *Writer's Digest* (September 1995).

Kolbert, Elizabeth, *The Sixth Extinction: An Unnatural History,* New York: Picador, 2004.

McPhee, John, "Atchafalaya," *The New Yorker,* (February 23, 1987).

Newman, Nic, Reuters Institute Digital News Report, 2020 accessed Oct. 1, 2021 at https://reutersinstitute.politics.ox.ac.uk/sites/default/files/2020-06/DNR_2020_FINAL.pdf

Schulte, Bret, "Temperature Rising," *U.S. News & World Report* (April 10, 2006).

Readership Institute, "The Power to Grow Readership," Northwestern University, 2001.

Rosenstiel, Tom, Sonderman, Jeff, Loker, Kevin, Ivancin, Maria, Kjarval, Nina, "Twitter and the News: How People Use the Social Network to Learn about the World," *American Press Institute,* September 1, 2015, accessed Sept. 10, 2021 at https://www.americanpressinstitute.org/publications/reports/survey-research/how-people-use-twitter-news/

Stoneman Douglas, Marjory, *Voice of the River, an Autobiography with John Rothschild*, Sarasota, FL: Pineapple Press, Inc., 1987.

Stoneman Douglas, Marjory. *The Everglades: River of Grass,* Sarasota, FL: Pineapple Press, Inc., 1947.

Quammen, David, *The Song of the Dodo,* New York: Scribner, 1996.

The Naturalist as Literary Journalist: David Attenborough's Sixty Years of Documentary Film-Making

Isabel Soares

> Zoology had changed since I was a student at Cambridge. Then [...] there was no suggestion that we might ultimately, as qualified zoologists, watch elephants in Africa or crouch in a hide in the depths of a tropical forest watching some rare bird at its nest. That was what naturalists did. Not scientists. But that was what I really wanted to do.
>
> —*(Attenborough, 2010: 326)*

Introduction

No (hi)story of documentary film-making can ever be validated without reference to David Attenborough (b. 1926). Natural science documentaries, in particular those relating to Earth, Nature, and Man's place in the

I. Soares (✉)
ISCSP, CAPP, Universidade de Lisboa, Lisboa, Portugal
e-mail: isoares@iscsp.ulisboa.pt

P. Calvi (ed.), *The Literary Journalist as a Naturalist*, Palgrave Studies in Literary Journalism,
https://doi.org/10.1007/978-3-031-56634-9_13

biosphere of our planet have evolved alongside Attenborough's remarkably long career as a broadcaster. A household name, his documentaries have captivated generations of audiences to the point that "in our fractured age, Attenborough is the closest we have to a universally beloved public figure."[1] Even A. A. Gill, late British writer and columnist known for his acerbic, controversy-prone style, enshrined Attenborough in a sort of Olympus beyond the reach of mere mortals: "If you ask [the English] what they think the Almighty will be like when they get to heaven, most of them will say David Attenborough."[2] As he has aged, so has his popularity increased. Debuting on Instagram on September 24, 2020, a coincidence to mark the release of his biopic *A Life on Our Planet*, his account reached one million followers in a swift 4 hours and 44 minutes thus breaking the world record previously held by actress Jennifer Aniston.[3] Enjoying pop stature and status, Attenborough can just as easily be seen among world leaders as making appearances before roaring crowds at music festivals. Whether interviewed by both President Obama in 2015, and HRH the Duke of Cambridge, in 2019, or acting as keynoter at the World Economic Forum at Davos, in 2019, Attenborough also has the rock star appeal that took him to the main stage at Glastonbury on June 2019 to promote his series *Seven Worlds, One Planet* and talk about the need to cut down on plastic use. His stellar credentials need no introduction. However, a biography of his life and work is yet to be written. As a prolific writer in the general category of non-fiction, much autobiographic documentation abounds from which to draw facts, anecdotes, dates, and milestones regarding Attenborough's career. Critical biography not entangled in the memories, feelings, and perceptions of self and his personal worldview are lacking and for that matter, Attenborough's biographic construct from an outside perspective needs to be erected on a varied corpus, including not just his books of an autobiographic nature but also interviews to media outlets, speeches before world organizations, reviews of his shows and books, dispersed articles relating to his work. Having been a public figure for over six decades, it means that, on the one hand,

[1] Patrick Barkham, "The Real David Attenborough," *The Guardian*, 22 October 2019. https://www.theguardian.com/tv-and-radio/2019/oct/22/david-attenborough-climate-change-bbc, accessed April 16 2020.

[2] A. A. Gill *The Angry Island: Hunting the English* (Phoenix: London, 2005), 140.

[3] Justine Calma, "David Attenborough Broke a World Record on Instagram," *The Verge*, 29 September 2020. https://www.theverge.com/2020/9/29/21493467/david-attenborough-instagram-guinness-world-record, Accessed June 6, 2021.

research pertaining to his work confronts the challenge of finding a path through the enormity of written or audiovisual material from whence to gather information. On the other hand, it is a striking finding worthy of mention that in these six odd decades no comprehensive, diachronic study of his biography has ever been attempted.[4] A caveat in our study is hence the unavoidable basis on much autobiography when addressing matters of Attenborough's life as a broadcaster.

Methodologically, our study is qualitative in nature and based mainly on two of the principles laid by Morse (2006): insight and inference. By insight it is meant that the researcher "must have much knowledge about what they are observing or hearing, know and be able to link it to relevant literature and be able to think conceptually and to link seemingly unconnected events, representations and ideas."[5] A lifelong viewer and reader of Attenborough's series and books and a specialist in literary journalism, the researcher in this particular instance is equipped with knowledge providing leverage to the research's objective: to show Attenborough's documentaries as instances of literary journalism. This aprioristic knowledge was strengthened by furthering investigation on bibliography (written and audiovisual) regarding Attenborough's documentarist work. Cumulatively, inference, as postulated by Morse is what "allows us to make linkages; [...] to recognize new instances of a case, inference allows us to categorize—to put somewhat similar things together—in other words, to create categories and themes [...], to link concepts, to create theory, to apply our research results, and move forward."[6] If looking at Naturalism as a possibility of literary journalism is incremental to the studies of both areas, so is a "move forward" connecting Attenborough's naturalist documentaries to the category of literary journalism by substantiating evidence to that statement.

In that vein, the epigraph, used to start a conversation debating David Attenborough not as the famed wildlife documentary presenter but as a literary journalist does not purport, as so usual in epigraphic transcript, to bring color or awaken curiosity to the topic. Attenborough's words are a

[4] On purpose we are excluding a biography of David Attenborough written for children: Sally Morgan *The Incredible Life of Sir David Attenborough* (Harper Collins: London, 2017).

[5] Janice M. Morse, "Insight, Inference, Evidence, and Verification: Creating a Legitimate Discipline," *International Journal of Qualitative Methods* Vol. 5, No. 1, 2006: 95. https://journals.sagepub.com/doi/pdf/10.1177/160940690600500108, accessed August 31 2021.

[6] Morse, "Insight," 96.

direct, first-person acknowledgment that his career was intent at shattering conventional or expected roles. From his perspective, naturalists are scientists. From our added discussion, naturalists can be literary journalists.

* * *

Addressing the UN Climate Summit in 2018 as representative of the People of the World, a 92-year-old David Attenborough urged global decision-makers to take effective action against climate change. The moment was solemn. Attenborough's standing as the "inventor" of the natural world documentary as we know it since the 1950s, and the genre's epitome ever since, make him a respected authority on matters concerning the environment and life on Earth. His career longevity further allows him a unique position of privileged witness to the ways nature has changed in consequence of anthropogenic risks. Revealingly, the subtitle to the book that pairs with the film documentary *A Life on Our Planet* is "My Witness Statement and a Vision for the Future" (2020). The firsthand vision of the "I," Attenborough's, despite mediated by an external element to the eye of the "I," the camera, is thus the premise on which truthfulness of the account reported on both book and film rests: "The natural world is fading. The evidence is all around. It has happened during my lifetime. I have seen it with my own eyes. It will lead to our destruction."[7] The somber warning in these words is given not just credibility but also pungency by the use of first person. Alastair Fothergill, who produced several of Attenborough's series, directed *A Life on Our Planet*, and is former head of the BBC Natural History Unit, noted in an interview with Patrick Barkham that Attenborough's professional activity spans two distinct geological eras. When he started making documentaries in the mid-twentieth century we were in the Holocene, whereas now the dominance of the *Homo sapiens* species over the planet has brought forth a new era, the Anthropocene. No other human being in history has "seen more of the natural world" or seen so much change as Attenborough.[8] A conclusion from these words is that hardly is there a more validated witness to the perils facing life on Earth than Attenborough.

[7] David Attenborough, *A Life on Our Planet: My Witness Statement and a Vision for the Future* (Witness Books: London, 2020), 7.

[8] Barkham, "The Real."

Aware of the fact that David Attenborough's work is not immune to criticism, that he has often been accused of airbrushing "human beings from his landscapes,"[9] using anthropomorphic characteristics to describe animals,[10] of ignoring homosexuality in the animal kingdom,[11] or sidestepping environmental problems,[12] we analyze the genesis of the natural history documentary and the part played by Attenborough in creating its narrative grammar. In a pioneering way, we also locate Attenborough's output in the field of literary journalism given the fact that natural history documentaries are narrative in mode and take hold of literary techniques, such as scene construction and character composition, which are ubiquitous in literary journalism, delivering what Joseph calls the "empathic" quality of journalism.[13] The standpoint we take is to look at Attenborough, the naturalist, as literary journalist and not the other way around. Attenborough does not come from the ranks of journalism and, for all intent and purpose, that premise should be acknowledged. In this approach, naturalism is linked to journalism while emphasis is placed on Attenborough's role in the dissemination of "nature as a story for the masses." Finally, we explore Attenborough's natural history journalism diachronically, so as to see its evolution from "conquest" of the natural world to "understanding" the need for its preservation.

[9] Laurie Taylor, "Watching David Attenborough: Laurie Taylor Talks to the Renowned Naturalist," *New Humanist*, 4 January 2008. https://newhumanist.org.uk/articles/1673/watching-david-attenborough-laurie-taylor-talks-to-the-renowned-naturalist, Accessed April 16, 2020.

[10] Taylor, "Watching" and Hale, Mike, "David Attenborough is Back, And He Brought the Walruses," *The New York Times,* 15 January 2020. https://www.nytimes.com/2020/01/15/arts/television/review-seven-worlds-one-planet.html, Accessed May 4, 2020.

[11] Brett Mills, "The Animals Went in Two by Two: Heteronormativity in Television Wildlife Documentaries," *European Journal of Cultural Studies* Vol. 16, No. 1, 2013: 100–114 and Tom Jeffreys, "Representation for Animals: An interview with Dr. Brett Mills," *The Journal of Wild Culture*, 2013, https://www.wildculture.com/article/representation-animals-interview-dr-brett-mills/1142, accessed April 16, 2020.

[12] George Monbiot, "David Attenborough Has Betrayed the Living World He Loves," *The Guardian*, 7 November 2018. https://www.theguardian.com/commentisfree/2018/nov/07/david-attenborough-world-environment-bbc-films, accessed April 27, 2020.

[13] Sue Joseph *Behind the Text: Candid Conversations with Australian Creative Nonfiction Writers* (Hybrid Publishers: Melbourne, 2016), xvii.

Serendipity

Attenborough's personal and professional odyssey of bringing the wonders of the natural world to millions of bewildered TV viewers had modest beginnings in an era when television was dawning as a medium for communication. Born in the inter-war period, David Frederick Attenborough majored in natural sciences from Clare College, Cambridge. Summoned for national service in the Royal Navy after the end of World War II, he took a job editing children's science books after his commission ended. Reading galley proofs lacked excitement and intellectual challenge[14] which led him to apply for a position as producer of radio talk shows with the BBC in 1950. He did not get the job but his vita found its way to Mary Adams, the head of the Talks Department, the BBC's branch for non-fiction. Adams thought the young applicant could be offered a training course at the Corporation's still infant Television Service.[15] At the time, in the early 1950s, Attenborough did not own a television set having only once watched a televised program at his in-laws', the only people he knew who did own a TV.[16] This was all his knowledge of the new medium when he accepted the three-month training course with no guarantee of getting a contract.

Attenborough's first assignment with the BBC was as co-interviewer at an afternoon show, targeting a mostly-female audience, called *Joan Gilbert's Weekend Diary.* His interviewee was a non-cooperative Olympic long-distance runner named Gordon Pirie, whose answers were, Attenborough recalls, a prodigal monosyllabic "Yeah."[17] Attenborough's contribution to the show was abruptly left at that. It would be decades before he found out the reasons for his abrupt dismissal from the show on an old note written by Adams: "David Attenborough is intelligent and promising and may well be producer material, but he is not to be used again as an interviewer. His teeth are too big."[18] After this short-lived début, he made a second appearance on the small screen, this time on a show about human ethnicities. Production needed "an example of a

[14] Attenborough, *A Life*, 9.

[15] David Attenborough *Life on Air: Memoirs of a Broadcaster* (BBC Books: London, 2010), 10.

[16] Attenborough, *Life*, 10.

[17] Attenborough, *Life*, 13.

[18] Attenborough, *Life*, 13.

Caucasian racial type"[19] to be pictured full-faced and in profile and Attenborough was the one at hand. These were not the most auspicious beginnings for someone who would rise up to be one of the most instantly recognizable faces on television, big teeth or not.

The serendipitous opportunity presented itself when a pre-historic fish (the *Coelacanth*), thought to have been long-extinct, was discovered in the sea depths off the Comoro Islands. Adams quickly devised a ten-minute program to mark the moment. As the only person around knowledgeable in natural sciences, Attenborough was assigned to help out on the show and assist Sir Julian Huxley, the then leading authority on Evolutionary Biology. Sir Julian had been invited to explain viewers why the newly discovered species was so important for the history of vertebrates.[20] Mary Adams was pleased with the final result and Attenborough had proven his worth for shows on the natural world. At the end of his training course, he was offered a six-month contract as assistant producer. It was 1952 and he was now a staff member of the BBC.

Creating a "Visual Grammar"

To trust his Biography Channel page, Attenborough holds, among his many records, those of the most travelled human in world history and of the "undisputed father of the modern nature documentary."[21] Watched by audiences of millions, his cutting-edge shows are the paradigm used by global networks including the BBC, Discovery Channel, or National Geographic, and, more recently, by streaming platforms such as Netflix, when catering for the public of natural history documentaries. Central in our argument is, however, that David Attenborough's overall work as a broadcaster can be regarded as that of a committed journalist. More precisely, a literary journalist.

When he landed on the BBC early in the 1950s, the Talks department was, within the Corporation, as the BBC was also known, the only exclusively televisual department. Televised image was in its infancy and everything was still to be invented. As Attenborough recalls on the problems of creating a language for televised contents:

[19] Attenborough, *Life*, 14.

[20] Attenborough, *Life*, 14–15.

[21] Biography.com Editors, "David Attenborough Biography." *Biography Channel*, 2020, https://www.biography.com/personality/david-attenborough, accessed April 21 2020.

> We felt it was up to us to invent a new kind of visual grammar – and we took the task seriously. Our discussions over coffee in the canteen were vigorous and never-ending. What, visually, was the best way to change shots …? Was it dishonest to mix film sequences shot earlier with live action without making it explicit we were doing so?"[22]

The BBC's path toward a paramount position within the audiovisual medium was beginning much thanks to the Talks department. In fact, a lot of unorthodox experimenting with televised material was coming from Alexandra Palace, the headquarters of Talks. If the press called the BBC "Auntie," Attenborough agrees it behaved as a stern parenting figure toward the department. Even if, as far as the news was concerned, "Auntie" did not quite know what to make of it:

> If Auntie did exist, she lived in Broadcasting House. She tended to regard the fashions and moral attitudes of yesterday as being eternal. She was dignified in language and manners and certainly knew what was best for people – even if they didn't. She regarded her young offspring at Alexandra Palace as feckless, irresponsible, and occasionally inclined to naughtiness.… Auntie had been very generous in giving us amusing toys to play with – our studios …
>
> Auntie also had a view on what and when viewers should view. News, it was felt, was not really suitable for television. Pictures were not to be trusted.… And then there was the knotty problem of what a news-reader was to do. If he […] read from a piece of paper in front of him, it would be very boring. But … if he tried to engage the viewer by looking at the camera, his words might lack authority.[23]

This quote speaks volumes about the struggles of journalism to adapt to the new audio-visual medium that came to dominate the twentieth century. It was against this backdrop of uncertainty and experimentation that Attenborough would pioneer the change within the way the BBC broadcast image. Firstly, however, there needed be a defining moment that set the change in motion.

In the 1950s, BBC shows about wildlife consisted mostly of having specimens brought to studios from zoos and have a specialist talk about them. Confounded by the lights and the hostile environment of the

[22] Attenborough, *Life*, 15–16.
[23] Attenborough, *Life*, 17–18.

studio, animals did all but behave naturally and many incidents happened, including the escape of an African squirrel which managed to live for days in the ventilation system making occasional *impromptu* appearances on "dramas, variety shows and epilogues."[24] If the BBC was to have any socially relevant pedagogical role, a change in approach was needed. As zoos frequently needed to replenish their stocks of wild animals, conservation issues not being a priority at the time, there came an opportunity for Attenborough to join a London Zoo expedition in search for wild animals. The BBC would send a filming crew along to Sierra Leone, a British colony at the time, to shoot the quest for a rare bird, *Picathartes gymnocephalus*, and a show was designed. It was called *Zoo Quest* and would be the *paterfamilias* of wildlife documentaries as we know them today.

It was 1954; filming in the damp, dense jungles of Western Africa was daunting, and there were no guiding models to follow. The forested environment was very dark for the type of film available at the time, and there was thought of cutting down trees to allow light to shoot in color. Further hindrances included bulky equipment which needed to be swapped by lighter one, at the time considered amateurish and below the standards of the BBC's filming, but Attenborough and his crew were confident they would succeed in delivering a finished product to the Corporation.[25] After two months' shooting on-site and six weeks production, *Zoo Quest* was aired. On episode two, Attenborough found himself unexpectedly in front of the camera because Jack Lester, London Zoo's Curator of Reptiles and original show host, fell ill with malaria and, proverbially, the rest is history. The episodes depicting the efforts to find *Picathartes* and to catch animals in their habitat for the London Zoo's collections received good reviews and audiences grew. When a new series was commissioned, Attenborough had the idea of going to British Guiana, another one of the then colonies of the Empire, in South America.[26] His time of intense traveling, script writing, and presenting had started. So too had the modern documentary.

With *Zoo Quest*, wildlife shows inaugurated a format still recognizable today. A narrative is built so that a story can be told about a certain

[24] Attenborough, David *Adventures of a Young Naturalist* (London: Two Roads: London, 2017), ix.

[25] Attenborough, *Adventures*, xii–xiv.

[26] British Guiana became independent from Britain in 1966, eleven years after David Attenborough visited it for his second *Zoo Quest* series.

species, a behavior or any feature of the natural world. In the first season of *Zoo Quest*, the main story was that of finding a rare bird, and the scenes during the "quest" showed a presenter guiding and involving the viewer in that chase. Meanwhile, other parallel narratives were constructed to talk about animals and peoples who were also encountered on the journey. The format would prove enduring: a presenter ventures in the natural habitats to show nature and its exoticism from a closer perspective to the audiences at home. In hindsight, it is possible to say this was a time of "conquest." It was the conquest of the natural world documentary and conquest of Nature, now visually dissected, specimens filmed as being retrieved from their environments for the enjoyment and the education of viewers. For Attenborough, "jumping on animals" stands as the greatest regret in his career,[27] but in the 1950s, environmental issues were not immediate worries.

Despite the absence of conservation concerns, a grammar for the documentary had been created and its mode was that of visual narrative, a story to be told: the quest for an animal, the explanation of a behavior or of an adaptation of an organism to the environment. *Zoo Quest*, which ran for seven seasons between 1954 and 1963, was the experiment that allowed the later groundbreaking releases of *Civilization* (1969) and *The Ascent of Man* (1973), the documentaries which trail-blazed color television and the "spectacular series."[28] Still, if any major subject could be treated "spectacularly" in a documentary fashion, it was the natural world. *Life on Earth* premiered in 1979. For Attenborough, who had devoted three labor-intensive years to the project, narrative was the grabbing element of documentaries because it "carries people from one sequence to another."[29] Simply, just like literary journalism is a modal genre of narrative[30] so too documentary emerged as a visual narrative genre. In the case of *Life on*

[27] Robin McKie, "David Attenborough: Force of Nature," *The Guardian*, 28 October 2012. https://www.theguardian.com/tv-and-radio/2012/oct/26/richard-attenborough-climate-global-arctic-environment, Accessed May 4, 2020.

[28] Attenborough, *Life*, 212. It was David Attenborough who, while controller of BBC2, commissioned the documentary series *Civilisation* (1969) with Kenneth Clark and Jacob Bronowski's *The Ascent of Man* (1973). As an aside, it was also during his tenancy as controller of BBC2 that Attenborough took a chance on a crackpot comedy by the name of *Monty Python's Flying Circus* (1969).

[29] Jacquelin Burgess and David Unwin. 1984. "Exploring the Living Planet with David Attenborough," *Journal of Geography in Higher Education* Vol. 8, No. 2, 1984: 95.

[30] John C. Hartsock *A History of American Literary Journalism: The Emergence of a Modern Narrative Form* (University of Massachusetts Press: Amherst, 2000), 11.

Earth the overarching narrative was that of the evolution of living creatures from their emergence as unicellular organisms to the complexity of vertebrates, and the ways they found to populate the planet in different geographies and conditions to form ecosystems. As an example, in the 1990 series *The Trials of Life*, narrative can be perceived in the very thread of the episode's titles which tell the stories of living creatures in today's geological era from birth (episode titled "Arriving"), to the stages in an animal's life until it dies. Within the episodes there are also shorter, more specific narratives that can focus on a developmental stage, a courting process leading to mating, a hunt with a killer and a victim, a fight for territory with winners and losers.

Mediation and Storytelling

In the documentary format inaugurated with *Zoo Quest* and perfected on *Life on Earth*, as a presenter, David Attenborough assumed not only the role of the knowledgeable entity on a subject but also that of the seeker of information, an attitude not too different from that of a journalist. This, however, is a simplistic generalization based on what is viewed on screen. Firstly, there are myriad forces at work in transforming the natural world into a visual commodity with information/documentary value. Attenborough calls the process a "complicated jigsaw"[31] involving putting pieces together which have sometimes been shot in different continents, perhaps even years apart.[32] Secondly, and more relevant for our analysis, is the fact that Attenborough is not solely a presenter of documentaries. He combines presenting with film-making and writing,[33] investing considerable detail in writing the scripts for his series from "precise turns of phrase in the narration" to "precise camera moves."[34] In these complexities, it is possible to locate the naturalist as literary journalist.

One of the most remarkable segments in *The Trials of Life* is the sequence showing chimpanzees hunting monkeys. Firstly, such behavior had never been shown to the public before. Secondly, these hunting chimpanzees in the Ivory Coast behaved in a manner distinct from the way popularized by the more amiable chimpanzees studied by Jane Goodall,

[31] Attenborough, *Life*, 294.
[32] Attenborough, *Life*, 294.
[33] Burgess and Unwin, "Exploring," 93.
[34] Burgess and Unwin, "Exploring," 95.

and their killer instincts if and when presented to audiences could "be only too easy to sensationalize."[35] Narrative provided a background and a plot complete with climax and a lesson to be had from watching primates closely connected to our own species engaged in a hunt that involved cunning and strategy to be successful. In this segment, here used for purposes of narrative illustration, a band of chimpanzees are seen as a group engaged in grooming activities or picking up the larvae from an ant colony. Then, a troop of colobus monkeys[36] is seen feeding in the canopy not far from the group of chimpanzees. Quickly, the behavior of the chimpanzees changes and a strategy is mounted to circle and ambush the monkeys with younger male chimps and a senior one playing different roles. When the monkeys notice the presence of the chimpanzees, fleeing chaos ensues. The female chimpanzees catch up with the rest of the group hooting and the male chimpanzee lying in ambush grabs a monkey and kills it by biting it in the back. Afterwards the carcass is dismembered and shared among the group without the senior chimpanzee having any privileges for having made the kill. Studio editing enabled the telling of the story, the killing part having actually been filmed on the sixth day of shooting with that particular group of chimpanzees which, on that day, were on a different corner of their territory than when they had been filmed eating ants. For the naturalist-literary journalist presenting the episode, this sequence is a staunch reminder that although the blood-smeared faces of the chimpanzees may "horrify us, [...] we might also see in them the faces of our long-distant ancestors. And if we are appalled by their mob-violence and blood-lust, we might also see in their actions the origins of the team-work that has [...] brought human beings some of their greatest achievements."[37] Like all stories, this one builds up to a moment of tension which, in its conclusion, entails some educational function with broader meanings.

This particular segment, which we choose to transcribe with parenthetical contextual annotations for script clarification, illustrates how narrative supports the scene Attenborough wants to bring to viewers. Also, it is a point in case of the naturalist being an investigative, immersive reporter as Attenborough is on location, witnessing and closely monitoring the hunt and then verbalizing to his audience how events are unfolding, who are the characters involved, and what roles they are playing. As a literary

[35] Attenborough, *Life*, 329.

[36] Family of arboreal monkeys native to eastern, central and western Africa.

[37] Attenborough, *Life*, 332.

journalist, Attenborough brings point of view to the narrative, addresses viewers directly, uses first-person narration, and makes informed judgments and conclusions to what he is reporting. For our transcript, we used the full six-minute, thirty-one-second clip as can be watched on the BBC Two website for *The Trials of Life* titled "Chimps on the hunt."[38] Punctuation, based on Attenborough's inflections of speech, is of our own making. It should also be noted that, because natural history documentaries, like journalism, are based on true happenings, in this case a hunt ending in a killing, this clip comes with the warning that it "contains scenes which some viewers may find disturbing."[39] As a further layer of warning, one has to click on the option "I am 16 years old or above" and also on the option that turns on parental guidance or not[40] in order to proceed to watching the clip.

> (Full screen image of trees where a group of primates can be seen on branches of the canopy grooming one another. Attenborough speaks in voice over).
>
> "Here, in the thick forests of the Ivory Coast in West Africa, live animals that are our closest relatives... chimpanzees. A peaceful scene of jungle harmony."
>
> (At second 29, Attenborough appears on screen. He is dressed in khaki, crouching at some distance from the group of chimpanzees, whispering so as not to disturb them).
>
> "You don't know me or think of them as hunters, more as gentle vegetarians munching food and picking leaves. But, if you follow them for any length of time in their true home, these forests in West Africa, you discover that they are hunters. What's more, they hunt in teams and have a more complex strategy than any other hunting animal except, except [*sic*], of course, Man." (Images of chimpanzees eating leaves alternating with the image of the narrator). "And one of the hunters, the experienced male, is sitting right there." (Focus shifts to the group of primates and then attention is drawn to a close-up of one of them yawning with its mouth wide open and teeth, specially large canines, showing).
>
> "This is the time they hunt, the wet season. Their regular prey are monkeys but they're very selective." (Image changes to a different primate leaping from branch to branch before the attentive gaze of the chimpanzees).

[38] BBC Two *The Trials of Life: A Natural History of Behaviour*, 2009, https://www.bbc.co.uk/programmes/p004hd8g, accessed June 29 2022.

[39] BBC Two, *Trials*.

[40] BBC Two, *Trials*.

"The diana monkey,[41] a big species, and one they seldom tackle." (More images of other primates). "A spot-nosed monkey,[42] a red colobus, they're much better jumpers than chimps and being half their weight, they can go on to much thinner branches. So, in theory, a chimp can't catch them. The only way they can do so is by working in team and there are half a dozen experienced males in this scattered group of around sixty who regularly do so." (Sequence shows chimpanzees walking at ground level, some of which carrying infants on their back. Attenborough is now filmed starting to follow the moving group with the camera shooting him from behind). "This is one of them." (Close-up of the mentioned experienced male walking amongst the foliage followed by other members of his troop).

"From the purposeful way in which he is walking, it's clear that the search for prey has started. The other members of the team are not far away. They have been steadily following the monkeys for about twenty minutes looking for an opportunity." (Camera changes angle and now focuses directly on Attenborough from the front while he, perspired, his clothes drenched from the forest's dampness, narrates the start of the hunt).

"The technique they will almost certainly use is that one of them will be driving the colobus ahead of him and there will be others that go up on either side who are blockers, who will make any attempt to catch the monkeys and then there are chasers who go and grab a monkey, if they can. And, finally, there's one male who will go up ahead and ambushes so bringing the whole trap close." (This part of the narration is only visually focused on Attenborough, who, for clarity gesticulates the acts of grabbing, ambushing and closing a trap. After the narration, the frames pass on to a colobus running in the canopy while followed by two chimps).

"The monkeys are now getting alarmed." (Shrieks are heard). "A driver's going up to prevent the group from settling and to drive them towards an area where they're more easily trapped." (Images show a chimpanzee climbing a tree trunk, followed by images of other chimpanzees monitoring the monkeys' movements and Attenborough walking fast to keep up pace with the events he is describing).

"That's, that's [*sic*] one of the blockers that's now [indistinct] come ahead of the colobus. He's half way up the tree now. He's deliberately making himself conspicuous. Now it looks as though they're all in position." (Shown on screen a group of chimps working in close ranks).

[41] *Cercopithecus diana*, a monkey species found in the canopy forests of Sierra Leone, Liberia, the Ivory Coast.

[42] Attenborough is not specific whether the images show a greater spot-nosed monkey or a lesser spot-nosed monkey.

"The drivers got them up, the blockers have gone up and now the one who's going to make the ambush and close the ring, he's gone up too. The colobus will be very lucky if they escape." (Loud, piercing shrieks, fast images of the hunt intermingling with short sequences of Attenborough's efforts of keeping up with what is going on).

"They've got one!" (Excitement in Attenborough's voice and images of what looks a frightened colobus screeching and secured beneath the feet of multiple chimpanzees). "The chimps are tearing it apart! Everyone, the hunters in the trees and the spectators on the ground, are screaming with excitement." (Shown a monkey's carcass being dismembered). "And now the kill is brought down so that the females and others share it." (Images reveal chimpanzees eating the meat of the kill). "And there's a reward for that long chase: the divided body of a colobus monkey." (Attenborough seen crouching again not far from the feeding group and watching them. He then turns his face to the camera while, in the background, the chimpanzees keep eating morsels of the colobus they have just killed). "These blood-stained faces may well horrify us but we might also see in them the faces of our long distant hunting ancestors and if we are appalled by that mob-violence and blood-lust, we might also see in that too, perhaps, the origins of the team work that have, in the end, brought human beings many of their greatest triumphs."

In this scene, Attenborough uses the *crescendo* devices of narrative. There is a contextualization showing a group of chimpanzees peacefully enjoying what seems time out up in the canopy. Then monkeys arrive in their territory, some of which will be targeted as prey. A set up for the hunt ensues when calm is replaced by the attentive eyes and expressions of the chimpanzees discerning the species of monkey they are going to decide to trap. The hunt is subsequently set in motion and all chimpanzees are aware of the parts they will play. Rapid pursuit of the prey follows until climax is reached when a colobus is ambushed and killed. Distressing images of dismembering a still alive monkey follow. Calm is resumed when the chimpanzees divide the meat spoils among themselves and eating leads to having their faces smeared with fresh blood in a suggestive change of paradigm that chimpanzees kill and eat warm-blooded animals. As a literary journalist, Attenborough constructed a plot, alive with scenes, characters, details, and a logical sequence from beginning to end. Also, as a literary journalist, he used perspective as in: "it's clear that the search for prey has started" or in "[h]e's deliberately making himself conspicuous." The narrator's/literary journalist's point of view is also present as, for example, in "[t]he colobus will be very lucky if they escape" or in the adjectivation used throughout ("a peaceful scene," "gentle vegetarians," "long chase"). Point of view, alongside the use of first person as exemplified in "you don't know me," "our long distant hunting ancestors," are indicators that Attenborough does not assume a neutral stance in what he is narrating, which further strengthens his inclusion in the realm of literary journalistic reporting. Finally, it is worthy of stressing that, just like an immersion literary journalist, Attenborough is there where events are unfolding. Dressed in a way as to blend in with the environment, suffering the heat and moist of the climate, panting because of the effort to pursue the chimpanzees in the midst of thick forest, he is nothing short of a reporter on site. Most importantly, perhaps, is his attempt at bringing to viewers information that, until the time of the broadcast of this scene, was unknown. Attenborough had news to share with his audience and the news on which he focuses is true and faithful to the truth because it is factual: this video shows chimpanzees on the hunt leaving no doubts as to the purposeful strategies they use and as to the fact that they are meat-eaters.

When venturing into the wildernesses of the far reaches of the Earth, Attenborough's position can also be likened to that of the guide, the intermediary between the place of the observer *Us* and the foreign, exotic observed *Other*, whether animal, vegetable, or human. His shows can hence be regarded as attempts at interpreting unfamiliar or unknown realities all the while letting audiences be put in tele-contact with otherness: awe-inspiring landscapes, exotic animals, strange plants, even different peoples. Otherness is made (more) comprehensible while being shown and explained by a guide unraveling or translating (a) reality:

> Attenborough adopts an "innocent" view of his own role, one which presumes that the documentary form does portray "reality" without intervention of technical and professional conventions; that the environment and nature do have "intrinsic worth" and that his own role is that of guide.[43]

As a guide, Attenborough is the translator of what is alien, hidden from sight, or new to us. In leading the public to places not easily accessible, he wants to show reality, even if it is the reality as apprehended by the

[43] Jacquelin Burgess, "The Production and Consumption of Environmental Meanings in the Mass Media: A Research Agenda for the 1990s," *Transactions of the Institute of British Geographers* Vol. 15, No. 2, 1990: 153.

camera's selective eye, his scripts, and post-production edition. The portrayal of different realities not readily available to the public without the mediation of the guide or presenter, serves the double purpose of entertaining and instructing.

In this guise, "Attenborough acts as the typical explorer-authority, the audience's representative in distant lands."[44] As in conventional literary journalism, he is the observing *Other*, the interpreter of other realities and the bridge between *Us* and *Them*. Whether in wildlife documentaries or in the ones of a more anthropological/ethnographic nature as *The Tribal Eye* (1975) or *The First Eden* (1987),[45] Attenborough acts just as any investigative literary journalist immersing into unknown universes and realities while detaching momentarily from their world. Indeed, documentaries, as the ones inaugurated by Attenborough, "are a key part of the apparatus through which our increasingly urbanized societies obtain their knowledge of the natural world."[46] This carries the responsibility of accountability as regards the information conveyed.

Gathering the knowledge compiled in a documentary is a monumental endeavor in research. If for both journalism and literary journalism "accuracy reigns as the supreme clause in the symbolic contract with readers,"[47] the same premise applies for documentaries. Investigation and fact-checking are cornerstones of documentary film-making. Describing the minutiae invested in the factuality and correction of the information brought forth in his documentaries, Attenborough explains that, after he finishes writing a script, the copy is sent for validation to the program director who then sends it to a researcher, usually a biologist, for fact-checking. At this stage, the script will be shown for comments to specialists on the subjects being written about. Inaccuracies will be corrected, new research will be incorporated and Attenborough will rewrite the script

[44] Paul Henley, "British Ethnographic Film: Recent Developments," *Anthropology Today* Vol. 1, No. 1, 1985: 6.

[45] David Attenborough's extensive filmography does not solely concern the wonders of wildlife. His interest in Man led him to a postgraduate course in social anthropology with the London School of Economics in the early 1960s and triggered documentaries related to the life and culture of tribal peoples and their relation to the natural world: *A Blank on the Map* (1971), *Eastwards with Attenborough* (1973), *Tribal Eye* (1975), *The First Eden* (1987) or *The Lost Gods of Easter Island* (2000).

[46] Jean-Baptiste Gouyon *BBC Wildlife Documentaries in the Age of Attenborough* (Palgrave Macmillan: London, 2019), 2.

[47] Norman Sims *True Stories: A Century of Literary Journalism* (Northwestern University Press: Evanston, 2007), 6.

in accordance with the suggestions and information received.[48] The emphasis on having the facts right is thus a common denominator between documentary and literary journalism, the genre against which we are positioning Attenborough's work.

Academically, the link connecting documentary to literary journalism has been, so far, woven mostly in relation to documentary about the social rather than the natural world. There seems to be an agreement that documentary has taken "major steps toward literary journalism"[49] and that documentaries "follow the lead of literary reportage" or "cinematic new journalism."[50] These assumptions support our interpretation that documentaries on the natural world are as related to literary journalism as are those on social matters. Literary journalism is appealing because, Hartsock argues, it "derives from the fact that the human mind is wired to engage in inquiry into the world by telling stories [...] At the heart of 'storytelling' is the symbiosis of narrative and descriptive modalities."[51] Likewise, wildlife documentaries are "well-crafted story-telling devices"[52] requiring, just as literary journalism does, investment in investigation and informative accuracy. Connections, however, do not end here.

Immersion and First-Person

As early as *Zoo Quest*, it is possible to verify that the perspective taken by the documentary's narrative is that of the first-person. Attenborough does not showcase the natural world and its wonders from either a neutral, or a third-person viewpoint. "I" is the stance adopted by the presenter to refer to what and how he sees. It is his apprehension of the world around that is taken to the viewers be it through Attenborough himself, collecting fragments of giant eggs in Madagascar or catching a Komodo Dragon in Indonesia. Similarly in literary journalism, and differently from journalism at large, there are no qualms about letting the journalist's voice be heard. This means that "in sharp contrast to the 'objectivity' that the reporter

[48] Burgess and Unwin, "Exploring," 95–96.

[49] Sims, *True*, 285.

[50] Kevin Kerrane, "Making Facts Dance," *The Art of Fact: A Historical Anthology of Literary Journalism*) (Kerrane, Kevin and Ben Yagoda eds.) (Touchstone: New York, 1997), 20.

[51] John C. Hartsock, "Note from the Editor," *Literary Journalism Studies* Vol. 1, No. 1, 2009: 5.

[52] Gouyon, *BBC*, 5.

strives for in the standard news article, the voice of the new journalist is frankly subjective; it bears the stamp of his personality."[53] Although Hollowell refers to the literary journalist as new journalist, one of the terminologies used for the genre,[54] the implication is that the personal/subjective element of the journalist's take on the facts enriches, rather than diminishes, the informative value of a piece. Besides, this allows the establishment of a closer bond between author/public and the recipient of the message, public, is invited to take part in the experience of the journalist, author.[55]

When we look at Attenborough's documentaries, audiences are similarly invited to accompany the presenter on journeys across the globe and likewise partake in his awe at new discoveries and the wonders of the planet. In the opening scenes of *The Tribal Eye's* episode "Behind the Mask," we see Attenborough nearing a Dogon village in Mali.[56] The filmic sequence shows the presenter as a path-breaking figure for the audiences he is bringing along and whom he invites to see what his eyes see. Both presenter and viewer share the experience:

> Attenborough, accompanied by a Dogon man, walk[s] down a steep path towards a village. Clearly, the members of the audience are supposed to feel that they too are approaching the village. Once in the village, the audience's contact with the Dogon themselves is continuously and obviously mediated by Attenborough's presence, either in person, standing on location directly in front of the camera, or in the form of his familiar voice in the almost perpetual narration, identifying and explaining the significance of what is on the screen.[57]

Attenborough is thus the catalyst facilitating viewers access to *Other* geographies and *Other* realities of life on Earth. Because his presence is not muted, audiences of his shows are given not an aseptic version of reality

[53] John Hollowell *Fact and Fiction: The New Journalism and the Nonfiction Novel* (University of North Carolina Press: Chapel Hill, 1977), 22.

[54] Looking at the evolution of literary journalism, two periods are often referred as "New Journalism": the late 19th century and literary journalists like British W.T. Stead, American Jacob Riis or Portuguese Eça de Queirós, and the 1960s with names such as that of Tom Wolfe.

[55] Hartsock, *History*, 71.

[56] See *The Tribal Eye* (1975), episode 1 "Behind the Mask," 1'35" to 2'50" at https://www.youtube.com/watch?v=6eHD8-Di0Kk. Accessed April 29, 2020.

[57] Henley, "British," 6.

but a personal one. Part of the enticing power of his television documentaries is not related to the "spectacular" but to the feeling that he is "genuinely excited about what he discovered. In a medium otherwise saturated with bogus enthusiasm, he [is] the genuine article."[58] If personalization is what attracts readers to literary journalism, so it is with Attenborough's documentaries.

Personalization is also found in immersion. Literary journalists go deep into their topics, sometimes to the point of dressing down to better blend in the environments they want to study. Jack London spent weeks as a vagrant in London's East End to write *The People of the Abyss* (1903); Ted Conover took a job as correction officer in Sing Sing Prison, publishing *Newjack: Guarding Sing Sing* (2000) as a testament to that experience, and countless other examples could be mentioned. Pivotal to literary journalism, immersion, or the coming into close contact with the object of reporting, was underlined by Wolfe when describing the coming of age of literary journalism in the 1960s: "It seemed all-important to *be there* when dramatic scenes took place, to get the dialogue, the gestures, the facial expressions, the details of the environment."[59] A hallmark of documentary, immersion is one of the genre's prerequisites which Attenborough has taken to its fullest, and sometimes most dangerous, meaning.

Frequently has Attenborough immersed in the surrounding environments of his documentaries, sometimes for weeks and months on end. The all-importance of being there referred by Wolfe was for Attenborough a brush with death on many occasions. Shipwrecked, bitten, stung, lost in impenetrable jungles, taken ill, awoken by presumably-cannibal indigenous peoples (who had never seen Westerners before) was a price Attenborough has paid on countless occasions so that audiences can appreciate the world in all its glory from the safety of their homes. Dressing down has also, in his case, literally meant to undress, to get rid of the clothes that made him an outsider, the paradoxical *Other* in the environment of the *Other*. A mediator with alterity, Attenborough has been himself a paradigm of alterity: the stranger coming from afar who needed to assume the trappings of the other *Other* and thus be allowed to film in their space. In Makaruka, Solomon Islands, Attenborough and his crew

[58] Taylor, "Watching."

[59] Tom Wolfe, "Like a Novel," *The New Journalism: With an Anthology* (Wolfe, Tom and E. W. Johnson, E. W. eds.) (Harper & Row: New York and London, 1973), 21. Italics in original.

had to film stark naked in order to be accepted by the autochthonous people.[60] It was in New Guinea, trying to find birds of paradise to film for *Zoo Quest*, where he realized, as in a moment of epiphany, his status as the *other* in the alien landscape he had come to observe and scrutinize. Looking at a Papuan chief dressed in traditional apparel which included an impressive headpiece adorned with feathers of birds of paradise, Attenborough recalls:

> I found myself looking at him as I would view a circus exhibit or a tourist attraction. Yet [...] it was I who was in the circus, and the man outside, together with thousands of his compatriots in the forests beyond, who was the audience.[61]

Immersion of this kind and his awareness about the fact that we are onlookers of other realities, is what permits Attenborough's documentaries to constantly surprise viewers, and it is what allows him to translate those other realities and guide us into the "spectacular." Being the stoic presenter shooting few feet away from prides of lions, surrounded by mountain gorillas or in the midst of Arctic blizzards is another form of creating the literary journalist's empathy with the subject when the literary journalist is the naturalist.

Full Circle

Attenborough's incursions into, and his reflections on the manifold varieties of environments and life on Earth are—we have been arguing—an unexpected form of literary journalism. Their pillars are erected on a "humanistic approach to culture as compared to the scientific, abstract, or indirect approach taken by much standard journalism,"[62] Attenborough's documentaries can be interpreted as literary journalism because literary journalism is "not immutable, self-defining, and non-transgressive."[63] If transgression is at the core of our argument and if literary journalism is transgressive, there is ground to locate the documentary genre pioneered

[60] Attenborough, *Life*, 271.

[61] David Attenborough, David. 2019. *Journeys to the Other Side of the World: Further Adventures of a Young Naturalist* (Two Roads: London, 2019), 10.

[62] Sims, *True*, 12.

[63] William Dow, "Reading Otherwise: Literary Journalism as an Aesthetic Narrative Cosmopolitanism," *Literary Journalism Studies* Vol. 8, No. 2, 2016: 120.

by David Attenborough, itself born out of transgression from the BBC's way of televisual broadcasting, within the borders of literary journalism.

As a literary journalist, Attenborough is not so "innocent" a guide as quoted above from Burgess. His voice is not neutral and his *persona* "has come to assume [a] heroic role in the battle for conservation in the political arena."[64] Here too lies another bond between literary journalism and the documentary: a form of activism.

From its genesis, literary journalism has been about "access, attention to ordinary lives, and the special qualities of a writer's connection to the subjects."[65] It has meant raising awareness for issues often neglected by conventional journalism. Central to the work of a literary journalist is the perpetual questioning of "Is there conflict, is there challenge? Is there urgency, are there links to larger issues? Is there a way for me to meaningfully participate in that world?"[66] In his work, David Attenborough has had the capacity of transforming the ordinary into the extraordinary. Plants, which no one thought had the star potential of vertebrates, featured as main characters in one of Attenborough's series, *The Private Life of Plants* (1995), as did bugs in *Life in the Undergrowth* (2005). For Attenborough, the "challenge" of a series like the latter was that invertebrates are "more unfamiliar [...], more unattractive" and therefore, "in television terms, largely unexposed."[67] The "urgency" was in the fact that "they constitute half of all known animal species."[68] Invertebrates are the least known, yet the most populous animal kingdom. In this resided the need for the series. And it was that paradox what fascinated Attenborough.

Despite bringing to light the secret life of lesser-known species, or making appealing ones not usually regarded as particularly interesting, much of the criticism targeting Attenborough's documentaries focuses on the assumption that his productions eschew conservation worries and draw a pristine image of a planet struggling against such threats as mass extinction.[69] Faced with the accusations of portraying a biosphere that no longer exists, Attenborough, who says his documentaries have long called

[64] Burgess, "The Production," 153.

[65] Sims, *True*, 12.

[66] Ted Conover, "Immersion and the Subjective: Intentional Experience as Research." *Literary Journalism Studies* Vol. 10, No. 2, 2018: 170.

[67] Attenborough, *Life*, 386.

[68] Attenborough, *Life*, 386.

[69] Caroline Mortimer, "David Attenborough's *Planet Earth II* Is a 'Disaster for Wildlife', Says Rival Producer," *The Independent*, 2 January 2017. https://www.independent.co.uk/

attention to the wrongdoings of Man over the natural world as was the case on Easter Island,[70] does not dismiss the menaces facing the environment. In his view, too much alarmism being invested in his series can be counter-productive and off-putting for viewers.[71] Instead, emphasis on the beauty of the natural world allows audiences to find in his shows "a respite from their concerns about the future of the planet."[72] Nevertheless, in more recent years, his documentaries have, more straightforwardly and increasingly, highlighted the plights of the globe's environments. Critique for his 2019, Emmy-winning, eight-part Netflix series, *Our Planet*, produced in a partnership with the World Wildlife Fund, underlined it was "full of sobering reminders about how climate change is threatening significant parts of the natural world, coupled with the hope that humans might find the collective willpower to avert the most catastrophic consequences."[73] Similarly, his other mammoth series of 2019, *Seven Worlds, One Planet*, a BBC production, also received reviews such as:

> Watching *Seven Worlds, One Planet*, it's hard to know what to worry about most: the future of wildlife – catastrophically imperiled by our fondness for fossil fuels, long-haul travel and convenience food – or Attenborough himself, now 93 and one of the few people that the world will listen to about the impending apocalypse.[74]

arts-entertainment/tv/news/planet-earth-ii-disaster-for-world-wildlife-david-attenborough-martin-hughes-games-extinction-a7504996.html, accessed April 29, 2020.

[70] Brady Dennis, "David Attenborough Gave the Natural World a Voice. Now He's Talking About Climate Change Like Never Before," *The Washington Post*, 11 April 2019. https://www.washingtonpost.com/climate-environment/2019/04/11/natural-world-is-under-attack-heres-why-david-attenborough-is-still-hopeful/, Accessed May 4, 2020.

[71] Jonathan Watts, "David Attenborough: Too Much Alarmism on Environment a Turn-off," *The Guardian*, 4 November 2018. https://www.theguardian.com/environment/2018/nov/04/attenborough-dynasties-ecological-campaign, Accessed April 29, 2020.

[72] Esther Addley, "*Planet Earth II* a Form of Therapy for Viewers, Says Attenborough," *The Guardian*, 1 December 2016. https://www.theguardian.com/tv-and-radio/2016/dec/01/planet-earth-ii-a-form-of-therapy-for-viewers-says-attenborough, Accessed May 5, 2020.

[73] Dennis, "David Attenborough."

[74] Fiona Sturges, "*Seven Worlds, One Planet* Review: Dazzling Series Is Proof David Attenborough's Existence Is Vital for Us All," *The Independent*, 27 October 2019. https://www.independent.co.uk/arts-entertainment/tv/reviews/seven-worlds-one-planet-bbc-review-david-attenborough-episode-1-watch-iplayer-a9173221.html, Accessed May 4, 2020.

The reception of these series shows growing acknowledgement that Attenborough's documentaries have become more directly aligned with addressing manmade disruption in the natural world. Apart from critique, a study using *Planet Earth II* to measure Twitter reaction and Wikipedia traffic in relation to the species featured on each episode concluded that, despite the series not focusing on conservation worries, interest on those species peaked during the broadcast and remained high for the following months.[75] These results "suggest that natural history films focused on entertainment can generate vicarious connections with the natural world and increase public interest in nature at both immediate and longer-time scales."[76]

In the case of the naturalist as literary journalist, Attenborough's vast body of work comes full-circle from "conquest" of the natural world and the art of the documentary genre to the "understanding" that natural history shows play no lesser role in the dissemination of conservation messages at a time when mass extinction, climate change and overall habitat destruction are upon us. More than a retrospective of Attenborough's life as a broadcaster of documentaries on the natural world, *A Life on Our Planet*, intercuts grim images of destruction made on the environment with spectacular ones of (still) thriving habitats so as to better call attention to the need to act fast to thwart the disaster facing Nature and, ultimately, Man. In this light, "such cinematic juxtapositions are persuasive: A dying planet is an ugly one, while healthy ecosystems please the eye and the earth."[77] The "visual grammar" that people in the Talks department were looking for in the 1950s was meantime created and has gained fluency and momentum through the epic series of David Attenborough. That his work touches on literary journalism or may be considered a form of literary journalism is a cause worthy of future, more consistent, study and a reason to further consider the stretching boundaries of literary journalism and its seemingly infinite possibilities.

[75] Darío Fernández-Bellon and Adam Kane, "Natural History Films Raise Species Awareness: A Big Data Approach," *Conservation Letters: A Journal of the Society for Conservation Biology* Vol. 13, 2020.

[76] Fernández-Bellon and Kane, "Natural History," 6.

[77] Natalia Winkelman, "*David Attenborough: A Life on Our Planet* Review: Ruin and Regrowth," *The New York Times,* 4 October 2020. https://www.nytimes.com/2020/10/04/movies/david-attenborough-a-life-on-our-planet-review.html, Accessed June 9, 2021.

References

Addley, Esther. 2016. "*Planet Earth II* a Form of Therapy for Viewers, Says Attenborough." *The Guardian* (1 December). https://www.theguardian.com/tv-and-radio/2016/dec/01/planet-earth-ii-a-form-of-therapy-for-viewers-says-attenborough, Accessed May 5, 2020.

Attenborough, David. 2010. *Life on Air: Memoirs of a Broadcaster*. London: BBC Books.

Attenborough, David. 2017. *Adventures of a Young Naturalist*. London: Two Roads.

Attenborough, David. 2019. *Journeys to the Other Side of the World: Further Adventures of a Young Naturalist*. London: Two Roads.

Attenborough, David. 2020. *A Life on Our Planet: My Witness Statement and a Vision for the Future*. London: Witness Books.

Barkham, Patrick. 2019. "The Real David Attenborough." *The Guardian* (22 October). https://www.theguardian.com/tv-and-radio/2019/oct/22/david-attenborough-climate-change-bbc, accessed April 16 2020.

BBC Two. *The Trials of Life: A Natural History of Behaviour*. https://www.bbc.co.uk/programmes/p004hd8g, accessed June 29, 2022.

Biography.com Editors. 2019. "David Attenborough Biography." *Biography Channel*. https://www.biography.com/personality/david-attenborough, accessed April 21 2020.

Burgess, Jacquelin and David Unwin. 1984. "Exploring the Living Planet with David Attenborough." *Journal of Geography in Higher Education* 8, no. 2: 93–113.

Burgess, Jacquelin. 1990. "The Production and Consumption of Environmental Meanings in the Mass Media: A Research Agenda for the 1990s." *Transactions of the Institute of British Geographers* 15, no. 2: 139–161.

Calma, Justine. 2020. "David Attenborough Broke a World Record on Instagram." *The Verge* (29 September). https://www.theverge.com/2020/9/29/21493467/david-attenborough-instagram-guinness-world-record, Accessed June 6, 2021.

Conover, Ted. 2018. "Immersion and the Subjective: Intentional Experience as Research." *Literary Journalism Studies* 10, no. 2: 162–173.

"David Attenborough Biography." *Biography Channel*, http://www.biography.com/people/david-attenborough, accessed April 21 2020.

Dennis, Brady. 2019. "David Attenborough Gave the Natural World a Voice. Now He's Talking About Climate Change Like Never Before." *The Washington Post* (11 April 2019). https://www.washingtonpost.com/climate-environment/2019/04/11/natural-world-is-under-attack-heres-why-david-attenborough-is-still-hopeful/, Accessed May 4, 2020.

Dow, William. 2016. “Reading Otherwise: Literary Journalism as an Aesthetic Narrative Cosmopolitanism.” *Literary Journalism Studies* 8, no. 2: 118–136.
Fernández-Bellon, Darío and Adam Kane. 2020. “Natural History Films Raise Species Awareness: A Big Data Approach.” *Conservation Letters: A Journal of the Society for Conservation Biology* 13:e12678. https://doi.org/10.1111/conl.12678.
Gill, A. A. 2005. *The Angry Island: Hunting the English*. London: Phoenix.
Gouyon, Jean-Baptiste. 2019. *BBC Wildlife Documentaries in the Age of Attenborough*. London: Palgrave Macmillan.
Hale, Mike. 2020. “David Attenborough is Back, And He Brought the Walruses.” *The New York Times* (15 January). https://www.nytimes.com/2020/01/15/arts/television/review-seven-worlds-one-planet.html, Accessed May 4, 2020.
Hartsock, John C. 2000. *A History of American Literary Journalism: The Emergence of a Modern Narrative Form*. Amherst: University of Massachusetts Press.
Hartsock, John C. 2009. “Note from the Editor.” *Literary Journalism Studies* 1, no. 1: 5.
Henley, Paul. 1985. “British Ethnographic Film: Recent Developments.” *Anthropology Today* 1, no. 1: 5–17.
Hollowell, John. 1977. *Fact and Fiction: The New Journalism and the Nonfiction Novel*. Chapel Hill: University of North Carolina Press.
Jeffreys, Tom. 2013. “Representation for Animals: An interview with Dr. Brett Mills.” *The Journal of Wild Culture*. https://www.wildculture.com/article/representation-animals-interview-dr-brett-mills/1142, accessed April 16 2020.
Joseph, Sue. 2016. *Behind the Text: Candid Conversations with Australian Creative Nonfiction Writers*. Melbourne: Hybrid Publishers.
Kerrane, Kevin and Ben Yagoda, eds. 1997. *The Art of Fact: A Historical Anthology of Literary Journalism*. New York: Touchstone.
McKie, Robin. 2012. “David Attenborough: Force of Nature.” *The Guardian* (28 October).https://www.theguardian.com/tv-and-radio/2012/oct/26/richard-attenborough-climate-global-arctic-environment, Accessed May 4, 2020.
Mills, Brett. 2013. “The Animals Went in Two by Two: Heteronormativity in Television Wildlife Documentaries.” *European Journal of Cultural Studies* 16, no. 1: 100–114.
Monbiot, George. 2018. “David Attenborough Has Betrayed the Living World He Loves.” *The Guardian* (7 November). https://www.theguardian.com/commentisfree/2018/nov/07/david-attenborough-world-environment-bbc-films, accessed April 27 2020.
Morse, Janice M. 2006. “Insight, Inference, Evidence, and Verification: Creating a Legitimate Discipline.” *International Journal of Qualitative*

Methods 5, no. 1: 93–100. https://journals.sagepub.com/doi/pdf/10.1177/160940690600500108, accessed August 31 2021.
Mortimer, Caroline. 2017. "David Attenborough's *Planet Earth II* Is a 'Disaster for Wildlife', Says Rival Producer. *The Independent* (2 January). https://www.independent.co.uk/arts-entertainment/tv/news/planet-earth-ii-disaster-for-world-wildlife-david-attenborough-martin-hughes-games-extinction-a7504996.html, accessed April 29 2020.
Sims, Norman. 2007. *True Stories: A Century of Literary Journalism.* Evanston: Northwestern University Press.
Sturges, Fiona. 2019. "*Seven Worlds, One Planet* Review: Dazzling Series Is Proof David Attenborough's Existence Is Vital for Us All." *The Independent* (27 October). https://www.independent.co.uk/arts-entertainment/tv/reviews/seven-worlds-one-planet-bbc-review-david-attenborough-episode-1-watch-iplayer-a9173221.html, Accessed May 4, 2020.
Taylor, Laurie. 2008. "Watching David Attenborough: Laurie Taylor Talks to the Renowned Naturalist." *New Humanist* (4 January). https://newhumanist.org.uk/articles/1673/watching-david-attenborough-laurie-taylor-talks-to-the-renowned-naturalist, Accessed April 16, 2020.
Watts, Jonathan. 2018. "David Attenborough: Too Much Alarmism on Environment a Turn-off." *The Guardian* (4 November). https://www.theguardian.com/environment/2018/nov/04/attenborough-dynasties-ecological-campaign, Accessed April 29, 2020.
Winkelman, Natalia. 2020. "*David Attenborough: A Life on Our Planet* Review: Ruin and Regrowth." *The New York Times* (4 October). https://www.nytimes.com/2020/10/04/movies/david-attenborough-a-life-on-our-planet-review.html, Accessed June 9, 2021.
Wolfe, Tom and E. W. Johnson, eds. 1973. *The New Journalism: With an Anthology.* New York and London: Harper & Row.

Interviewing Nature: The Dangers and Delights of the Pathetic Fallacy and Anthropomorphism in Nature Writing and Environmental Journalism

Susan E. Swanberg

Introduction

The first time I heard about Henry David Thoreau was when—at the age of eight—I lost my newly acquired birthstone ring in Walden Pond. I searched and searched at the edge of the famous kettle hole where I'd been wading, hoping I might detect the sparkle of an asparagus-green peridot in the slimy alluvion between my toes. I never found the ring, but I recall that, in an attempt to console me, somebody present at the scene explained who Thoreau was and the significance of Walden Pond as the inspiration for his classic nonfiction work. Despite his misanthropic

S. E. Swanberg (✉)
University of Arizona, Tucson, AZ, USA
e-mail: swanberg@arizona.edu

P. Calvi (ed.), *The Literary Journalist as a Naturalist*, Palgrave Studies in Literary Journalism,
https://doi.org/10.1007/978-3-031-56634-9_14

tendencies, New Englanders generally seem to approve of Thoreau, his civil disobedience, and his experiment in the woods.

Years later, as a high school student in California, I acquired a copy of Thoreau's *Walden*.[1] Much of the book's transcendentalism escaped me and I thought the tone was too acerbic in its judgment of humanity's follies, but I was drawn to Thoreau (pronounced a "sometime narrative literary journalist" by John Hartsock)[2] because Thoreau wrote true stories about the New England environs I had known so well and missed so much.

Thoreau—the erstwhile hermit of Walden Pond—wrote about the flora and fauna he observed in a way I could feel to the marrow of my bones. Among the many entertaining natural vignettes recorded in *Walden* were Thoreau's charmingly sentimental depiction of the antics of a female partridge and her precocial young[3] and his harrowingly graphic description of an epic battle between red and black "Myrmidons."[4]

I'd seen similar dramas play out in my New England backyard—a yard which merged with the woods behind our house on Edgell Road in Framingham, Massachusetts and was less than fifteen miles from Walden Pond.[5] I fancied myself a female Thoreau as I ventured alone into those woods, unearthing treasures like lady's slippers and salamanders. It wouldn't be until Rachel Carson's *Silent Spring* (1962) was published that I discovered a female nature writer with whom I could identify.

Silent Spring, which I also read in high school, had a tremendous impact on my life. From my science fair project on the impact of DDT to a summer school class in ecology, I felt myself more and more inclined toward studying the natural world. Now, as my understanding of Thoreau and Rachel Carson has deepened and I have become aware of subspecies within the nature writing genre, I ask myself "What is nature writing, and

[1] Henry David Thoreau, *Walden: Return to Nature: An American Classic* (New York: The New American Library, 1953).

[2] John Hartsock, *A History of American Literary Journalism: The Emergence of a Modern Narrative Form* (Amherst, University of Massachusetts Press, 2000), 210.

[3] Ralph Waldo Emerson and Henry David Thoreau, "Walden" in *Transcendentalists Collection: Walden, Walking, Self-Reliance and Nature* (Los Angeles: Carousel Books, 2020), 191–192.

[4] "Myrmidons" is the term Thoreau used to describe the armies of red and black ants whose battle he recounted in *Walden*. In Greek mythology, Myrmidons were the loyal followers of Achilles. Myrmecology is the scientific study of ants. Henry David Thoreau, "Walden" in *Transcendentalists Collection* (Los Angeles: Carousel Books, 2020), 193–195.

[5] A glimpse of my old neighborhood via a satellite image shows that, sadly, the woods behind the home my parents built have shrunken to a fraction of their former size.

what are its canons (if it has them)?" and "Does nature writing fit into the genre we call literary journalism?"

Although there are many books and articles about nature writers and nature writing, there is very little, if any, scholarly work on nonfiction nature writers as literary journalists. Some literary journalism anthologies include representative excerpts from the works of nature writers, suggesting that the authors of those pieces are, *sine dubio*, literary journalists. Robert Alexander's essay "Literary Journalism and Ecocriticism," published in *The Routledge Companion to American Literary Journalism*[6] discusses environmental writing as seen through the lens of critical theory but mentions nature writing only in passing. While Mark Neuzil notes in his book, *The Environment and the Press* (2008)—that "no tributary of the river that is environmental journalism has had more attention from scholars than nature writing,"[7] I suggest that nonfiction nature writing merits focused attention from scholars of literary journalism, in particular.

The descriptor *nonfiction* is used herein to modify nature writing because "nature writing defines an expansive genre which, according to David Rains Wallace (1984) also includes fiction, poetry, drama, and even speculative fiction.[8] This essay includes a discussion of the works of a small but germane sample of nonfiction nature writers (past and present) who have either been identified as literary journalists by scholars and aficionados of the genre or who might yet be deemed literary journalists.

The Reality Boundary and Nonfiction Nature Writing

In an essay published in the inaugural issue of *Literary Journalism Studies*, the late, prominent literary journalism scholar, Norman Sims, described what he called the *reality boundary* at the core of literary journalism. Literary journalism, Sims suggested, could be envisioned as a central literary form bordered by other forms including autobiography, fiction,

[6] Robert Alexander, "Literary Journalism and Ecocriticism," in the *Routledge Companion to American Literary Journalism* (2020), 482–97.

[7] Mark Neuzil, *The Environment and the Press: From Adventure Writing to Advocacy*, (Evanston: Northwestern University Press, 2008), 96.

[8] David Rains Wallace, "The Nature of Nature Writing," *The New York Times*, July 22, 1984, Section 7, Page 1, www.nytimes.com/1984/07/22/books/the-nature-of-nature-writing.html; Ursula K. le Guin, *The Word for World is Forest* (New York: Harper Collins), 1980.

science writing, conventional journalism, and history. "Sometimes a writer can stray over a border without damage—say into science writing or history," wrote Sims, "but when the writer crosses the border into fiction, it triggers a hunt by the guardians of journalism."[9]

Several paragraphs later, Sims referred to John McPhee, calling him a "realist literary journalist."[10] It should be noted that John McPhee's eclectic body of work includes nonfiction nature writing from "The Woods From Hog Wallow," to "Fire in the Pines" and other selections from McPhee's *The Pine Barrens* not to mention, *Annals of the Former World* (1980–1982), *The Control of Nature* (1989), and *Coming into the Country* (1976).[11] So why has so much nonfiction nature writing by so many other authors gone largely unrecognized as (at the very least) a sometime subgenre of literary journalism? Could it be that too many early nature writers strayed too far over the reality boundary?

A Brief Natural History of the Nature Writing and Literary Journalism Traditions

Nonfiction Nature Writing

The title of this collection is *The Literary Journalist as Naturalist.* This chapter flips that script, however, by examining *nonfiction nature writers as literary journalists.* A nonfiction nature writer, whether a literary journalist or not, must be something of a naturalist in the first instance. That is, the writer must have been sufficiently immersed in the natural world to become knowledgeable about its features—or at least the features of which the author intends to write. Rachel Carson was a professional biologist and an author.[12] Henry David Thoreau was a writer and a

[9] Norman Sims, "The Problem and the Promise of Literary Journalism Studies," *Literary Journalism Studies* 1, no. 1 (2009), 11–12.

[10] Sims, "The Problem and the Promise…," 12.

[11] John McPhee, *The Pine Barrens* (New York: Farrar Straus & Giroux, 1968); *Annals of the Former World* (New York: Farrar Straus & Giroux, 1980–1982); *The Control of Nature* (New York: Farrar Straus & Giroux, 1989); *Coming into the Country* (New York: Farrar Straus & Giroux, 1976).

[12] David Pimentel, "Silent Spring, the 50th Anniversary of Rachel Carson's Book," *BMC Ecology* 12: (2012): 20.

naturalist—among other things.[13] Both Carson and Thoreau immersed themselves in nature before they put pen to paper.

"[N]aturalists are those who study nature," wrote David J. Schmidly in the 2005 issue of the *Journal of Mammalogy*.[14] According to Peter R. Grant, the roots of natural history—how he defines what naturalists study—originated in a branch of the natural sciences that dates back before Aristotle.[15] Natural history, Grant thinks, tends to unify the natural world rather than splitting it into multiple parts, which necessitates multiple disciplines to study one world. Each issue of *The American Naturalist* announces that the journal is "Devoted to the Conceptual Unification of the Biological Sciences."[16] Some scholars say that the ancient discipline we call "natural history" is destined to be supplanted by the newer scientific disciplines that have steadily gained adherents.[17] Despite this dire prediction, many well-known nonfiction nature writers have taken and continue to take a naturalist's approach to their subject matter.

In a *Natural History of Nature Writing*, Frank Stewart noted the following regarding nature writing.

> Whatever form it takes, nature writing at its best is a literary art as rigorous as natural science, with a similar allegiance to verifiable fact. As literary artists, nature writers interpret and vivify their observations through aesthetic language. More important, they are mindful of the role that storytelling and dramatic narration play in our psychic and cultural well-being. This awareness makes their best writing as expressive, as rigorous, and as carefully structured as a novel or poem.[18]

Stewart stressed the fact that nature writers are aware of the importance of accuracy as they pursue both the literary and the rational.

[13] Diane Timblin, "Thoreau as Naturalist: A Conversation with Four Authors," *American Scientist*, https://www.americanscientist.org/article/thoreau-as-naturalist.

[14] David J. Schmidly, "What It Means to be a Naturalist and the Future of Natural History at American Universities," *Journal of Mammalogy* 83, no. 3 (2005): 449

[15] Peter R. Grant, "What Does It Mean to Be a Naturalist at the End of the Twentieth Century?" *The American Naturalist* 155, no. 1 (2000): 1.

[16] Grant, "What Does It Mean to Be a Naturalist…," 1–2.

[17] Schmidly, 449, citing D.S. Wilcove and T. Eisner, "The Impending Extinction of Natural History," *Chronicle Review, Chronicle of Higher Education* 47, no. 3 (2000): B24.

[18] Frank Stewart, *A Natural History of Nature Writing* (Washington, D.C.: Island Press, 1995), xv–xvi.

> From their own direct experiences [nature writers] are aware of the limits of both objectivity and subjectivity in giving accurate accounts of nature that will grip our emotional as well as rational understanding. They pursue this understanding with an avidity for fact accessible only through the scientific method and with a passion for metaphors, patterns, feelings, and self-awareness accessible only through poetry and art.[19]

The term "nature writing" covers much ground and the genre's tent has been especially expansive during the twentieth and twenty-first centuries.[20] Some nonfiction nature writers are explainers—providers of detailed scientific information about the world in all its "elemental" splendor.[21] Other nonfiction nature writers paint portraits of the natural world in lyrical essays bordering on poetry.[22] Still others stand on the edge of the abyss issuing a clarion call to action in the face of global catastrophe.[23] In other words, nonfiction nature writers play a variety of roles within their genre.

Some early-twentieth-century nature writers engaged in what zoologist and conservationist, William T. Hornaday called "nature-faking"—writing about "wild-animal imaginings under the stamp of truth."[24] Most nature writers, however, tell true stories about what they saw, heard, touched, or smelled as they were immersed in their observations of the

[19] Stewart, *A Natural History of Nature Writing*, xix. I personally disagree that the scientific method must be followed in nature writing, because the pursuit of natural history is not an experimental science, but a descriptive science based upon observation. A naturalist/nature writer's observations must, however, be honestly and accurately conveyed and many nonfiction nature writers do feature the findings of experimental science in their work.

[20] John Elder and Robert Finch, eds., "Introduction" in *The Norton Book of Nature Writing* (New York: W.W. Norton & Company, 2002), 15–20.

[21] Diogo Rodriguez, "How to Create a Good Explainer, Explained," *Medium*, May 6, 2019, https://medium.com/notes-from-the-classroom/how-to-create-a-good-explainer-explained-4e4236870b24.

[22] Aldo Leopold, *A Sand County Almanac: With Essays on Conservation from Round River* (New York: The Random House Publishing Group, 1966; Oxford University Press, Inc., 1949).

[23] Rachel Carson, *Silent Spring* (Boston: Houghton Mifflin Company, 1962); Bill McKibben, *The End of Nature* (New York: Random House Paperbacks, 2006).

[24] William Hornady, *Hornaday's American Natural History, Fourteenth Edition - Revised* (New York: Charles Scribner's Sons, 1927), xxii–xxiv; Jack London's, *The Call of the Wild* (New York: The McMillan Company, 1903), which described the life of Buck, a domestic dog who ultimately joins a wolf pack in the Klondike Region, might have been one of the books Hornaday described as containing "wild-animal imaginings under the stamp of truth." Interestingly, Hornaday noted at p. xxiii that "The newspapers of the day were abundantly stocked" with such stories.

natural world. Some, but not all, nature writing fits the description of literary journalism put forth by Sims or Kramer, which demand, above all, that its practitioners write truthfully and accurately. It is the ***nonfiction*** nature writers whose works are examined here and whom, I suggest, we should consider accepting as bona fide members of the literary journalism genre, not the nature writer/novelists who write, for example, about the inner thoughts of a domestic dog gone feral.[25]

In his *American Natural History* (1927), Hornaday expressed a concern that presaged debate about the reality boundary. Recounting his evolution from a scientist who disdained "nature fakery and exaggeration" to an author who "registered cordial approval of the wild-animal-hero tale of imagination that is true to life so far as the personality and daily life habits of the hero are concerned" Hornaday was describing his own struggles with nature writing's reality boundary. The debate continues in part because, like literary journalism, nature writing is not as well-defined a genre as one might think.[26]

John Elder and Robert Finch refer to nature writing as a "particular form of the personal essay," while also recognizing that many outstanding nature writers do *not* always choose the personal essay as their genre of choice.[27] In his 1984 *New York Times* essay, "The Nature of Nature Writing," David Rains Wallace described the confusion surrounding the term "nature writing." According to Wallace, the roots of nature writing include "hunting stories, fables, herbals, bestiaries, pastorals, lyrics and traveler's tales."[28] Sprouting from these roots, noted Wallace, were works as disparate as *Walden*, nature fiction, nature poetry, nature reporting, and nature drama, all of which had in common "appreciative esthetic responses to a scientific view of nature."[29] How habitually these "appreciative esthetic responses" can be invoked in nature writing without the genre becoming

[25] Jack London, *The Call of the Wild*.

[26] In Richard Lance Keeble's "Literary Journalism As A Discipline: Tom Wolfe and Beyond," *Brazilian Journalism Research* 14, no. 3(2018): 871, the author describes literary journalism as "at its core a messy term" which "has at its essence a provisional quality that captures many of the uncertainties and contradictions of the writer's predicament today."

[27] John Elder and Robert Finch, *The Norton Book of Nature Writing* (New York: W.W. Norton & Company, Inc., 1990, 2002), 17.

[28] David Rains Wallace, "The Nature of Nature Writing," *The New York Times*, July 22, 1984, Section 7, Page 1.

[29] Wallace, "The Nature of Nature Writing,"

overly effusive, resulting in an inaccurate view of the natural world is an issue worth considering.

Literary Journalism

According to Norman Sims, Edwin H. Ford—a teacher of journalism at the University of Minnesota—was likely the first to use the term *literary journalism* in its contemporary scholarly meaning.[30] A revolutionary genre that gained attention in the second half of the twentieth century and is kin to literary journalism was named ("however ungracefully," noted Tom Wolfe) "the New Journalism."[31] Wolfe delineated the following characteristics as his essential New Journalism canons:

1. Scene-by-scene construction, or as Wolfe put it "telling the story by moving from scene to scene and resorting as little as possible to sheer historical narrative;"
2. Dialog recorded in full as a scene was reported in real time;
3. A third person point of view enhanced by interviewing the subject about their thoughts and emotions;
4. Recording the subject's "status life" including everyday gestures, habits, manners, customs, styles, and modes of living—characteristics that symbolize the subject's relationship within their environment.

Much nonfiction nature writing does not (and, by its nature, cannot) adhere to all of Wolfe's essential New Journalism rules. For example, nature writers can describe nature's flora and fauna as well as its geologic and astronomical characteristics using scene-by-scene construction and nature writers can examine the behavior, life histories, relationships, and ecologies of many of nature's plants and animals—their status lives, in essence. One cannot, however, interview nature or its nonhuman inhabitants nor can one record nature's "dialogs"—at least not literally.[32] Observers of nature can, of course, be interviewed regarding their obser-

[30] Tom Wolfe, "Preface," in *The New Journalism* (New York: Harper and Row, 1973), 11.

[31] Tom Wolfe, "Why They Aren't Writing the Great American Novel Anymore?" *Esquire*, December 1972; Tom Wolfe and Edward Warren Johnson, eds., *The New Journalism* (New York: Harper and Row, 1973), 46–47.

[32] Observers of nature can be interviewed as to their observations and the conclusions they have drawn from them, but that is quite different from interviewing Nature which, I suggest, is the primary protagonist in much nonfiction nature writing. If nonfiction can be said to have a "protagonist."

vations and the conclusions they have drawn from them. In addition, many nonfiction nature writers use first person quite effectively in their writing without damaging the authenticity of the story they are telling.

In *Literary Journalism*, Norman Sims characterized the source of the literary journalism genre's continuing influence—condensing and clarifying Wolfe's rules in the process. "[Literary journalism's] appeal has grown from the solid foundations of the form—immersion reporting, narrative techniques that free the voice of the writer, and high standards of accuracy."[33] Clearly one can immerse oneself in nature and—using a high standard of accuracy and a lyrical narrative voice that frees the writers from the constraints of old forms—convey the essence of the natural world.

In "Breakable Rules for Literary Journalists," Mark Kramer reinforced the revised literary journalism norms Sims had delineated.[34] A condensed inventory of Kramer's defining traits of literary journalism include immersion in a subject's world; background research; adherence to a covenant with readers and sources regarding accuracy and candor; avoiding acts such as creating composite characters, enhancing quotations, or misleading sources as to the nature of the writer/source relationship; writing in a plain style and an intimate, informed voice; assuming a digressive, mobile stance from which primary and secondary scenes are described; and building upon the readers' reactions to a progressing yet digressive narrative.[35]

Notably, both Sim's and Kramer's descriptions of the essence of literary journalism are compatible with Frank Stewart's depiction of nature writing.[36] Nonfiction nature writing—examined both retrospectively as earlier examples of the genre are identified and categorized, as well as with a fresh perspective as more nonfiction writers flock to the genre—is a burgeoning phenomenon to which literary journalism has attended only sporadically for several decades.

[33] Norman Sims, "The Art of Literary Journalism," in *Literary Journalism: A New Collection of the Best American Nonfiction*, Sims and Kramer eds. (New York: Random House, 1995), 3.

[34] Mark Kramer, "Breakable Rules for Literary Journalists," in *Literary Journalism: A New Collection of the Best American Nonfiction* (New York: Ballantine Books, 1995), 21–34.

[35] With apologies to Mark Kramer for condensing and combining his list of eight "breakable rules." Mark Kramer, "Breakable Rules for Literary Journalists," in *Literary Journalism: A New Collection of the Best American Nonfiction* (New York: Ballantine Books, 1995), 22–33.

[36] See notes 18 and 19 and the corresponding text.

The Use of Figurative Literary Devices in Nature Writing

Both science and literary journalism hold truth and accuracy in high regard. Neither tradition finds it acceptable to make things up. It is not uncommon, however, for nonfiction nature writers to use figurative literary devices that are frowned upon or considered obsolete by scholars of the humanities and others.[37] Astrophysicist and science writer, Caleb Scharf, noted that the use of metaphor, analogy, anthropomorphism, and the pathetic fallacy in descriptive science writing is controversial.[38] Use of certain literary devices including the pathetic fallacy,[39] personification,[40] anthropomorphism,[41] and others could be said to contradict both literary journalism *and* scientific traditions regarding accuracy and truth because they are not literal expressions of reality. Should use of these and other emotive, figurative literary devices disqualify a nonfiction nature writer from inclusion among the ranks of literary journalists?

The pathetic fallacy, in which inanimate nature "reacts" to the human situation with human-like emotion, is exemplified by the phrase, "the skies *wept bitter tears* upon us during the brave soldier's funeral." Personification is the figurative attribution of human traits or attributes to a non-human animal or an inanimate object as when humans attribute a warrior's attitude, motivation, or behavior to animals in conflict or ascribe human

[37] Bryan L. Moore, "Part 1 Personification in Practice and Theory," *Ecology and Literature: Ecocentric Personification from Antiquity to the Twenty-first Century* (New York: Palgrave Macmillan – a division of St. Martin's Press LLC., 2008), 1; John Ruskin in *Modern Painters, Volume III. Containing Part IV. Of Many Things Chapter XII "Of the Pathetic Fallacy."*

[38] Caleb Scharf, "In Defense of Metaphor in Science Writing," *Scientific American*, July 19, 2013, https://blogs.scientificamerican.com/life-unbounded/in-defense-of-metaphors-in-science-writing/

[39] The "pathetic fallacy" is a term originated by John Ruskin in *Modern Painters, Volume* III, where he debated the meaning of the objective and the subjective which he described as "the ordinary, proper, and true appearances of things to us; and the extraordinary, or false appearances, when we are under the influence of emotion, or contemplative fancy; false appearances … as being entirely unconnected with any real power or character in the object, and only imputed to it by us." Ruskin cited as examples of the pathetic fallacy: "the spendthrift crocus" and "the cruel, crawling foam." See Frederick William Roe, ed., *John Ruskin: Selections and Essays*, (Mineola: Dover Publications, Inc., 2013), 116–117.

[40] Bryan L. Moore noted that *The Oxford English Dictionary* defines personification as the "[a]ttribution of personal form, nature, or characteristics; the representation of a thing or abstraction as a person," (29).

[41] Moore also noted that anthropomorphism is "spontaneous, less consciously or artistically used as than personification, which is a more temporal and deliberate metaphor," (29).

characteristics to non-human animals as in "the loon ***danced*** across the lake." Anthropomorphism also involves the attribution of human traits or attributes to a non-human animal but is a more robust, literal attribution of human characteristics as when an animal is named and its role in nature described in human terms. Walt Disney often engaged in this form of anthropomorphism in his nature films.[42]

As these and other figurative literary devices arguably go against the grain of literary journalism by presenting a subjective description of the relationship of humans and nonhumans to the natural world and vice versa, one might suppose that using these devices in nature writing could result in nature writing being excluded from the literary journalism genre—if literary journalism's tenets were rigidly enforced. A number of nature writers, however, have used these figurative literary devices without suffering a diminution in their status as literary journalists. In fact, use of these devices by nature writers recognized as literary journalists has often gone unnoticed—or at least not commented upon.

In *Walden*, for example, Thoreau made use of the pathetic fallacy, personification, and anthropomorphism. Despite Thoreau's use of these literary devices, a number of authors and scholars include Thoreau as a member or the literary journalism pantheon, whether implicitly or explicitly. For example, *Hot Zone* author, New Yorker contributor, and literary journalist Richard McCann Preston wrote in his PhD dissertation that Thoreau's Walden is "a brilliant piece of literary journalism, an accurate (although intensely imagined) account of some things that its author witnessed."[43] Scholar Nancy L. Roberts, wrote the following regarding Thoreau: "In 1854, Henry David Thoreau (1917–1862), produced Walden as a kind of literary journalistic call to action for environmental preservation."[44] Thoreau is also referenced a number of times in John L. Hartsock's *A History of American Literary Journalism: The Emergence of a Modern Narrative Form* (2000).[45]

[42] Margaret J. King, "Audience in the Wilderness: The Disney Nature Films," *Journal of Popular Film and Television* 24, no. 2 (1996): 60–68.

[43] Richard McCann Preston, "The Fabric of Fact: The Beginnings of American Literary Journalism," Ph.D. Princeton, 1983, 34.

[44] Nancy Roberts, "Literary Journalism and Social Activism," in *The Routledge Companion to Literary Journalism* (New York: Routledge, Taylor and Francis, 2020), 264.

[45] John L. Hartsock, *A History of American Literary Journalism: The emergence of a Modern Narrative Form*, (Amherst: University of Massachusetts Press, 2000).

One notable example of Thoreau's attribution of human qualities to non-human creatures is found in his description of the battle between red and black ants that took place near Thoreau's woodpile. An "internecine war" Thoreau called it, "the red republicans on the one hand, and the black imperialists on the other....and human soldiers never fought so resolutely.... It was evident that their battle-cry was 'Conquer or die.'"[46] The ant battle, upon which Thoreau spent several pages, was remarkable in its detail and its tragedy. Thoreau humanized the ant soldiers so well that the reader is tempted to mourn the fallen individuals he depicted.

Thoreau also occasionally invoked the pathetic fallacy, as he did with his description of nature assisting a loon trying to avoid the humans hunting him.

> ...rustling through the woods like autumn leaves, at least ten men to one loon. Some station themselves on this side of the pond, some on that, for the poor bird cannot be omnipresent; if he dive here he must come up here. But now the *kind October wind rises, rustling the leaves and rippling the surface of the water, so that no loon can be heard or seen.... The waves generously rise and dash angrily, taking sides with all water-fowl,* (emphasis provided) and our sportsmen must beat a retreat to town and shop and unfinished jobs.[47]

Thoreau thus adorned his work with figurative language; but did his use of these literary devices diminish his candid approach to Nature's inhabitants? There seems to be very little disagreement among literary journalism scholars that at least some of Thoreau's nonfiction nature writing should be considered literary journalism.

According to Mark Canada of the *Observer*:

> [Thoreau] framed "Walden" as a kind of dispatch from nature, a more reliable source of the truth that lay beneath the superficial details of the ordinary world. He was, he said, "trying to hear what was in the wind, to hear and carry it express," adding that "If it had concerned either of the political parties, depend upon it, it would have appeared in the Gazette with the earliest intelligence." In short, because traditional newspapers were more

[46] Henry David Thoreau, "Brute Neighbors," from *Walden* in *The Norton Book of Nature Writing* Finch and Elder, eds. (New York: W.W. Norton & Company, 1990, 2012), 174–176.

[47] Thoreau, "Brute Neighbors," 177–178.

interested in the trivial details of politics, Thoreau tasked himself with reporting truths through his own form of journalism. We call it "literature."[48]

David Quammen and John Mcphee—both of whose works appear in Sims and Kramer's anthology, *Literary Journalism* (1995)—also use figurative literary devices. Quammen occasionally uses devices that resemble personification as when he described the living results of nature's eccentric experiments, "[m]any of the world's gaudiest life forms, both plants and animals, occur on islands. There are giants, dwarfs, *crossover artists, nonconformists* of every sort.... The point is that islands are havens and breeding grounds for the unique and anomalous. *They are natural laboratories of extravagant evolutionary experimentation.*"[49] The terms "crossover artists" and "nonconformists," as well as the image of nature as a scientific laboratory run by its flora and fauna, suggest that the behavior of nature's inhabitants might be perceived by an observer as demonstrating omniscience or a level of consciousness—sometimes called theory of mind—that scientists hypothesize does not exist in non-human animals or in plants.[50]

Quammen likes his metaphors and analogies and sometimes combines them. For example, his extended metaphorical analogy of the carved-up Persian rug as a way to explain that a fragmented habitat is less than the sum of its parts, is very effective.[51] And the phrase referenced above—"*They are natural laboratories of extravagant evolutionary experimentation*"—metaphorically juxtaposes the idea of a traditional human laboratory where conscious experimentation occurs, to the natural stew that gives rise to much animal and plant diversity through natural selection.[52]

John McPhee has demonstrated the excellence of his research repeatedly. He also uses immersion, scene setting, dialog, and reporting of status

[48] Mark Canada, "Henry David Thoreau's views of 19th-century media resonate today," *The Conversation*, https://theconversation.com/henry-david-thoreaus-views-of-19th-century-media-resonate-today-81926.

[49] David Quammen, *The Song of the Dodo: Island Biogeography in an Age of Extinctions* (New York: Simon & Schuster, 1996), 18.

[50] Theory of mind is the ability to represent and understand that other individuals have their own independent mental states, including their thoughts and beliefs. Jane R. Conway, Caroline Catmur, and Geoffrey Bird, "Understanding individual differences in theory of mind via representation of minds, not mental states," *Psychonomic Bulletin and* Review 26 (2019): 798–812 and citations therein.

[51] Quammen, *The Song of the Dodo*, 18.

[52] Quammen, *The Song of the Dodo*, 18.

details—all to great advantage. His writing is entertaining and informative. Who else could have made reading not one, not two, but four books about geology and geologists so enjoyable? His writing style is spare when it comes to literary devices. He relies primarily on dialog and detailed description to sculpt his narratives. On the rare occasion when he seizes upon a literary device he likes, he just might amplify it in a subsequent story as he did in *Basin and Range*[53] after writing the following in "Travels in Georgia."

> [I]t was a journey that tended to mock the idea of a state—as an unnatural subdivision of the globe, as a metaphor for the human ego sketched on paper and framed in straight lines and in riparian boundaries behind an unalterable coast. Georgia. A state? Really a core sample of a continent, a plug in the melon, a piece of North America. Pull it out and wildcats would spill off the high edges. Alligators off the low ones.[54]

In a 2019 interview with John McPhee, Tyler Malone wrote the following about the author:

> John McPhee, in particular, borrows Ernest Hemingway's affinity for distilled grandeur and Robert Frost's understanding of the power of a pure image. He finds in simplicity both sublime beauty and profound depth. "Don't slather one verbal flourish on top of another lest you smother them all," one student remembers him saying in his Princeton course. A true master can make the plain feel pyrotechnic without unleashing a barrage of syntactical fireworks.[55]

Other literary journalists have relied more heavily on figurative literary devices than McPhee. Rachel Carson, author of *Silent Spring*—an environmental classic by an author who has been identified by many scholars

[53] John McPhee, *Basin and Range*, (New York: Ferrar, Straus, and Giroux, 1981), 15. "… a journey that tends to mock the idea of a nation, of a political state, as an unnatural subdivision of a globe, as a metaphor of the human ego sketched on paper and framed in straight lines and in riparian boundaries between unalterable coasts. The United States: really a quartering of a continent, a drawer in North America. Pull it out and prairie dogs would spill off one side, alligators off the other…"

[54] John McPhee, "Travels in Georgia," excerpted in Norman Sims, ed., *The Literary Journalists* (New York: Ballantine Books, a division of Random House, 1984), 47.

[55] Tyler Malone, "John McPhee: Seven Ways of Looking at a Writer," *Lithub*, January 17, 2019. https://lithub.com/john-mcphee-seven-ways-of-looking-at-a-writer/

as a literary journalist—occasionally used personification as when she described environmental pollution by dangerous chemicals as if contamination was a living being.

> The most alarming of all man's assaults upon the environment is the contamination of air, earth, rivers, and sea with dangerous and even lethal materials. This pollution is for the most part irrecoverable, the chain of evil it initiates not only in the world that must support life but in living tissues is for the most part irreversible. In this no universal contamination of the environment, *chemicals are the sinister and little-recognized partners of radiation* in changing the very nature of the world—the very nature of its life.[56]

and

> Strontium 90, released through nuclear explosions into the air, comes to earth in rain or drifts down as fallout, *lodges* in soil, *enters* into the grass or corn or wheat grown there, and in time *takes up its abode* in the bones of a human being, there to remain until his death.[57]

A Scientific Critique of Anthropomorphism

Of the rhetorical devices reflected upon in the foregoing sections, anthropomorphism is the one science most despises.[58] As a matter of fact, science actively discourages anthropomorphism in scientific literature.[59] Epley et al. define anthropomorphism as "the tendency to imbue the real or imagined behavior of nonhuman agents with humanlike characteristics, motivations, intentions, or emotions," a characterization resembling the definition of personification.[60]

[56] Rachel Carson, *Silent Spring* (Boston: Houghton Mifflin Company, 1962), 6.

[57] Carson, *Silent Spring*, 6.

[58] In *Ecology and Literature: Ecocentric Personification from Antiquity to the Twenty-First Century* (2), Bryan L. Moore notes that, "...among scientists, the attribution of human qualities to the nonhuman became abhorrent after the rise of modern science in the seventeenth century."

[59] Clive Wynn writes "...anthropomorphism must be resisted. Its drawbacks remain the same as they have always been: mentalistic folk-psychological accounts of animal psychology have no useful role to play in a modern objective science," in "What are animals? Why anthropomorphism is still not a scientific approach to behavior," *Comparative Cognition and Behavior Reviews* 2 (2007): 134.

[60] Nicholas Epley, Adam Waytz, and John Cacioppo, "On seeing human: A three-factor theory of anthropomorphism," *Psychological Review* 114, no. 4 (2007): 864–886; John

The scientific bias against anthropomorphism might stem from the etymology of the word itself. According to Clive Wynne, who cautions against the perils of anthropomorphism, the term originally meant ascribing human qualities to God and the angels.[61] Use of the term "anthropomorphism" increased in 1858 when Darwin himself wrote about the similarities between animal and human behavior.[62] Scientist George Herbert Lewes, a peer of Darwin's, wrote the following in an 1858 study of mollusks:

> [...] we are incessantly at fault in our tendency to anthropomorphize, a tendency which causes us to interpret the actions of animals according to the analogies of human nature.[63]

In "The perils of anthropomorphism," Wynne wrote:

> But anthropomorphism is not a well-developed scientific system. On the contrary, its hypotheses are generally nothing more than informal folk psychology and may be of no more use to the scientific psychologist than folk physics to a trained physicist.[64]

Frans de Waal, primatologist and professor of psychology at Emory University, would argue that Wynne's concerns about anthropomorphism are a sign of Wynne's own anthropodenial.[65] According to de Waal, "[a]nthropodenial is a blindness to the human-like characteristics of ani-

Kennedy, *The New Anthropomorphism* (The Cambridge University Press, 1992); Clive Wynne, "What are animals?"; Robin L. Murray and Joseph K. Heumann, Robin L. *Monstrous Nature: Environment and Horror on the Big Screen* (Lincoln: University of Nebraska Press, 2013).

[61] Clive Wynne, "What are animals? 125–135.

[62] Charles Darwin, *The Origin of Species: 150th Anniversary Edition* (New York, N.Y.: Signet Classics, 2008); Robert A. Boakes, *Darwin and Animal Behavior: Psychology and the Minds of Animals* (Cambridge: Cambridge University Press, 1984).

[63] Lewes, George (1858), *Sea-side Studies at Ilfracombe, Tenby, The Scilly Isles, & Jersey* (London: William Blackwood and Sons, 1858), https://archive.org/details/seasidestudiesat00lewe; Clive Wynne, "What are animals?: 125–135.

[64] Clive Wynne, "Concept: The perils of anthropomorphism," *Nature* 428 (2004): 606.

[65] Frans B.M. de Waal, "Anthropomorphism and anthropodenial: Consistency in our thinking about humans and other animals," *Philosophical Topics* 27, no. 1 (1999): 255–280, https://www.jstor.org/stable/43154308; Frans B.M. de Waal, "What I learned from tickling apes," *New York Times*, April 8, 2016, https://www.nytimes.com/2016/04/10/opinion/sunday/what-i-learned-from-tickling-apes.html.

mals, or the animal-like characteristics of ourselves."[66] An increasing number of researchers, however, suggest that anthropomorphism has a place in science.

Alvin Chan wrote that animals likely share analogous human cognitive characteristics and emotions, therefore anthropomorphism can and should be used to promote empathy for animals and thoughtful conservation practices.[67] Chan cautions that anthropomorphism "should be cast in a conservative light and used in a controlled manner."[68] This can be accomplished by avoiding: overzealous interpretations of animal intelligence and behavior, and taking care not to create fictional animal personas or unjustifiably project human qualities onto non-human animals.

If we can do this, notes Chan, anthropomorphism could be a useful mechanism "to create increased empathy toward target species and populations among the general public that will lead to actions that are more considerate toward them." [69] In other words, anthropomorphism, when applied appropriately, might be an effective way to communicate the plight of both non-human animals and humans.[70] Could an argument be made that the judicious use of anthropomorphism/personification, the pathetic fallacy, and other figurative literary devices in nonfiction nature writing doesn't violate literary journalism's emphasis on accuracy, and that writers do not cross the reality boundary if and when they use these devices?

In his book *Ecology and Literature: Ecocentric Personification from Antiquity to the Twenty-first Century*, Bryan L. Moore notes that writers who personify nature do so "not reflexively, as a prettifying cliché, or to assert a childish understanding of nonhuman nature, but consciously, to account for interrelationships, human and otherwise."[71] Should we not, then, relax literary journalism's criteria to allow nature writers who use personification-like devices, metaphor, analogy, and other less-than-literal techniques into the club, as long as those writers pay attention to literary

[66] de Waal, "Anthropomorphism and anthropodenial, 258.

[67] Alvin Chan, "Anthropomorphism as a conservation tool," *Biodiversity and Conservation*, 21 (2012): 1889–1892.

[68] Chan, "Anthropomorphism, 1890.

[69] Chan, "Anthropomorphism, 1890.

[70] Fredrik Karlsson, "Critical anthropomorphism and animal ethics," *Journal of Agricultural and Environmental Ethics*, 25 (2012): 707–720.

[71] Bryan L. Moore, "Prefatory Note," *Ecology and Literature: Ecocentric Personification from Antiquity to the Twenty-first Century* (New York: Palgrave Macmillan – a division of St. Martin's Press LLC., 2008), ix.

journalism's core tenets by displaying a rigorous allegiance to verifiable fact? Whether consciously or not, literary journalism's canons were relaxed when authors like Henry David Thoreau, Rachel Carson, John McPhee, and David Quammen—all of whom sometimes relied upon reality-bending literary language—became standard bearers for the genre. Let's lend cordial approval to other nonfiction nature writers who meet the proposed standards set forth below.

Toward a New Literary Journalism Canon

Based upon the discussion above, I suggest that the literary journalism community *explicitly* modify the tenets of literary journalism to be more amenable to nonfiction nature writers. Starting with Mark Kramer's "breakable rules," the following is proposed as a modern canon for nonfiction nature writers who might be deemed literary journalists—a canon that would incorporate nonfiction nature writers who are scrupulous in their methods and their accuracy:

1. Nonfiction nature writers who are literary journalists immerse themselves in their subjects' worlds. *Many nature writers already do this when they carefully observe and report what happens in the natural world.*
2. Nonfiction nature writers who are literary journalists immerse themselves in background research. *Again, this is what many nature writers already do.*
3. Nonfiction nature writers who are literary journalists work out implicit covenants concerning accuracy and candor with readers and sources. This covenant requires the natural world to be portrayed accurately, even when emotive or figurative literary devices are used by the writer.
4. Nonfiction nature writers who are literary journalists write primarily about routine events. *The lives of non-human animals as well as other entities in the environment are often routine. However, it should be fair game for the literary journalist/nature writer to write about natural catastrophes (including potential future catastrophes as Rachel Carson and others have done) so long as the truth of the matter is documented.*
5. Nonfiction nature writers who are literary journalists write in "intimate voice" –informal, frank, human, and ironic.

6. Style counts and tends to be plain and spare. *Here I would suggest that the judicious use of literary devices including the pathetic fallacy, personification, anthropomorphism, metaphor, analogy, and others should be allowed so long as the way these devices are used is consistent with literary journalism's implicit covenants concerning accuracy and candor.*
7. Nonfiction nature writers who are literary journalists *can write from a disengaged stance, they may turn and address readers directly, or they can write from a personal stance (as Rachel Carson did in* Silent Spring*) so long as the covenants concerning accuracy and candor are not violated.*
8. Nonfiction nature writers who are literary journalists develop meaning by building upon the readers' sequential reactions.[72]

The adoption of these or similar guidelines would go a long way toward breaking down barriers within the literary journalism genre and enriching it by including more nonfiction nature writers, even though their works might not conform in every detail to previously established canons. Literary journalism should not be an "I know it when I see it" genre, but an inclusive genre if writers adhere to the overarching requirements of immersion, candor and accuracy, and an intimate author's voice as they write about the status lives of non-humans including but not limited to flora, fauna, rivers, streams, woods, desert, and bighorn sheep.

Recommended Additions to the Literary Journalism Pantheon

There are innumerable nonfiction nature writers, past and present, who might be accorded status as literary journalists. Many members of the Chicago school of nature writers whose works appear in Joel Greenberg's *Of Prairie, Woods, and Water: Two Centuries of Chicago Nature Writing* might be considered. For example, Gene Stratton-Porter's "Moths of the Limberlost," Jens Jensen's tale of a vanishing midwestern landscape in "Sifting," and Leonard Dubkin's "My Secret Places: One Man's Love Affair with Nature in the City," are worth considering.[73] There are many

[72] With more apologies to Mark Kramer for reframing his list of eight "breakable rules." Mark Kramer, "Breakable Rules for Literary Journalists," in *Literary Journalism: A New Collection of the Best American Nonfiction* (New York: Ballantine Books, 1995), 22–33.

[73] Gene Stratton-Porter, "Moths of the Limberlost," Jens Jensen, "Sifting," and Leonard Dubkin, "My Secret Places: One Man's Love Affair with Nature in the City," in Greenberg, Joel ed., *Of Prairie, Woods, and Water: Two Centuries of Chicago Nature Writing* (University of Chicago Press, 2008).

other nonfiction nature writers whose works are worth examining through the literary journalism lens. One of these writers is Gary Paul Nabhan, a knowledgeable naturalist who chronicles the flora and fauna of the Southwest. Nabhan writes the following in his introduction to the edited collection, *Twenty Ways of Seeing Desert Bighorn*:

> This book is more than a collection of essays about the bighorn sheep that live in the desert along the U.S.-Mexico border. It is an attempt to broaden what the natural history essay can be, for many of the following offerings build upon but diverge from the narrative journal style and didactic tone of the early twentieth-century scientist-explorers who searched exotic lands for wildlife and wisdom.[74]

Finally, the writings of the French naturalist and entomologist, J. Henri Fabré, have much in common with the works of recognized literary journalists. Fabré, an autodidact who became acquainted with science via the creatures he found in his immediate environment, immersed himself in the ordinary lives of his subjects—including the insects and arachnids of rural Provence. Fabré disdained scientists who dissected and anatomized the bodies of the creatures he studied. His goal was to uncover the larger scientific truth of his subject's lives and he did so by intensely studying their behavior—largely in natural settings. Fabré's powers of observation, united with his ability to translate his observations using a literary style, revealed his admiration for the overlooked or undervalued members of the natural world. Fabré transported his readers into a vivid domain inhabited by beetles, caterpillars, and the Languedocian scorpion.

His aesthetically pleasing tale of the life cycle of the Languedocian scorpion, written in 1923, begins with an account of the very dilemma nature writers encounter when their work centers around creatures, plants, and features of the earth, wind, and waters that do not speak and therefore cannot be interviewed.

Although, as Fabre so accurately notes, he cannot interview a scorpion, he is able to observe the creature, record the creature's habits, behaviors, environment, and his interactions with other living creatures. In other

[74] Gary Paul Nabhan, "Introduction: Diversity in Desert Wildlife Writing," in *Counting Sheep: 20 Ways of Seeing Desert Bighorn* (Tucson, AZ: The University of Arizona Press, 1993), xi. Nabhan's narrative nonfiction contribution to the volume, "When the Spring of Animal Dreams Run [sic] Dry," (7–26) informs the reader about Arizona's diminished herds of desert bighorn sheep using techniques including natural history (accurate reporting of past and present status details concerning the sheep), dialog, an intimate voice, immersion in his subject, and emotive, figurative language.

words, Fabre can record the "status life" of an extraordinary, yet humble, non-human being. As a substitute, perhaps, for the limitations of Nature's mute creatures, Fabre personified the creature's secretive, taciturn nature in sparkling prose.

> The Scorpion is an uncommunicative creature, secret in his practices and disagreeable to deal with, so that his history, apart from anatomical detail, amounts to little or nothing. The scalpel of the experts has made us acquainted with his organic structure; but no observer, as far as I know, has thought of interviewing him, with any sort of persistence, on the subject of his private habits.[75]

Conclusion

Nature writing has a long history—as journalism and as literature. Although it might be challenging because of that long history and the diffuse, multivalent character of the nature writing genre, why shouldn't we explore nature writing in a systematic, scholarly way, debate the issue of inclusion, and make conscious decisions to include in the literary journalism community those nature writers who create truth-telling narratives about the natural world using the techniques of the literary journalism genre—including immersion, dialog (when possible), intimate voice, and reporting on Nature's status details? Literary journalism scholarship has much to gain and little to lose by expanding the ranks of those who—like Thoreau, Carson, McPhee, and Quammen—have brought the realities of the natural world home to us with grace and style.

References

Alexander, Robert, "Literary Journalism and Ecocriticism," in *The Routledge Companion to American Literary Journalism* (New York: Routledge, Taylor and Francis, 2020), 481–497.

Boakes, Robert A., *Darwin and Animal Behavior: Psychology and the Minds of Animals* (Cambridge: Cambridge University Press, 1984).

Canada, Mark, "Henry David Thoreau's Views of 19th-century media resonate today," *The Conversation*, https://theconversation.com/henry-david-thoreaus-views-of-19th-century-media-resonate-today-81926

[75] J. Henri Fabre, *The Life of the Scorpion* (Honolulu: University Press of the Pacific, 2002, reprinted from 1923 edition), 3.

Carson, Rachel, *Silent Spring* (Boston: Houghton Mifflin Company, 1962).
Chan, Alvin, "Anthropomorphism as a conservation tool.' *Biodiversity and Conservation*, 21 (2012): 1889–1892.
Darwin, Charles, *The Origin of Species: 150th Anniversary Edition* (New York, N.Y.: Signet Classics, 2008).
de Waal, Frans B.M., "Anthropomorphism and anthropodenial: Consistency in our thinking about humans and other animals," *Philosophical Topics* 27 no. 1 (1999): 255–280.
de Waal, Frans B.M., "What I learned from tickling apes," *New York Times*, April 8, 2016. https://mobile.nytimes.com/images/100000004315815/2016/04/10/opinion/sunday/what-i-learned-from-tickling-apes.html.
Dubkin, Leonard, "My Secret Places: One Man's Love Affair with Nature in the City," in Greenberg, Joel ed., *Of Prairie, Woods, and Water: Two Centuries of Chicago Nature Writing* (University of Chicago Press, 2008), 278–288.
Epley, Nicholas, Adam Waytz, and John Cacioppo, "On seeing human: A three-factor theory of anthropomorphism," *Psychological Review* 114 no. 4 (2007): 864–886.
Elder, John and Robert Finch, eds. "Introduction," in The Norton Book of Nature Writing (New York: W.W. Norton and Company, 2002).
Fabre, J. Henri, *The Life of the Scorpion* (Honolulu: University Press of the Pacific, 2002, reprinted from 1923 edition).
Grant, Peter R., "What Does It Mean to Be a Naturalist at the End of the Twentieth Century?" *The American Naturalist* 155 no. 1 (2000): 1–12.
Hartsock, John, *A History of American Literary Journalism: The Emergence of a Modern Narrative Form* (Amherst, University of Massachusetts Press, 2000).
Hornady, William, *Hornady's American National History, Fourteenth Edition – Revised* (New York: Charles Scribner's Sons, 1927), xxii–xxiii.
Jensen, Jens, "Sifting," in Greenberg, Joel ed., *Of Prairie, Woods, and Water: Two Centuries of Chicago Nature Writing* (University of Chicago Press, 2008), 398–402.
Karlsson, Fredrik, "Critical anthropomorphism and animal ethics," *Journal of Agricultural and Environmental Ethics*, 25 (2012), 707–720.
Keebles, Richard Lance, "Literary Journalism As A Discipline: Tom Wolfe and Beyond," *Brazilian Journalism Research* 14, no. 3 (2018).
Kennedy, John, *The New Anthropomorphism* (The Cambridge University Press, 1992).
King, Margaret J., "Audience in the Wilderness: The Disney Nature Films," *Journal of Popular Film and Television* 24 no. 2 (1996), 60–68.

Kramer, Mark, "Breakable Rules for Literary Journalists," in *Literary Journalism: A New Collection of the Best American Nonfiction* (New York: Ballantine Books, 1995), 21–34.

Leopold, Aldo, *A Sand Country Almanac: With Essays on Conservation from Round River*, (New York: The Random House Publishing House, 1966; Oxford University Press, Inc., 1949).

le Guin, Ursula K., *The Word for World is Forest* (New York, NY: Harper Collins), 1980.

London, Jack, *The Call of the Wild* (New York: The McMillan Company, 1903).

Lewes, George (1858), *Sea-side Studies at Ilfracombe, Tenby, The Scilly Isles, & Jersey* (London: William Blackwood and Sons, 1858), https://archive.org/details/seasidestudiesat00lewe.

McPhee, John, Basin and Range, (New York: Farrar Straus & Giroux, 1981).

McPhee, John, *Annals of the Former World* (New York: Farrar Straus & Giroux, 1980–1982).

McPhee, John, *Coming into the Country* (New York: Farrar Straus & Giroux, 1976).

McPhee, John, *The Control of Nature* (New York: Farrar Straus & Giroux, 1989).

McPhee, John, *The Pine Barrens* (New York: Farrar Straus & Giroux, 1968).

McPhee, John, "Travels in Georgia," excerpted in Norman Sims, ed., *The Literary Journalists* (New York: Ballantine Books, a division of Random House, 1984), 47.

Moore, Bryan L., "Part 1 Personification in Practice and Theory," *Ecology and Literature: Ecocentric Personification from Antiquity to the Twenty-first Century* (New York: Palgrave Macmillan – a division of St. Martin's Press LLC., 2008a), 1–41.

Moore, Bryan L., "Prefatory Note," *Ecology and Literature: Ecocentric Personification from Antiquity to the Twenty-first Century* (New York: Palgrave Macmillan – a division of St. Martin's Press LLC., 2008b), ix–xi.

Murray, Robin L., and Heumann, Joseph K. *Monstrous Nature: Environment and Horror on the Big Screen* (Lincoln: University of Nebraska Press, 2013).

Nabhan, Gary Paul, "Introduction: Diversity in Desert Wildlife Writing," in *Counting Sheep: 20 Ways of Seeing Desert Bighorn* (Tucson, AZ: The University of Arizona Press, 1993a), xi–xx.

Nabhan, Gary Paul, "When the Spring of Animal Dreams Run [sic] Dry," in *Counting Sheep: 20 Ways of Seeing Desert Bighorn* (Tucson, AZ: The University of Arizona Press, 1993b), 7–26.

Neuzil, Mark and Russell Train, *The Environment and the Press: From Adventure Writing to Advocacy* (Northwestern University Press, 2008).

Pimentel, D. Silent Spring, the 50th anniversary of Rachel Carson's book. *BMC Ecolology* 12, 20 (2012). https://doi.org/10.1186/1472-6785-12-20

Preston, Richard McCann, "The Fabric of Fact: The Beginnings of American Literary Journalism," Ph.D. Dissertation, Princeton, 1983.

Quammen, David, *The Song of the Dodo: Island Biogeography in an Age of Extinctions* (New York: Simon & Schuster, 1996).

Rodriguez, Diogo, "How to Create a Good Explainer, Explained," *Medium*, May 6, 2019, https://medium.com/notes-from-the-classroom/how-to-create-a-good-explainer-explained-4e4236870b24

Roberts, Nancy, "Literary Journalism and Social Activism," in *The Routledge Companion to Literary Journalism*, (New York: Routledge, Taylor and Francis, 2020).

Roe, Frederick William, ed., *John Ruskin: Selections and Essays*, (Mineola: Dover Publications, Inc., 2013).

Ruskin, J. (2010) 'OF THE PATHETIC FALLACY', in E.T. Cook and A. Wedderburn (eds.) *The Works of John Ruskin*. Cambridge: Cambridge University Press (Cambridge Library Collection - Works of John Ruskin), pp. 201–220.

Scharf, Caleb, "In Defense of Metaphor in Science Writing," *Scientific American*, July 19, 2013. https://blogs.scientificamerican.com/life-unbounded/in-defense-of-metaphors-in-science-writing/

Schmidly, David J., "What It Means to be a Naturalist and the Future of Natural History at American Universities," *Journal of Mammalogy* 86, no. 3 (2005): 449–456. https://doi.org/10.1644/1545-1542(2005)86[449:WIMTBA]2.0.CO;2

Sims, Norman, "The Art of Literary Journalism," in *Literary Journalism: A New Collection of the Best American Nonfiction*, Sims and Kramer eds. (Random House, 1995).

Sims, Norman, "The Problem and the Promise of Literary Journalism Studies," *Literary Journalism Studies* 1, no. 1 (2009), 7–16.

Stewart, Frank, *A Natural History of Nature Writing* (Washington, D.C.: Island Press, 1995).

Stratton-Porter, Gene, "Moths of the Limberlost," in Greenberg, Joel ed., *Of Prairie, Woods, and Water: Two Centuries of Chicago Nature Writing* (University of Chicago Press, 2008), 213–220.

The Routledge Companion to American Literary Journalism (New York: Routledge, Taylor and Francis, 2020).

Thoreau, Henry David, "Walden" in *Transcendentalists Collection* (Los Angeles: Carousel Books, 2020), 7–277.

Thoreau, Henry David, *Walden: Return to Nature: An American Classic* (New York: The New American Library, 1952).

Thoreau, Henry David, "Brute Neighbors," from *Walden* in *The Norton Book of Nature Writing* Finch and Elder, eds. (New York: W.W. Norton & Company, 1990, 2012), 172–180.

Timblin, Diane, "Thoreau as Naturalist: A Conversation with Four Authors," *American Scientist*, https://www.americanscientist.org/article/thoreau-as-naturalist

Wallace, David Rains, "The Nature of Nature Writing," *The New York Times*, July 22, 1984, Section 7, Page 1, https://www.nytimes.com/1984/07/22/books/the-nature-of-nature-writing.html.

Wolfe, Tom, "Preface," in *The New Journalism* (New York: Harper and row, 1973).

Wynne, Clive, "The perils of anthropomorphism," *Nature* 428 (2004): 606, https://doi.org/10.1038/428606a

Wynne, Clive, "What are animals?" Why anthropomorphism is still not a scientific approach to behavior," *Comparative Cognition and Behavior Reviews* 2 no. 1 (2007): 125–135.

Thanatos Syndrome: Literary Forms in Domosławski's *Death in the Amazon*

Aleksandra Katarzyna Wiktorowska

Introduction

They set up the ambush by a small bridge across a stream. They've been hiding among the trees since early morning—and they're lying in wait. They know that José Cláudio and Maria will have to slow down here. That's when the first shot is fired.

Discharged from a hunting rifle, the bullet hits both of them at once […]. When they fall off their motorbike, they are still alive. The killers finish them off at close range, drag the bodies off the highway, and drive away toward the town.

But first they sever José Cláudio's right ear.[1]

[1] Artur Domosławski, *Śmierć w Amazonii* [Death in the Amazon] (Warsaw: Wielka Litera, 2013), 9. All translations from Polish to English, unless otherwise noted, are my own.

A. K. Wiktorowska (✉)
University of Lower Silesia (DSW), Wrocław, Poland
e-mail: aleksandra.wiktorowska@dsw.edu.pl

P. Calvi (ed.), *The Literary Journalist as a Naturalist*, Palgrave Studies in Literary Journalism,
https://doi.org/10.1007/978-3-031-56634-9_15

Thus begins the story recounted by Artur Domosławski on the ecological disasters brought to the human habitat in Brazil, Peru, and Ecuador, and on the responsibility for the destruction that rests with the citizens of developed countries. By following the trail of such raw materials as wood, charcoal, and beef in Brazil, and the persecution of indigenous people together with the destruction of the environment by major extraction industries in Peru and Ecuador, the author of *Śmierć w Amazonii. Nowe Eldorado i jego ofiary* [Death in the Amazon: The New Eldorado and its Victims] (2013) depicts the food chain of the world economy, which begins with crimes committed against local people and ends with consumers from the global north.

Born in 1967, Domosławski is a Polish reporter who specializes in Latin American affairs and has spent a large part of his career in South America. His books include *Gorączka latynoamerykańska* [Latin American Fever] (2004) on the twentieth-century revolts and revolutions in Argentina, Brazil, Chile, Colombia, Cuba, Mexico, and Venezuela, told through the personal accounts of local people. Among his other works are: *Ameryka zbuntowana* [Rebel America] (2007), a collection of interviews with 17 American left-wing intellectuals, such as Noam Chomsky, Eric Foner, Richard Rorty, and Howard Zinn; *Kapuściński non-fiction* (2010) [*Ryszard Kapuściński: A Life*, 2012], a biography of the famous Polish reporter that sparked controversy by revealing that Kapuściński fictionalized some of what he had presented as factual; and *Wykluczeni* [Excluded] (2016), a book about groups of people from Latin America, South East Asia, and the Middle East, who for social or political reasons are deprived of their rights and in many cases also their lives.[2]

The aim of this chapter is to present Domosławski as the author of book-length reportage that depicts nature as a human habitat in crisis. After a brief introduction on *Death in the Amazon* and the history of its publication, I demonstrate the narrative devices and techniques Domosławski uses in his reportage, as well as the theories he implements to back up his findings and engage his readers. I conclude by showing how this piece of environmental journalism is unique among books dealing with ecology and responsibility for the environmental crisis.

[2] Only one of Domosławski's books, Kapuściński's biography, has appeared in English.

The History of the Publication

Like Rachel Carson's 1962 environmental classic *Silent Spring*, which grew out of her *New Yorker* series, Domosławski's *Death in the Amazon* developed from his reportage for *Gazeta Wyborcza* (2005). There he covered the coal mafia's assassination of the American nun Dorothy Stang, who had fought for the rights of the landless workers in Brazil and against the destruction of the rainforest by timber traders, charcoal burners, and cattle breeders.

Domosławski came up with the idea for his book in 2011: "I had a hunch that such a story could say something not only about South America but also about the mechanisms that govern the modern world."[3] He had just read on *The New Yorker* website an article by Jon Lee Anderson on the murder of two Brazilian ecologists, José Cláudio Ribeiro da Silva and Maria do Espírito Santo da Silva. This information—as Domosławski recalls—unsettled him. He had visited the region some years before and knew it first hand, so he decided to go back and research this crime.[4] If in 2005, when writing on Dorothy Stang, he had been interested mostly in the landless workers, his main focus now was the environment. Domosławski remembered that the American nun had been killed by the coal mafia responsible for the destruction of the rainforest, whereas José Claudio and his wife Maria were environmental activists killed "in the trademark style of contracted executions"[5] on the orders of people with a vested interest in stopping them. Maria do Espírito Santo da Silva—as Domosławski discovered with surprise—appeared on the hit list he had quoted in his reportage six years earlier. The story started to write itself.

Domosławski's book-length—over 300 pages—journalism was published in the autumn of 2013 by the Polish publishing company Wielka Litera. Strangely enough, it attracted little attention, flying under the radar of the general audience, save for a few journalists who interviewed the author and a few critics whose reviews appeared mainly in the Polish independent media. Domosławski was already an acclaimed writer, so this

[3] Artur Domosławski, interview with the author, June 24, 2020.
[4] Ibid.
[5] Jon Lee Anderson, "Murder in the Amazon." *The New Yorker*, June 15, 2011.

lack of interest was surprising. His previous publication, *Ryszard Kapuściński: A Life*, had been widely debated and in 2010 he was named in Poland the Journalist of the Year; his subsequent book-length reportage, *Excluded*, received comprehensive coverage in major Polish newspapers and magazines, and was nominated for the Nike Literary Award in 2017. How is it, then, that his book on the climate and global markets was almost completely ignored? One possible reason is that Domosławski moved to Argentina for ten months just as the book was published, so he could not be involved in its promotion. Another possible explanation is a rather marginal interest in the topic, as the fate of the Amazon and climate change were not widely discussed in Poland at the time. A third eventuality is that the ongoing trial against Domosławski—he was sued by Ryszard Kapuściński's wife and daughter for infringement of personal rights, a case he finally won in 2016—might have influenced his credibility as a reporter.[6]

Even though it sold relatively well in Poland, the book interested just one foreign publisher despite the best efforts of Antonia Lloyd-Jones, an English translator, and Agata Orzeszek, a Spanish translator.[7] The reasons remain unclear, as the book market plays by its own inexplicable rules. The fact that the topic was of limited interest and that after writing Kapuściński's biography Domosławski had become the enfant terrible of Polish reportage might have had something to do with the cold reception the book received abroad. Another potential factor is that the book openly criticizes big corporations, who—just like as tyrants and occupiers—usually prefer to move in the silence of shadows:

> Silence is a signal of unhappiness and, often, of crime. It is the same sort of political instrument as the clatter of weapons or a speech at a rally. Silence is necessary to tyrants and occupiers, who take pains to have their actions accompanied by quiet. Look at how colonialism has always fostered silence; at how discreetly the Holy Inquisition functioned; at the way Leonidas Trujillo avoided publicity.[8]

[6] Artur Domosławski, interview with the author, June 24, 2020.

[7] Although the book has had two editions and has sold over 10,000 copies in Poland, when "unpopular" books rarely sell more than 2000–3000 copies, so far it has only one foreign translation. In 2017 it was published by Hungarian publishing house Typotex.

[8] Ryszard Kapuściński, *The Soccer War*, trans. William Brand (London: Granta Books, 2007), 189.

The book's atypical genre—literary journalism written with a strong thesis that goes beyond classical reportage and is rooted directly in socio-economic and sociocultural theory—might have also influenced its fate.

Narrative Techniques in the Service of Nature

The main subject of Domosławski's book is death:

> Death was very important, because this time it was not just about a community being expropriated from the land, but about a war for money and access to raw materials in which people are being killed. It was totally shocking. For a man like me, brought up on the myth of capitalism and North America, even if I went to the other side of the mirror and saw that I had grown up in some false, hypocritical myth, it was unbelievable that big business works like the mafia in the countries of the South. I went there several times and I kind of knew about it, but full realization that shooting a man is not a problem came only while working on this book. It is "part of the business".[9]

In his account, death functions as a narrative hook; some critics have even argued that *Death in the Amazon* reads like a thriller. The first part is set in northern Brazil and starts with the assassination of the two ecologists, José Cláudio Ribeiro da Silva and his wife Maria do Espírito Santo da Silva. Domosławski then recounts the killing of Dorothy Stang and of the first hero of the jungle, Chico Mendes, who fought to protect the environment against cattle farmers and was assassinated in 1988 on the orders of a landowner. The author continues with a hit list that names potential victims whose personal stories he relates: "Their history is characterized by cruel repetition. Scorched earth surrounds everyone: murdered comrades, friends, relatives."[10] In the second part, set in north-western Peru, Domosławski follows the same pattern, focusing on the methods used to terrorize the Yanacocha–Newmont gold mine's critics, including Marco Arana and his friends from the local organization Grufides who are campaigning against the persecution of indigenous people and the destruction of the environment in the Cajamarca region by the mine. The activists are followed, bugged, issued death threats, and at

[9] Artur Domosławski, interview with the author, June 24, 2020.
[10] Domosławski, *Śmierć w Amazonii*, 52.

worst beaten up or killed. Domosławski reflects: "The most remarkable thing about Marco Arana's life is that he is still alive."[11]

In part three, set in north-eastern Ecuador, Domosławski narrates the struggle of 30,000 inhabitants of Amazonia led by local lawyer Pablo Fajardo against the oil giant Texaco–Chevron, which is responsible for polluting water and land in the region. The central murder related here is that of Pablo Fajardo's brother. There are also other victims, such as Nicolás Soto, who died of cancer after entering contaminated rivers without protective clothing in an effort to remove oil spills. "There are many like him, his wife and sons. In every village, community, settlement,"[12] the author remarks, and quotes testimonials about dozens of children and adults who died of cancer and other diseases caused by the toxic air and water.

The list of fatalities—a register of the direct and indirect victims of big business—seems endless. Nonetheless, death is an integral element in the process of meaning-making bound up with the concepts of character, space, and plot, as it concerns not only humans. In his narrative, Domosławski also shows the gradual demise of nature: the rainforest that is being destroyed in Brazil, land and water that are being polluted by gold mining in Peru and by oil extraction in Ecuador. The habitat dies gradually, along with its people. This direct connection between nature and human life allows the author to relate the ecological disasters in the Amazon from the perspective of its inhabitants, by paying what Sims describes as "attention to ordinary lives,"[13] a core characteristic of literary journalism. In the same vein, Domosławski's style of writing reflects the humanist tradition, as opposed to the positivist approach typical of 5Ws journalism.[14] His message is, therefore, similar to the point made by Robinson in *Mother Country* that "abuse of the natural environment involves contempt for the health and the life of human beings."[15]

To describe nature in danger Domosławski employs images–symbols, a popular device in socioenvironmental writing. In Rachel Carson's

[11] Ibid., 128.

[12] Ibid., 229.

[13] Norman Sims, *True Stories: A Century of Literary Journalism* (Evanston, IL: Northwestern University Press, 2007), 12.

[14] Ted Conover, "Immersion and the Subjective: Intentional Experience as Research," *Literary Journalism Studies* Vol. 10, No. 2 (Fall 2018), 168.

[15] David Dowling, "Banned in Britain: Marilynne Robinson's Environmental Literary Journalism," *Literary Journalism Studies* Vol. 11, No. 2 (December 2019), 76.

"Undersea" published in *The Atlantic*, it is the ocean with "the foam and surge of the tide that beats over the crab hiding under the seaweed of his tide-pool home; or the lilt of the long, slow swells of midocean, where shoals of wandering fish prey and are preyed upon, and the dolphin breaks the waves to breathe the upper atmosphere…."[16] In *Silent Spring* it is the silent garden that the birds no longer visit, killed by insecticide from crop spraying.[17]

Domosławski presents two powerful images in *Death in the Amazon*. The first is a tree, the biggest chestnut tree José Cláudio ever saw in his life. Domosławski quotes the following interview:

Q: When a tree as fine as this one is cut down, doesn't it make you think of bloodshed?
A: My friend, when a tree like this is cut down, it gives off an odor that you can smell. When it's about to topple, you can hear its groan: the creaking of its trunk. And you can see the leaves falling from it, like a farewell gesture …
Q: And does it feel to you as if someone were dying?
A: It feels as if someone were being killed. It's a living creature.[18]

The author returns to the chestnut tree further on when rendering the lives of the assassinated ecologists. He describes a photograph: "There is beauty and strength in this image: José Cláudio spreads his arms widely against a background of a giant chestnut tree, like Christ the Redeemer at the top of Corcovado Mountain in Rio de Janeiro. It is His Majesty—Sua Majestade—as he called it."[19]

The tree is a symbol of life, strength, stability, growth, and of the Amazon rainforest. The chestnut tree also symbolizes a bygone time when chestnut farming provided a source of income for small-scale local peasant farmers. The fact that for José Cláudio cutting a tree was like murder, and knowing that he and his wife were murdered because they were fighting against the *madeireiros*, the timber dealers who have been systematically

[16] Alex Lockwood, "Recruiting the Imagination: the environmental literary journalism of Rachel Carson," in *Global Literary Journalists: Exploring the Journalistic Imagination*, ed. Richard Lance Keeble and John Tulloch (New York: Peter Lang, 2011), 335.

[17] See Rachel Carson, *Silent Spring* (Cambridge: The Riverside Press, 1962).

[18] Domosławski, *Śmierć w Amazonii*, 13.

[19] Ibid., 33.

devastating the region, complements this powerful symbolic image. The second image is a description of pollution:

> Animals were the first to die.
> Nocturnal birds, curious about the plumes of fire from the burners rising above the ground, flew too close and turned into burning spooks. The last bird scream, then the unpleasant stench of burnt feathers.
> People found dead snakes, lizards, wild pigs—all stuck in a dark, oily slush. Cows that drank the water from the streams pissed with blood-colored urine and then died. Pigs too. [...] They say that when they cut them up to grill them, the guts stank like never before. The roasted meat tasted strange. Kind of chemical. So did the fried dead fish they caught from rivers and streams. The dogs didn't want to eat them. Besides, dogs also died. Then the children began to fall ill.
> First, they were swimming in the Aguarico River, playfully trying to catch a rainbow floating on the surface. In special reservoirs, resembling ponds or small lakes, where the remains of kerosene mixed with water and slush were amassed, children practiced diving. They collected the oily slush in their hands—whoever had the most won.[20]

The idyllic rainforest in Ecuador is defiled by the fires that accompany oil extraction. Water, the universal symbol of life and purity, is contaminated, endangering all living creatures that drink it. Domosławski even invokes the leitmotif of innocent children threatened by contamination. The fearsome chronology of death helps the author to depict the sense of the rising danger that step by step threatens human health, and life itself.

If Carson needed to animate nature to evoke empathy and involve her readers in her campaign, Domosławski's images–symbols serve the same purpose, but they are always in close relation to humans. Unlike Carson, who believed that "if the behavior of the fish is to be understandable to us, we must describe it in the words that most properly belong to human psychological states,"[21] and thus imbued her actors—birds, insects and fish—with feelings, Domosławski's nature remains silent and inanimate. His actors are people, and he establishes a parallel between human life and the fate of their habitat, that is, nature. Without the human component,

[20] Ibid., 225.

[21] Lockwood, "Recruiting the Imagination: the environmental literary journalism of Rachel Carson," 337.

his powerful images lose their metaphorical meaning—the chestnut tree would be just a big tree, the polluted rainforest merely a contaminated area.

Overall, even if *Death in the Amazon* is focused on nature and ecological disasters, humans form the core of Domosławski's narrative. The author presents his story through the eyes and lives of his characters: "I can't tell my stories—this or any other one—differently. They are lifeless until someone recounts them. And we can't tell a story without a hero."[22]

Each part of *Death in the Amazon* is constructed around a main character. Of the assassinated ecologists Domosławski writes: "Today they are called heroes of the forest, martyrs of the Amazon. This is how history is written retrospectively. A medal, a commemorative plaque, a monument. But what were these Amazonian villagers like in real life? Where did they come from? How did they live? Why did they risk their lives?"[23]

But even if each part of the book has its main character whose life and fearless campaign the author reconstructs, we are in fact dealing with a collective protagonist. This becomes clear especially in the Ecuadorean part of the book when Domosławski quotes Pablo Fajardo's protest:

> People in America and Europe—says Pablo—like to have a hero, preferably someone weak, who unexpectedly defeats a bigger, more powerful player, someone who doesn't stand a chance at the start but wins in the end. It's very American, very Hollywood. You are individualists. For us, community and collective action are more important. The collective hero here is thirty thousand people who filed a lawsuit and took up the fight for their rights. Pablo Fajardo is only one of thirty thousand.[24]

Domosławski deploys what Hartsock calls "narrative literary journalism [that] embraces the more personal as revealing a different dimension to the cultural in the attempt to narrow the empathetic distance between the protagonists in the discourse, the author, and the readers."[25] Pablo Fajardo may be just one of 30,000, but Domosławski tells the lawyer's story because he knows readers would not be able to identify with an anonymous collective.

[22] Domosławski, *Śmierć w Amazonii*, 320.

[23] Ibid., 36.

[24] Ibid., 320.

[25] Dowling, "Banned in Britain: Marilynne Robinson's Environmental Literary Journalism," 69.

Domosławski's approach to his characters is worth separate analysis. Unlike some other Western writers who deal with colonialism, he is aware of postcolonial theory and his attitude is completely different from that condemned by Said in *Orientalism*.[26] His story is one of local people, and to steer his coverage in the right direction he digs deep into the historical context, reflecting on various forms of conquest, past and present. A perfect example is in the section about Peru, where some 500 years ago the conquistadores killed the Inca chief, Atahualpa. "The little shop offering locally crafted items sells two chess sets: one classical—the Spanish versus the Incas, and another contemporary—gold prospectors from mining companies versus the natives."[27] The topicality of the chief's murder and the local people's recourse to the history of conquest is supported by the book

Haine de l'Occident [Hate For the West] by Jean Ziegler. Domosławski explains that like individuals, societies can suffer from a crippling form of post-traumatic shock following an unexpected and extremely violent attack from the outside. He writes:

> Memories of traumatic events return after a long time. The victims, or in most cases their descendants, finally start to talk about the injustices, speak up for their own and their ancestors' dignity, for public remembrance of the crimes, and for compensation. [...]
>
> In Cajamarca, the trauma of the past is being reinforced by the injustices of the present. The native people's past is being played out again today, in the very same place, in the very same square.[28]

[26] In the introduction to *Orientalism* Said observes: "On a visit to Beirut during the terrible civil war of 1975–1976 a French journalist wrote regretfully of the gutted downtown area that 'it had once seemed to belong to ... the Orient of Chateaubriand and Nerval.' He was right about the place, of course, especially so far as a European was concerned. The Orient was almost a European invention, and had been since antiquity a place of romance, exotic beings, haunting memories and landscapes, remarkable experiences. Now it was disappearing; in a sense it had happened, its time was over. *Perhaps it seemed irrelevant that Orientals themselves had something at stake in the process, that even in the time of Chateaubriand and Nerval Orientals had lived there, and that now it was they who were suffering; the main thing for the European visitor was a European representation of the Orient and its contemporary fate*, both of which had a privileged communal significance for the journalist and his French readers" (1; emphasis added).

[27] Domosławski, *Śmierć w Amazonii*, 219.

[28] Ibid., 220.

Understanding the historical, material context of Native Latin American nations is what Pablo Calvi claims is "instrumental for understanding their plight and their mistrust while guiding the reporting in more fruitful directions."[29] Domosławski deconstructs the colonial narrative, showing how years of exploitation and the appropriation of territory by multinationals resonates deeply in the community. This strategy of "emplotment" (i.e. imposing the historical meaning on historical facts by choices made in storytelling) to use Hayden White's term,[30] enables Domosławski to place his text within a specific historic framework. He writes about Native Latin Americans with empathy and engagement, situating himself on the side of the oppressed.

The interconnectedness of life systems, as expressed, for example, in *Silent Spring*, gains an additional level in Domosławski's narrative, for he shows the deep correlation not only between different species and ecosystems but also between historical colonial abuse and modern industrial exploitation.

The author does not give the two sides of the conflict equal consideration. While the local people's plight is described in detail, the standpoint of the transnational corporations, local bosses, corrupt officials, and so on is not as well depicted. When asked about this, Domosławski explains:

> At one point, I understood that this is not a dialogue between differing political positions, where some, for example, believe taxes should be higher, and others the reverse, but that we are dealing with a realm in which the line between economic and criminal activity is completely blurred. It was naive of me to expect someone to tell me the truth on behalf of the company; after all, the criminal does not admit to murder.[31]

In the part dedicated to Peru, Domosławski openly admits: "Since arriving and learning many more details of this story, I have doubts to what extent— if at all—I am interested in the corporation's version of events."[32] After looking up the Newmont website with such blatant lies as "the corporation meets ecological standards" and "promotes the constant

[29] Pablo Calvi, "Reporting on Indigenous Issues: The Extractive Matrix of Journalism vis-à-vis Native Latin Americans," *Literary Journalism Studies* Vol. 10, No. 1 (Spring 2018), 54.

[30] Hayden White, *Tropics of Discourse: Essays in Cultural Criticism* (Baltimore: Johns Hopkins University Press, 1985), 83.

[31] Artur Domosławski, interview with the author, June 24, 2020.

[32] Domosławski, *Śmierć w Amazonii*, 187.

development of the region," he writes: "Enough!"[33] When describing the corporation he focuses on the fact that it is almost an abstract being, "it is virtual, though the money is real; it is a maze of laws and regulations, behind-the-scene pressures, mining machinery, chemicals.... And also leaked mercury, acid water or no water at all, dead fish."[34] The author explains that those who are responsible for a corporation's operations are usually invisible. The multinational is, therefore, like a ghost, like the monster from Zbigniew Herbert's poem *The Monster of Mr Cogito* published in 1974 that Domosławski cites. The only proof of the monster's existence is its victims.[35]

Even when the author finally meets with the PR manager from Yanacocha–Newmont, he is clear about his sympathies. He likens the manager to a smooth-talking guy from a commercial,[36] and later alternates this PR spiel, in related speech, with his own observations in brackets: "He begins the story as if he were a Grufides activist: he talks about the disastrous social climate that the company has created around its activities in the Cajamarca region. (A good start. First lower the nosy journalist's guard)."[37] In the next section Domosławski condemns the image of locals created by the corporation's PR specialists, showing its catastrophic consequences for local communities:

> This propaganda goes to "information consumers"—primarily middle-class—and makes them indifferent to the rebel persecution in the provinces. If to insults such as "narrow minded" or "anti-development" you add "left extremist," which is what the corporation's propaganda and the top media operating in cahoots with the government do, the portrait of the enemy is ready. Then, with the silent consent of the opinion-forming classes, you can beat, torture, shoot.[38]

By situating himself on the side of the oppressed, Domosławski not only demonstrates his empathy with voiceless people fighting for their habitat but also shows his engagement with their struggle. He writes: "Should a reporter, historian, writer, or researcher give a voice to the

[33] Ibid.
[34] Ibid., 188.
[35] Ibid., 172.
[36] Ibid., 200.
[37] Ibid.
[38] Ibid., 202.

other side of the conflict on a non-discriminatory basis when he has enough examples that the other side is lying, committing abuses, manipulations, and even more serious sins?"[39]

Domosławski's work resonates with Bell's call for a "journalism of attachment,"[40] and with what Lynas maintains about the journalist's job being just telling the story "unless you're a campaigning journalist, then you may seek to raise awareness."[41] Domosławski thus seems the perfect example of a caring and campaigning reporter.

Another device that helps Domosławski elicit empathy and relate the tragedy of his protagonists is the blending of various genres. His narrative choices shift from classical reportage—with its "knowledge by acquaintance," to use Code's term[42] based on first-hand experience, and "knowledge by description," acquired in this case through press articles, history books and testimonies—to critical essay, where the author not only expresses his personal views but also supports his findings with theory. It is as if the reporter's tools were insufficient to report the complex story of the Amazon.

Seen from an analytical perspective, we can observe how distinct modes of narration are employed to build up the meaning. This is especially visible at the end of the first part, where conversations with victims' family members are juxtaposed with theoretical analysis. "Will there be a night when I don't see that scene?"[43] asks Laísa, Maria's sister who saw her sister and brother-in-law right after they were killed. In response, Domosławski reflects:

> Does it really matter who pulled the trigger?
>
> After all, everyone—from those cutting down forests, buying timber and burning it for coal, producing pig iron in steel mills, exporting and importing meat, leather, wood, and steel, right up to those buying cars, floors, shoes and other goods…either finances the destruction of the Amazon or profits from it. Whether we like it or not, we have all contributed to or

[39] Ibid., 210.

[40] Martin Bell, "The Journalism of Attachment," in *Media Ethics*, ed. Matthew Kieran (New York: Routledge, 1998), 16.

[41] Lockwood, "Recruiting the Imagination: the environmental literary journalism of Rachel Carson," 340.

[42] Lorraine Code, *Epistemic Responsibility* (Hanover: University Press of New England, 1987), 206.

[43] Domosławski, *Śmierć w Amazonii*, 110.

earned a dime from the murder of two environmentalists from Nova Ipixuna. Indirectly. Unknowingly.[44]

Laísa's highly emotional first-hand account of events is followed by rational scrutiny. In the same chapter Domosławski explains in disengaged analytical language Bauman's sociological theory on the notion of moral blindness, before returning to José Cláudio's niece. Clara moved out of the village after the murder and has not been back since. Domosławski shows her despair: "She is afraid of anger, lust for revenge. Of her own emotions. She knows they are poisoning her—not the killers, not those who ordered the crime, but her. She tries to control them, but sometimes she loses. 'I want to leave, run away. So much life ahead, so much to see…' Tears."[45]

By shuttling back and forth between his protagonists' lives and his own personal essay rooted in critical theory, Domosławski manages to build an emotional crescendo that engages his readers.

'Moral Sleeping Pills' and Responsibility in Liquid Times

To analyze his findings and bring them into the global discourse, Domosławski draws on two theories: the notion of moral blindness as formulated by Polish sociologist Zygmunt Bauman, and the concept of moral responsibility as rendered by Swedish author and historian Sven Lindqvist. Both are interconnected, as they deal with the question of moral responsibility. However, while the first shows what generates moral blindness, the second denounces higher-income countries and their inhabitants and places the responsibility for atrocities against people and nature on them.

First, Domosławski tracks the final destination of the raw goods from the Amazon: "The initial links in the long chain of meat and leather production—as well as wood and steel production—are corrupted by land grabs, environmental destruction, human exploitation, and sometimes even homicide."[46] For example, tracing the illegally logged Amazon timber takes him to North American Lumber Liquidators, which installs

[44] Ibid., 110–111.
[45] Ibid., 115.
[46] Ibid., 93.

wooden floors in the houses of such celebrities as Angelina Jolie and Donald Trump and buys its timber from a dodgy Brazilian company, Vitória Regia, which was fined more than US$270,000 in 2013 for stocking and selling illegal timber.

Domosławski then tries to identify those responsible for the death of the two ecologists. He explains that the timber, charcoal, and cattle industries operate like syndicates, and the deaths of activist ecologists at the hands of local bandits are ordered from on high through a sort of informal chain of command that ensures total impunity for the original commissioner, who remains unconnected to the actual murder.

The interpretation is self-evident—the great physical and mental distance between the original commissioner of the crime and his victim is as great as that between the homicide and a chestnut floor. The theory of moral blindness shows that, as happened in the Holocaust, modern society can involve a far greater number of people in genocide, war or other atrocity, for they never face moral choices in the process and do not need to battle with their consciences.[47] A worker from Yanacocha–Newmont sums this up very aptly:

> When asked why he works for the corporation, a laborer replies: I earn a living. When asked if he knows that the company harms local people, he may not even understand the question. After all, he himself is doing nothing wrong. The same answer is given by a manager in an expensive suit who is driven to work from a gated community in a luxury car. After all, he just crunches numbers on his computer; he does not commission the stalking of activists.[48]

Moral blindness is possible only in a complex modern society, where the principle of "intermediation of action" holds sway.[49] There is a great distance between an intention and its actual accomplishment, and it is filled with a great many minor acts carried out by insignificant intermediaries. The middle man shields off the outcomes of the action from the actors' sight.[50] As a result, horrendous acts are committed that no one feels responsible for.

[47] See Zygmunt Bauman, *Liquid Modernity* (Cambridge: Polity, 2000), 24.
[48] Domosławski, *Śmierć w Amazonii*, 188.
[49] Ibid., 113.
[50] Zygmunt Bauman, *Liquid Modernity*, 24–25.

Still, the story does not end there for Domosławski. He brings in the idea of responsibility by drawing an analogy with another time and place. The author narrates the story of a Swedish non-fiction writer and journalist Sven Lindqvist, who traveled to Iceland aged 17 aboard a ship that docked where en route in Norway. The locals' initially warm welcome turned cool on learning of his nationality, which in their eyes made him complicit in Sweden's agreement to Nazi transits in 1942. Lindqvist was initially indignant at being hold responsible for the decision his country made in the past, for he was just ten at the time. But after years spent writing on colonialism and its victims, he saw things differently:

> The national debt is passed on from generation to generation. Just like the nation's assets [...]. Simply by being born Swedish, I was born rich. It wasn't my own efforts that had made me better off than a Congolese or an Indonesian. I was the heir to an undamaged society and a fully functioning economy; in short, it was as a Swede that I was well off. And having accepted the advantages of being Swede, how could I then disown the disadvantages?[51]

Domosławski makes the point that we are the ones responsible for the destruction of the Amazon, which functions here as the perfect paradigm. Earth's lungs filter the air we all breathe, so the fate of the rainforest determines the fate of the whole human race. By showing the food chain of the global economy, Domosławski positions consumers within the last link of that chain and shows how we are complicit in ecological disasters. This is what makes his environmental narrative powerful and innovative.

The literary journalist invariably looks for a story with a wider reach, one that encapsulates a broader phenomenon and explains a complex issue. In short, a story that can become a synecdoche. Domosławski goes further. To give his story another dimension, he applies a theoretical framework that allows him to bring the global issue down to a local level. But this time he is addressing his readers, telling us that we are the ones responsible for what is happening in the Amazon. By adopting this novel perspective the author demonstrates the existence of global responsibility for global issues. He shows that the environment is in fact a system of communicating vessels, and that specific actions in one part of the world always have far-reaching consequences somewhere else.

[51] Sven Lindqvist, *Terra Nullis,* trans. Sarah Death (New York: The New Press, 2007), 11–12.

Conclusion, or the Literary Journalist as a Naturalist

A literary journalist is—to use Plato's metaphor from *Phaedrus*—like a charioteer. The charioteer has to control two horses, one of which is like a writer who chases a story, with its plot, characters, setting, and dialogues, and who is concerned with the narrative axis and structure of his book. The other horse, a journalist, understands the reality, all the collected data, first-hand experience, and the mechanisms rooted in science, social sciences, or culture that back up and explain his findings. He too chases a story, but his is a scoop, a top story, a heated issue. For the chariot to reach its destination, the charioteer needs to control both horses, to find the right balance. The final destination is an account that reads like a novel but is set in reality and based on facts, so that it can raise the awareness of its readers and tell them something about the world.

Domosławski's chariot is drawn by four horses. There is the writer who composes his book using narrative strategies and literary devices, and the reporter who interviews the protagonists, gathers first-hand material, digs into the historical background, and looks for new informants and information. But there is also an essayist, who tries to theorize about his findings, and an engaged or perhaps even militant naturalist who denounces the ecological disasters. Domosławski drives his chariot using journalistic, literary, sociological, and even philosophical tools to establish a direct link between the lives of his readers and those of his characters. He then shifts to global responsibility, analyzing how local daily choices influence the world we all live in.

In the 1960s Marshall McLuhan spoke about the "global village." Many have criticized the term since then, including Ryszard Kapuściński, who pointed out that in a village, people know and help each other and are united by their common fate, while our modern society resembles the anonymous crowd at a big airport, where people are in a rush, mutually indifferent and unfamiliar.[52] By showing in *Death in the Amazon* what impact capitalism and consumerism have on nature and people, Domosławski makes at least some of the inhabitants of the global village aware of—to use Kapuściński's metaphor—the airport's layout.

Research for this article was partially developed as part of the three-year binational project, "CRONICA: Comparative Reportages: An Ontology of

[52] Ryszard Kapuściński, *Ten Inny* [The Other] (Kraków: Znak, 2007), 60–61.

French Narrative Journalistic Influences and Dialogue in Chile and Argentina,” funded by ECOS–Sud in France and CONICYT/ANID in Chile (N° C18H01/ECOS180053).

References

Anderson, Jon Lee. “Murder in the Amazon.” *The New Yorker*, June 15, 2011. https://www.newyorker.com/news/news-desk/murder-in-the-amazon

Bauman, Zygmunt. *Liquid Modernity.* Cambridge: Polity, 2000.

Bell, Martin. “The Journalism of Attachment.” In *Media Ethics*, edited by Matthew Kieran, 15–22. New York: Routledge, 1998.

Berning, Nora. 2011. “Toward a Critical Ethical Narratology for Literary Reportages: Analyzing the Story Ethics of Alexandra Fuller’s *Scribbling the Cat.*” *Interférences littéraires/Literaire interferenties*, Vol. 7 (November 2011): 189–221.

Calvi, Pablo. “Reporting on Indigenous Issues: The Extractive Matrix of Journalism vis-à-vis Native Latin Americans.” *Literary Journalism Studies* Vol. 10, No. 1 (Spring 2018): 46–69.

Carson, Rachel. *Silent Spring.* Cambridge: The Riverside Press, 1962.

Code, Lorraine. *Epistemic Responsibility.* Hanover: University Press of New England, 1987.

Conover, Ted. “Immersion and the Subjective: Intentional Experience as Research.” *Literary Journalism Studies* Vol. 10, No. 2 (Fall 2018): 163–173.

Domosławski, Artur. “Lista zapowiedzianych śmierci” [List of the deaths foretold]. *Gazeta Wyborcza*, August 22, 2005, 6–8.

Domosławski, Artur. *Śmierć w Amazonii* [Death in the Amazon]. Warsaw: Wielka Litera, 2013.

Dowling, David. “Banned in Britain: Marilynne Robinson’s Environmental Literary Journalism.” *Literary Journalism Studies* Vol. 11, No. 2 (December 2019): 56–87.

Hoeks, Henk. “The Vulnerability of Literary Non-Fiction.” *Publishing Research Quarterly*, Vol. 16, No. 1 (2000): 38–42.

Kapuściński, Ryszard. *Ten Inny* [The Other]. Kraków: Znak, 2007a.

Kapuściński, Ryszard. *The Soccer War.* Translated by William Brand. London: Granta Books, 2007b.

Lindqvist, Sven. *Terra Nullis.* Translated by Sarah Death. New York: The New Press, 2007.

Lockwood, Alex. “Recruiting the Imagination: the environmental literary journalism of Rachel Carson.” In *Global Literary Journalists: Exploring the Journalistic Imagination*, edited by Richard Lance Keeble and John Tulloch, 332–344. New York: Peter Lang, 2011.

McLuhan, Marshall. *The Gutenberg Galaxy*. Toronto: University of Toronto Press, 1962.

Plato. *Phaedrus*. Translated by Christopher Rowe. London: Penguin Books, 2005.

Said, Edward W. *Orientalism*. New York: Pantheon Books, 1978.

Sims, Norman. *True Stories: A Century of Literary Journalism*. Evanston, IL: Northwestern University Press, 2007.

White, Hayden. *Tropics of Discourse: Essays in Cultural Criticism*. Baltimore: Johns Hopkins University Press, 1985.

It's Personal: Women's Writing on Weather Disaster in the Context of Climate Crisis in Australia

Deb Anderson and Pepi Ronalds

> *Because our house is on fire, and this should come as no surprise.*
> (Naomi Klein 2020)

Bronwyn Adcock was surrounded by cremated trees when she saw the forest. The former SBS TV *Dateline* reporter, who last decade moved off grid to the coast south of Sydney, was working freelance on a story for *The Monthly* magazine. It was late 2019. A fire disaster was unfolding in northern New South Wales. Even before the scale and ferocity of Black Summer emerged, Adcock was considering a story frame of climate attribution. What caught her ear, she recalls, was a quote from leading fire ecologist David Bowman that acknowledged humans as an epoch-defining biophysical force: "He basically said what is happening and about to happen in

D. Anderson (✉) • P. Ronalds
Monash University, Melbourne, VIC, Australia
e-mail: deb.anderson@monash.edu; pepi.ronalds@monash.edu

P. Calvi (ed.), *The Literary Journalist as a Naturalist*, Palgrave Studies in Literary Journalism,
https://doi.org/10.1007/978-3-031-56634-9_16

Australia is going to be studied around the world for decades to come."[1] Then the inferno came to Adcock's neck of the woods.

The Currowan fire on the NSW South Coast burned for 74 days. Adcock phoned her editor to "download" about the chaos—"our house nearly burning down, losing most of our property, my husband nearly dying, our whole town [being] evacuated."[2] But having experienced the terror of those whose homes and loved ones were threatened underlined the failures of Australian climate leadership, and prompted a dramatic shift in civic agency. The story was *hers*. "I switched from being a dispassionate journalist to actually thinking … I've actually got lived experience now," Adcock says. "That's when it became, I guess, more personal."[3]

* * *

If, as environmental historian Libby Robin writes, "climate change is making Australia's experience with strange weather salutary for the rest of the world,"[4] then the lessons of Australian journalism could be cause for despair. While our scientists named the ten years to 2020 the "critical decade" to manage the increasing risks of weather disasters in light of climate change,[5] our government's "wilful political blindness on climate" has persisted even as the Intergovernmental Panel on Climate Change's Sixth Assessment Report warns that the country may warm by 4 degrees Celsius this century.[6] Certainly, the constitutive role of our hegemonic news media is a topic of scholarly concern.[7] Climate adaptation and risk management are central political matters that carry deep ideological

[1] Bronwyn Adcock, Personal Communication, July 27 2021.

[2] Adcock, Pers. Com.

[3] Adcock, Pers. Com.

[4] Libby Robin, "The Eco-Humanities as Literature: A New Genre?" *Australian Literary Studies* 23, no. 3 (2008): 302.

[5] Will Steffen and Lesley Hughes, "The Critical Decade: Climate Science, Risks and Responses," Australian Climate Commission, 2019, accessed November 14 2019, https://www.climatecouncil.org.au/resources/critical-decade-risks-responses/.

[6] Mark Kenny, "With the Release of a Terrifying IPCC report, Australia Must Face its Wilful Political Blindness on Climate," *The Conversation*, August 10 2021, https://theconversation.com/with-the-release-of-a-terrifying-ipcc-report-australia-must-face-its-wilful-political-blindness-on-climate-165868.

[7] Robert A. Hackett, Susan Forde, Shane Gunster and Kerrie Foxwell-Norton, *Journalism and Climate Crisis: Public Engagement, Media Alternatives* (New York: Routledge, 2017).

conflicts in Australia, divisions informed by the nation's reliance on coal.[8] In a news industry heavily influenced by media ownership, the conservative political allegiances of Nine, Australia's highest-rated TV network, and News Corp, with more than 140 newspapers, have delimited environmental coverage, to say the least.[9] Robert Hackett et al. put it bluntly in their 2017 review of the relevant scholarship, *Journalism and Climate Crisis*: most Australian news media are failing to mobilize public engagement or challenge the powerful interests vested in our high-carbon economy.[10] The problem of conventional media in this country is not a lack of information, they write. "Rather, the key deficit is one of agency, hope and efficacy."[11]

What might the kind of reporting that builds a bridge to political change look like? Celebrated poet and activist Judith Wright strove to remind us that Australian nature is "a living and creative force, capable of arousing, not merely receiving, the human imagination."[12] In this vein, we detect a direct response to climate crisis *and* the deficits of journalism that is both personal and distinctively Australian—and, in Wright's footsteps, we argue it is led by women.

This chapter proffers a lens onto thematic, if generic, change in long-form journalism on extreme weather as a harbinger of climate crisis in Australia. Here, weather disasters—each with their own time, space and mythology—have come to form "focusing events," that is, discursive sites of contention in which multiple actors interpret events as political opportunities or threats.[13] Our research is tracing the punctuated development of this work through the "critical decade": from the tail end of the 2001–2009 millennium drought, through the 2009 Black Saturday fires, Queensland's 2011 summer of disaster with flooding across an area the

[8] Lyn McGaurr and Elizabeth Lester, "Complementary Problems, Competing Risks: Climate Change, Nuclear Energy and *The Australian*," in *Climate Change and the Media*, ed. Tammy Boyce and Justin Lewis (Oxford: Peter Lang, 2009).

[9] Maxine Newlands, "Environmental Journalism in Australia and New Zealand," in *The Routledge Handbook of Environmental Journalism*, ed. David B. Sachsman and JoAnn Myer Valenti (New York: Routledge, 2020).

[10] Hackett et al., *Journalism and Climate Crisis.*

[11] Hackett et al., *Journalism and Climate Crisis.*

[12] Judith Wright, *Because I Was Invited* (Melbourne: Oxford UP, 1975), 71.

[13] Eitan Y. Alimi and Gregory M. Maney, "Focusing on Focusing Events: Event Selection, Media Coverage, and the Dynamics of Contentious Meaning-Making," *Sociological Forum* 33, no. 3 (2018).

size of France and Germany and "monster killer" Cyclone Yasi,[14] to Australia's worst bushfire season of 2019–2020 (or Black Summer) and the 2021 NSW floods. This is a dynamic subgenre that speaks to the local and the global. Published to audiences near and far—from *The Monthly* to *The Paris Review*, *The New York Times* to *Granta*—this work bears witness to distinct moments of the human experience of weather disasters that present cautionary tales of climate crisis. Its proponents shadow the visionary path of ecologist and journalist Rachel Carson: they illuminate how climate shapes culture *and* culture shapes climate.

Writers who identify as women are not the only ones doing this important work. Amid ongoing change in the wider relationship between Australians and their environment—which *Griffith Review* noted in 2019 has "shifted from one of innocence to one of anxiety"[15]—our project derives from a larger, concurrent, nationwide study with journalists on change in Australian disaster reporting.[16] But here, our critical intervention relates not only to an ongoing cultural narrative authorizing the plunder of the planet but also to gender. Last decade, Louise North concluded from Australia's largest, longitudinal survey of women journalists that our news media must confront the "uncomfortable truth" that gender bias is ongoing "at all levels of seniority across all media platforms, and in rural, regional and metropolitan areas."[17] Regrettably, in 2021 the situation is largely unchanged.[18] As North argued, researchers cannot go on merely documenting the inequity and harassment in an industry that is ultimately failing its workers; we must challenge it, too.[19]

[14] Tom Reilly and Aaron Cook, "North Queenslanders Batten Down for 'Monster Storm'," *The Sydney Morning Herald*, February 2 2011, https://www.smh.com.au/national/north-queenslanders-batten-down-for-monster-storm-20110201-1acfs.html.

[15] "Writing the Country." *Griffith Review* no. 63 (2019), https://www.griffithreview.com/editions/writing-the-country/.

[16] This project derives from a national study, *Covering Climate Disasters: The Professional and Ethical Challenges for Journalists*, for which Anderson received funding from the Journalism Education Research Association of Australia Grant for Excellence in Journalism Research (2019–2021) and Monash University Faculty of Arts.

[17] Louise North, "'Blokey' Newsrooms Still a Battleground for Female Journalists," *Australian Journalism Review* 34, no. 2 (2012).

[18] Our Watch, "Newsrooms and Gender Equality," Our Watch: Media Making Change, n.d., accessed August 21 2021, https://media.ourwatch.org.au/reporting-violence-against-women/newsrooms-and-gender-equality/.

[19] Louise North, "Damaging and Daunting: Female Journalists" Experiences of Sexual Harassment in the Newsroom," *Feminist Media Studies* 16, no. 3 (2015): 496.

In the task of grounding new knowledge in women's experiences and by shifting attention from men's lives as central, our approach is to explore not just the meaning of this work but how the meaning is created. We aim to examine and offer women writers the chance to reflect upon their work and what shaped its production, including power relations, social hierarchies, institutionalized dominance, climate knowledge and gendered lives. On the latter, we also invite considered reflection on the ways gender identities are socially constituted through struggles and experiences of environment, disasters, and climate change.[20] In Australia, research shows women are more concerned about environmental issues in general and about climate change in particular;[21] climate knowledge has a stronger association with anthropogenic climate change beliefs among women than it does among men.[22] Our line of questioning is not to rest on easy assumptions about women's writing or the oneness and benign nature of shared womanhood; indeed, so far, where some interviewees align their writing with an ecofeminist ethics of care, others have argued explicitly against the idea of gender having influenced their empathetic way of addressing nature. Rather, by exploring how these women have met the challenge of writing about extreme weather in the context of climate crisis, we call for a deeper understanding of women's engagement as a means of opening up dialogue on the ways environmental knowledge and voice interact with gender.

And we are uncovering a remarkable diversity of long-form journalistic voices on climate crisis. We deliberately say "journalistic," for our selection process is anchored in industry *and* guided by output. On the one hand, we have heeded Sue Joseph's finding that a prevailing discomfort exists among Australian writers with the term "literary journalist."[23] As Joseph writes, "some of the leading proponents of this type of long-form narrative may well be literary but definitely do not regard themselves as journalists, nor what they do as journalism."[24] On the other hand, we have accounted for shifts in the political economy of the Australian news media.

[20] Sherilyn MacGregor, "Gender and Environment: An Introduction," in *Routledge Handbook of Gender and Environment*, ed. Sherilyn MacGregor (New York: Routledge, 2017).

[21] Bruce Tranter, "Climate Scepticism in Australia and in International Comparative Perspective," in *Australian Social Attitudes IV: Australia in the Age of Insecurity*, ed. Shaun Wilson and Markus Hadler (Sydney: Sydney UP, 2018).

[22] Bruce Tranter, "Does Public Knowledge of Climate Change Really Matter in Australia?" *Environmental Communication* 14, no. 4 (2020).

[23] Sue Joseph, "Telling True Stories in Australia," *Journalism Practice* 4, no. 1 (2010).

[24] Joseph, "Telling True Stories in Australia," 95.

With newsrooms suffering at the hands of media mergers, redundancies and restructuring,[25] many of these women no longer work in a newsroom, or never worked there, but instead write from a university or home office.

* * *

There is no explicit culture of literary journalism or creative non-fiction on nature or environmental concern in Australia,[26] although prominent poet and scholar Mark Treddinick sought to spur the genre on with the 2004 edited anthology *A Place on Earth: Nature Writing in Australia*.[27] There are the profound voices of First Nations women whose storytelling tradition does not distinguish between human and natural worlds, each of whose distinctive work calls for an entire project devoted to the topic. They include Melissa Lucashenko, a Goorie writer of Bundjalung and European heritage, as well as authors Claire G. Coleman (see, for example, "Boodjar Ngan Djoorla/Country, My Bones"),[28] Natalie Cromb ("Water Is Life and Life Is Finite")[29] and Karen Wyld ("Thirteen Jetties and a Man on a Hill"),[30] each of whom is concerned not only with ecological collapse but also a wider relationship to Country—with "earthspeaking," as the acclaimed Lucashenko wrote in her 2004 essay, "Not Quite White in the Head"[31]—and invader-harms inflicted.

In contrast, unresolved issues of belonging have long pervaded non-Indigenous Australian writing on nature.[32] In this context, novelist Inga Simpson notes how nature writing has, over time, become a mode for examining evolving relationships with landscapes.[33] Here we detect the

[25] Newlands, "Environmental Journalism in Australia and New Zealand."

[26] Joseph, "Telling True Stories in Australia."

[27] Mark Treddinick, *A Place on Earth: Nature Writing in Australia* (Sydney: UNSW Press, 2004).

[28] Claire G. Coleman, "Boodjar Ngan Djoorla/Country, My Bones," *Griffith Review* no. 63/February (2018).

[29] Natalie Cromb, "Water is Life and Life is Finite," *Kill Your Darlings* June (2020), https://www.killyourdarlings.com.au/article/water-is-life-and-life-is-finite/.

[30] Karen Wyld, "Thirteen Jetties and a Man on a Hill," *Meanjin* Spring (2019), https://meanjin.com.au/essays/thirteen-jetties-and-a-man-on-a-hill/.

[31] Melissa Lucashenko, "Not Quite White in the Head," *Griffith Review* no. 2/December (2004), https://www.griffithreview.com/articles/not-quite-white-in-the-head/.

[32] Inga Simpson, "Encounters with Amnesia: Confronting the Ghosts of the Australian Landscape," *Griffith Review* no. 63/February (2019): 272.

[33] Simpson, "Encounters with Amnesia."

emergence of writing that balances the seriousness of climate crisis with a sense of wonder, while conjoining literary styles with journalistic beats. In the wider subgenre of women's writing on ecological collapse, we note the poetic essays of Annamaria Weldon (such as "Threshold Country[34]), the work of Rebecca Giggs, whose book *Fathoms: The World in the Whale* was awarded the 2021 Andrew Carnegie Medal,[35] and the writing of Anna Krien, who wrote of realizing she was on "the island of last things" in her intrepid book of investigative journalism, *Into the Woods: The Battle for Tasmania's Forests*.[36] But some have focused specifically on extreme weather. Half a decade ago, for instance, *Griffith Review* editor Ashley Hay, who has regularly produced literary science journalism, wrote a prized cover story for *Australian Book Review*, "The Forest at the Edge of Time," on what we know about bushfire-resilient eucalypts and climate change.[37] That same year Jane Rawson (who spent years editing "article after article about extinction" for *The Conversation*[38]) co-authored *The Handbook: Surviving and Living with Climate Change* to help people "prepare for increased droughts, floods, fires and heatwaves."[39]

It was perhaps former senior Fairfax science and environment reporter Jo Chandler who led the way in this subgenre in 2011, in the mainstream press. This followed a period of near-silence on climate attribution in mainstream media coverage after Black Saturday;[40] as the Fairfax press reported a week after those fires of February 2009: "no climate scientist will say that one event, no matter how extreme, is the manifestation of

[34] Annamaria Weldon, "Threshold Country," *Indigo: Journal of West Australian Writing* Summer (2011).

[35] Rebecca Giggs, *Fathoms: The World in the Whale* (London: Simon & Schuster, 2021).

[36] Anna Krien, *Into the Woods: The Battle for Tasmania's Forests* (Melbourne: Scribe Publishing, 2010), 297.

[37] Ashley Hay, "The Forest at the Edge of Time," *Australian Book Review* no. 375 (2015), https://www.australianbookreview.com.au/abr-online/current-issue/161-october-2015-no-375/2724-ashley-hay.

[38] Jane Rawson, "The Invisible Extinctions," *Meanjin* Spring (2018), https://meanjin.com.au/essays/the-invisible-extinctions/.

[39] Jane Rawson and James Whitmore, *The Handbook: Surviving and Living with Climate Change* (Melbourne: Transit Lounge, 2015).

[40] Deb Anderson, Philip Chubb and Monika Djerf-Pierre, "Fanning the Blame: Media Accountability, Climate and Crisis on the Australian 'Fire Continent'," *Environmental Communication* 12, no. 7 (2018).

climate change."[41] Entrenched, mediatized "climate debate" persisted also in the wake of the Brisbane floods of 2011, when even Germaine Greer wrote in the *Guardian* of some experts being "rather too quick to claim the extreme weather as a direct consequence of global warming."[42] Chandler's 2011 book, *Feeling the Heat*, took a literary and personal approach to the assignment, traveling with scientists through distinctive landscapes in Australia and Antarctica (and their associated weather extremes) as they tried to decode climate information that is critical to the future of the planet.[43] A decade later, having produced a wealth of climate-related storytelling on weather disaster, Chandler speaks directly to the "cracks in that fraught construct of journalistic objectivity"—comparing her most frightening journalistic posts, such as covering gender-related violence in tribal-war-torn Papua New Guinea, to the emotional impact of climate change.[44] The latter, she writes, has been testing her more: "Facing the climate emergency demands the courage of my convictions."[45]

The writer's responsibility rises to the fore in contemporary long-form non-fiction that homes in on weather disasters—notably on Black Summer, which affected nearly 80 per cent of Australians directly or indirectly and killed at least 1 billion animals, 33 of them human.[46] Lucy Treloar, for example, pondered the ways fellow novelists would create meaning from this bushfire season in her *Meanjin* essay, "Writing the Apocalypse."[47] "Now we have smelt it in the air, we have breathed it into our lungs," she writes. "The things we have read about have become manifest."[48] Charlotte

[41] Michael Bachelard and Melissa Fyfe, "Lessons from the Ashes," *The Sydney Morning Herald*, February 15 2009, https://www.smh.com.au/national/lessons-from-the-ashes-20090215-8810.html.

[42] Germaine Greer, "Why Were We So Surprised?" *Guardian,* January 15 2011, https://www.theguardian.com/environment/2011/jan/15/australian-floods-queensland-germaine-greer.

[43] Jo Chandler, *Feeling the Heat* (Melbourne: University of Melbourne Press, 2011).

[44] Jo Chandler, "Tales from the Frontline: The Emotional Impact of Climate Change," *Griffith Review* no. 71/January (2021), https://www.griffithreview.com/articles/tales-from-the-frontline/.

[45] Chandler, "Tales from the Frontline."

[46] Lesley Hughes et al., "*Summer of Crisis,*" The Climate Council, 2021, accessed August 21 2021, https://www.climatecouncil.org.au/wp-content/uploads/2020/03/Crisis-Summer-Report-200311.pdf.

[47] Treloar, Lucy, "Writing the Apocalypse." *Meanjin* Winter (2020), https://meanjin.com.au/essays/writing-the-apocalypse/.

[48] Treloar, "Writing the Apocalypse."

Wood wrote in the *Guardian* of stages of petulance, patient reasoning then dread, brought on by a new physiological understanding of fear.[49] "It feels like karma. This is what the scientists have warned us about, ***begged*** us to think of, all these years. It's here. And it's going to get worse."[50] Novelist Jennifer Mills took her frustrations with Australian political inertia on climate to a European audience via Denmark's *Weekendavisen*,[51] also published down under on *Literary Hub*:

> One of the challenges for writers in the Anthropocene is that events like these fires belong at once to the everyday and to the unthinkable. In processing the shock of the fires, the images of death and destruction, the corpses of animals, the scale of the grief, writers must think in two directions at once: we must try to integrate these events into our understanding of the reality we live in, and we must refuse to accept them as normal.[52]

For Mills, weather disasters present the opportunity to interrogate the relationship of politics to memory and historical generalization: "I see it as a responsibility to write openly about what is happening to our world."[53]

* * *

Extreme weather has long served as a backdrop against which Australian culture defines itself.[54] Drought, in particular, has played a potent role in post-colonial interpretations of a harsh and unpredictable Australian climate.[55] Certainly, since the millennium drought, which scientists deemed

[49] Charlotte Wood, "From Disbelief to Dread: The Dismal New Routine of Life in Sydney's Smoke Haze," *Guardian,* December 7 2019, https://www.theguardian.com/australia-news/2019/dec/07/from-disbelief-to-dread-the-dismal-new-routine-of-life-in-sydneys-smoke-haze.

[50] Wood, "From Disbelief to Dread." Author's emphasis.

[51] Jennifer Mills, "The Disaster and the Literature," *Weekendavisen,* January 16 2020, https://www.weekendavisen.dk/2020-3/boeger/katastrofen-og-litteraturen.

[52] Jennifer Mills, "Battling the False Narratives around Australia's Devastating Bushfires," *Literary Hub,* January 22 2020, https://lithub.com/battling-the-false-narratives-around-australias-devastating-bushfires/.

[53] Mills, "Battling the False Narratives."

[54] Tom Griffiths, "The Nature of Culture and the Culture of Nature," in *Cultural History in Australia*, ed. Hsu-Ming Teo and Richard White (Sydney: UNSW Press, 2003).

[55] Deb Anderson, *Endurance: Australian Stories of Drought* (Carlton, Vic.: CSIRO Publishing, 2014); Brad West and Philip Smith, "Drought, Discourse and Durkheim: A Research Note," *Australian and New Zealand Journal of Sociology* 32, no. 1 (1996).

"the worst in 1,000 years,"[56] drought has been read as a harbinger of global warming-driven times ahead.[57] In this territory, we find journalists with a scholarly bent writing on crisis as a historical constant. Senior journalist and academic Margaret Simons, for instance, chronicled decades of political failure to meet the needs of drought-affected rural places in her 2020 *Quarterly Essay,* "Cry Me a River."[58] In Australia's troubled foodbowl, the Murray-Darling Basin, Simons uncovers "a society without being a community" and a natural state so altered as to exist "outside living memory, in the realm of dreaming and anecdote."[59] Linguist Christine Keneally has also examined a streak of self-reliance in the Australian character in her journalism on weather disaster. In Keneally's work, the story frame of climate attribution appears to have strengthened over time. Her *New Yorker* reportage on Black Saturday, for instance, referenced a link between catastrophic fire risk and a drought "widely thought to be a result of global warming."[60] A decade later, Keneally examined Black Summer squarely in terms of global temperatures having steadily increased and the continent having become drier—as "a case study in the realities of climate change."[61] In contrast, South Africa-born *Guardian Australia* reporter Helen Sullivan invoked a multi-generational sense of foreboding in a *Monthly* article on drought-stricken Gunnedah, a country town struggling with a loss of social capital due to rural restructuring.[62] Sullivan, who is studying for a Masters in Climate Change, says that she seeks "truth and the right metaphor" when writing explicitly about climate

[56] John Vidal, "Australia Suffers Worst Drought in 1,000 Years," *Guardian,* November 8 2006, www.guardian.co.uk/world/2006/nov/08/australia.drought.

[57] Anderson, *Endurance*; Melissa Fyfe, "It's Not Drought, it's Climate Change, say Scientists," *The Age,* August 30 2009, https://www.smh.com.au/national/its-not-drought-its-climate-change-say-scientists-20090829-f3cd.html.

[58] Margaret Simons, "Cry Me a River: The Tragedy of the Murray-Darling Basin," *Quarterly Essay* no. 77 (2020).

[59] Simons, "Cry Me a River."

[60] Christine Keneally, "The Inferno," *New Yorker,* October 26 2009, https://www.newyorker.com/magazine/2009/10/26/the-inferno.

[61] Christine Keneally, "When Fire Weather Becomes the Norm," *The Atlantic,* January 26 2020, https://www.theatlantic.com/ideas/archive/2020/01/when-fire-weather-becomes-norm/605470/.

[62] Helen Sullivan, "Formal Night in Gunnedah," *The Monthly* February (2019), https://www.themonthly.com.au/issue/2019/february/1548939600/helen-sullivan/formal-night-gunnedah.

change[63]—which she does regularly for an international audience (notably in *The New York Times* and *New Yorker*). Her aim is to make a complex topic not only easier to understand but also more memorable, or "more likely to stay with the reader."[64]

The deep and curious work of writers like these fills the annual anthology of *Best Australian Science Writing*, which not only shirt-fronts climate deniers but also strives to make readers feel something. Of course, some of these writers *are* scientists, contributing compelling first-person dramas of discovery and loss. Climate scientist Joëlle Gergis, for instance, inserted herself as an actor in her book of historical climatology, *The Sunburnt Country*,[65] and in regular essays she has connoted a wider appreciation of the many ways that readers, to cite literary scholar Mark Doty, derive pleasure "recognizing a described world."[66] Such work attempts to show readers the provenance of information; we gain insight into how it was gathered and, significantly, acknowledge the subjectivity and uncertainty that exist in factual discovery. Likewise, Tasmanian author and ecologist Nicole Gill engages in the art of thick description, if slow journalism,[67] to explore personal connections to drought-prone and fire-threatened places, and the task of choosing which battles to fight in a world morphing into new and terrifying forms. In a standout 2016 *Monthly* article, "Rupture in Tasmania," Gill centered on the concept of solastalgia: the mix of emotion that arises when the comfort of the lived and known home environment is "disfigured by the awareness that something is horribly wrong."[68] Despair forms a call to action, after ancient lineages of World Heritage-protected alpine and rainforest plants *burned* in bushfires. As Gill later reflected, under the conditions of climate change, "there is nowhere that is far enough from humanity to be safe."[69]

Indeed, colliding crises are taking center stage in the present. Take *Fire Flood Plague*, for instance: a 2020 anthology edited by Sophie Cunningham,

[63] Helen Sullivan, Personal Communication, June 29 2021.

[64] Sullivan, Pers. Com.

[65] Joëlle Gergis, *The Sunburnt Country: The History and Future of Climate Change in Australia* (Melbourne: Melbourne University Press, 2019).

[66] Mark Doty, *The Art of Description: World into Word* (Minneapolis: Graywolf Press, 2010), 11.

[67] Megan Le Masurier, "What is Slow Journalism?" *Journalism Practice* 9, no. 2 (2015).

[68] Nicole Gill, "Rupture in Tasmania." *The Monthly* April (2016), https://www.themonthly.com.au/issue/2016/april/1459429200/nicole-gill/rupture-tasmania#mtr.

[69] Nicole Gill, Personal Communication, July 21 2021.

one of few writers on nature to have considered the impact of past cyclones in the current context of evermore destructive weather.[70] Here she offered twenty-six Australian writers space to respond to a year in which environmental unraveling met with the global COVID-19 pandemic—and sixteen contributors are women.[71] Gergis, for instance, relates watching "the horror of planetary collapse" during Black Summer to that of the death of her father.[72] "We stood back helplessly," she writes, "knowing that nothing more could be done, that something vital had slipped away."[73] Speaking directly to the rationale behind the techniques deployed in this first-person essay, Gergis urges other scientists to acknowledge their devastation, their personal and collective grief, as a means to promote understanding. Indeed, she quotes Carson: "It is not half so important to know as to feel ..."[74] Meanwhile, *Guardian Australia* political journalist Gabrielle Chan contributes a recovery story on one family's micro food-production project in the Upper Murray.[75] It is a mission borne of "vegetable gardens next to burnt-out buildings," a metaphor for possibility in times of upheaval.[76] Here Chan, who is building a reputation for quality long-form journalism on rural Australia,[77] examines the value of that metaphor in rebuilding not only bushfire-devastated rural communities but also our national regard for food production. As Chan says, the local story forms a vital opportunity "to reflect on national themes."[78] And Jess Hill, an investigative journalist and author of the critically acclaimed 2019 book *See What You Made Me Do* on domestic abuse,[79] considers Black Summer specifically in terms of power.[80] There's no running from it, she writes:

[70] Sophie Cunningham, 2014, *Warning: The Story of Cyclone Tracy* (Melbourne: Text Publishing, 2014).

[71] Sophie Cunningham, ed., *Fire, Flood, Plague: Australian Writers Respond to 2020* (Melbourne: Vintage, 2020).

[72] Joëlle Gergis, "The Great Unravelling," in Cunningham, ed., *Fire, Flood, Plague*, 44.

[73] Gergis, "The Great Unravelling," 44.

[74] Gergis, "The Great Unravelling."

[75] Gabrielle Chan, "Our Daily Bread," in Cunningham, ed., *Fire, Flood, Plague*, 87.

[76] Chan, "Our Daily Bread," 87.

[77] Gabrielle Chan, *Why You Should Give a F*ck About Farming* (Melbourne: Vintage, 2021).

[78] Gabrielle Chan, Personal Communication, October 8 2021.

[79] Jess Hill, *See What You Made Me Do: Power, Control and Domestic Abuse* (Melbourne: Black Inc. Books, 2019).

[80] Jess Hill, "The Ghost in the Machine," in Cunningham, ed. *Fire, Flood, Plague*.

> "It" won't *ever* be over because "it" is not just one thing. *It* is not the catastrophic fires of last summer … *it* is not climate change … and *it* is not the coronavirus – though all are emblematic of the mess we're in … "It" is, in short, the shift we made from societies built on the principle of balance to what we have now: a dominant culture obsessed with power-over and control.
>
> There's a neat little term for this culture of power-over: patriarchy.[81]

* * *

Evidently, the "quiet revolution" of women's writing on climate crisis,[82] outlined here, contrasts sharply with the mainstream news media's fixation on the "shenanigans" that are Australia's climate policy,[83] including those politicians avoiding the link between extreme weather events and climate change.[84] In this context, we might question how well convention has served Australian journalism in the contemporary era of climate-related catastrophe. Indeed, it was not until Black Summer that a story frame of climate attribution took hold in mainstream reporting of weather disaster—that is, in half of their reporting.[85] Media scholars are not alone in calling for a fundamental rethinking of relationships between journalism, publics, democracy and climate crisis, or noting the facilitative potential of long-form and literary journalism to engage, motivate and mobilize publics.[86] While it is unlikely that long form will have the power or reach to challenge the dominant culture of fast news, many of the literary journalists-as-naturalists surveyed here welcome the chance to reflect upon it, too.

How have these women met the challenge of writing convincingly about climate crisis in Australia without undermining notions of journalistic objectivity or integrity? By writing from, and about, lived experience. United by a desire to "rescue" stories from the abstract, and by responding to the need to deal directly with climate crisis, this work operates on a

[81] Hill, "The Ghost in the Machine," 166.

[82] Simpson, "Encounters with Amnesia," 272.

[83] Newlands, "Environmental Journalism in Australia and New Zealand," 308.

[84] Tom Griffiths, "Savage Summer," *Inside Story*, 2020, accessed August 21 2021, https://insidestory.org.au/savage-summer/.

[85] Tahnee Burgess et al., *Black Summer: Australian Newspaper Reporting on the Nation's Worst Bushfire Season* (Melbourne: Monash Climate Change Communication Research Hub, Monash University, 2020).

[86] See Hackett et al., *Journalism and Climate Crisis.*

human scale. It stands out from daily journalism for its deep engagement with subjects.[87] It eschews the conventional ideal of dispassionate disaster reporting, bearing witness to but also creating a moral imaginary through "everyday" lived understandings of disaster, often giving a rare voice to the explicit political critique of ordinary citizens.[88] It does the necessary work of exploring the cultural dimensions of Australian climate literacy. It cannot help but expose the core interpretive problems of climate change, or its politicization. And it seeks to act upon a sense of lost trust in basic democratic institutions for their (in)capacity to act on climate, or to grapple deeply with how we live on this continent, which eminent historian Tom Griffiths has termed "the burning frontier of a warming world."[89]

The latter is at the heart of *Currowan*, the 2021 book Bronwyn Adcock wrote from that interminable fire of 2019–2020. "I think it's one thing to have the debate up at a high level, but I think that what we're seeing now, increasingly, is people living with it," she says.[90] Writing about weather disaster granted her the agency to talk about climate crisis differently: "tangibly."[91] For, as Adcock put it, the failures of Australian leadership are all too real when the distance between yourself and the trauma you cover "collapses under a climate-changed world."[92]

References

Adcock, Bronwyn. 2020. "Living Hell.' *The Monthly* (February). https://www.themonthly.com.au/issue/2020/february/1580475600/bronwyn-adcock/living-hell.

Adcock, Bronwyn. 2021. *Currowan: The Story of a Fire and a Community during Australia's Worst Summer.* Melbourne: Black Inc.

[87] Joseph, "Telling True Stories in Australia."

[88] Mervi Pantti, Karin Wahl-Jorgensen and Simon Cottle, *Disasters and the Media* (New York: Peter Lang, 2012); Justin Lewis, Sanna Inthorn and Karin Wahl-Jorgensen, *Citizens or Consumers? What the Media Tell us about Political Participation* (Buckingham, UK: Open UP, 2005).

[89] Tom Griffiths in Black Inc., "Books: Imprint: Black Inc.: True Stories: *Currowan: The Story of a Fire and a Community during Australia's Worst Summer*," Black Inc. Books, 2021, https://www.blackincbooks.com.au/books/currowan.

[90] Adcock, Pers. Com.

[91] Adcock, Pers. Com.

[92] Bronwyn Adcock, *Currowan: The Story of a Fire and a Community during Australia's Worst Summer* (Melbourne: Black Inc., 2021).

Alimi, Eitan Y. and Gregory M. Maney. 2018. "Focusing on Focusing Events: Event Selection, Media Coverage, and the Dynamics of Contentious Meaning-Making." *Sociological Forum* 33 (3): 757–782.

Anderson, Deb, Philip Chubb and Monika Djerf-Pierre. 2018 "Fanning the Blame: Media Accountability, Climate and Crisis on the Australian "Fire Continent"." *Environmental Communication* 12 (7): 928–941.

Anderson, Deb. 2014. *Endurance: Australian Stories of Drought*. Carlton, Vic.: CSIRO Publishing.

Bachelard, Michael and Melissa Fyfe. 2009. "Lessons from the Ashes." *The Sydney Morning Herald*, February 15. https://www.smh.com.au/national/lessons-from-the-ashes-20090215-8810.html.

Black Inc. 2021. "Books: Imprint: Black Inc.: True Stories: *Currowan: The Story of a Fire and a Community during Australia's Worst Summer*." Black Inc. Books. https://www.blackincbooks.com.au/books/currowan.

Burgess, Tahnee, James R. Burgmann, Stephanie Hall, David Holmes and Elizabeth Turner. 2020. *Black Summer: Australian Newspaper Reporting on the Nation's Worst Bushfire Season*. Melbourne: Monash Climate Change Communication Research Hub, Monash University.

Chan, Gabrielle. 2020. "Our Daily Bread." In *Fire, Flood, Plague: Australian Writers Respond to 2020*, edited by Sophie Cunningham, 84–91. Melbourne: Vintage.

Chan, Gabrielle. 2021. *Why You Should Give a F*ck About Farming*. Melbourne: Vintage.

Chandler, Jo. 2011. *Feeling the Heat*. Melbourne: University of Melbourne Press.

Chandler, Jo. 2021. "Tales from the Frontline: The Emotional Impact of Climate Change." *Griffith Review* 71 (January). https://www.griffithreview.com/articles/tales-from-the-frontline/.

Coleman, Claire G. 2018. "Boodjar Ngan Djoorla/Country, My Bones." *Griffith Review* 63 (February): 47–52.

Cromb, Natalie. 2020. "Water is Life and Life is Finite." *Kill Your Darlings* (June). https://www.killyourdarlings.com.au/article/water-is-life-and-life-is-finite/

Cunningham, Sophie. 2014. *Warning: The Story of Cyclone Tracy*. Melbourne: Text Publishing.

Doty, Mark. 2010. *The Art of Description: World into Word*. Minneapolis: Graywolf Press.

"Edition 63: Writing the Country.' 2019. *Griffith Review* 63 (February). https://www.griffithreview.com/editions/writing-the-country/.

Fyfe, Melissa. 2009. "It's Not Drought, it's Climate Change, say Scientists." *The Age*, August 30. https://www.smh.com.au/national/its-not-drought-its-climate-change-say-scientists-20090829-f3cd.html.

Gergis, Joëlle. 2019. *The Sunburnt Country: The History and Future of Climate Change in Australia*. Melbourne: Melbourne University Press.

Gergis, Joëlle. 2020. "The Great Unravelling." In *Fire, Flood, Plague: Australian Writers Respond to 2020*, edited by Sophie Cunningham, 44–52. Melbourne: Vintage.

Giggs, Rebecca. 2021. *Fathoms: The World in the Whale*. London: Simon & Schuster

Gill, Nicole. 2016. "Rupture in Tasmania." *The Monthly* (April). https://www.themonthly.com.au/issue/2016/april/1459429200/nicole-gill/rupture-tasmania#mtr.

Greer, Germaine. 2011. "Why Were We So Surprised?" *Guardian*, January 15. https://www.theguardian.com/environment/2011/jan/15/australian-floods-queensland-germaine-greer

Griffiths, Tom. 2003. "The Nature of Culture and the Culture of Nature." In *Cultural History in Australia*, edited by Hsu-Ming Teo and Richard White, 67–80. Sydney: UNSW Press.

Griffiths, Tom. 2020. "Savage Summer." *Inside Story*. Accessed August 21 2021. https://insidestory.org.au/savage-summer/.

Hackett, Robert A., Susan Forde, Shane Gunster and Kerrie Foxwell-Norton. 2017. *Journalism and Climate Crisis: Public Engagement, Media Alternatives*. New York: Routledge.

Hay, Ashley. 2015. "The Forest at the Edge of Time." *Australian Book Review* 375. https://www.australianbookreview.com.au/abr-online/current-issue/161-october-2015-no-375/2724-ashley-hay.

Hill, Jess. 2019. *See What You Made Me Do: Power, Control and Domestic Abuse*. Melbourne: Black Inc Books.

Hill, Jess. 2020. "The Ghost in the Machine." In *Fire, Flood, Plague: Australian Writers Respond to 2020*, edited by Sophie Cunningham, 163–171. Melbourne: Vintage.

Hughes, Lesley, Will Steffen, Greg Mullins, Annika Dean, Ella Weisbrot and Martin Rice. 2020. *Summer of Crisis*. The Climate Council. Accessed August 21 2021. https://www.climatecouncil.org.au/wp-content/uploads/2020/03/Crisis-Summer-Report-200311.pdf.

Joseph, Sue. 2010. "Telling True Stories in Australia." *Journalism Practice* 4 (1): 82–96.

Keneally, Christine. 2009. "The Inferno." *New Yorker*, October 26. https://www.newyorker.com/magazine/2009/10/26/the-inferno.

Keneally, Christine. 2020. "When Fire Weather Becomes the Norm." *The Atlantic*, January 25. https://www.theatlantic.com/ideas/archive/2020/01/when-fire-weather-becomes-norm/605470/.

Kenny, Mark. 2021. "With the Release of a Terrifying IPCC report, Australia Must Face its Wilful Political Blindness on Climate." *The Conversation*, August 10. https://theconversation.com/with-the-release-of-a-terrifying-ipcc-report-australia-must-face-its-wilful-political-blindness-on-climate-165868.

Klein, Naomi. 2020. "On Fire." In *All We Can Save: Truth, Courage and Solutions for the Climate Crisis*, edited by Ayana Elizabeth Johnson and Katharine K. Wilkinson, 59–69. New York: One World.

Krien, Anna. 2010. *Into the Woods: The Battle for Tasmania's Forests*. Melbourne: Scribe Publishing.

Le Masurier, Megan. 2015. "What is Slow Journalism?" *Journalism Practice* 9 (2): 138–152.

Lewis, Justin, Sanna Inthorn and Karin Wahl-Jorgensen. 2005. *Citizens or Consumers? What the Media Tell us about Political Participation*. Buckingham, UK: Open UP.

Lucashenko, Melissa. 2004. "Not Quite White in the Head." *Griffith Review* 2 (December). https://www.griffithreview.com/articles/not-quite-white-in-the-head/.

MacGregor, Sherilyn. 2017. "Gender and Environment: An Introduction." In *Routledge Handbook of Gender and Environment*, edited by Sherilyn MacGregor, 1–24. New York: Routledge.

McGaurr, Lyn and Libby Lester. 2009. "Complementary Problems, Competing Risks: Climate change, nuclear energy and *The Australian*." In *Climate Change and the Media*, edited by Tammy Boyce and Justin Lewis, 174–85. Oxford: Peter Lang.

Mills, Jennifer. 2020a. "The Disaster and the Literature." *Weekendavisen*, January 16. https://www.weekendavisen.dk/2020-3/boeger/katastrofen-og-litteraturen.

Mills, Jennifer. 2020b. "Battling the False Narratives around Australia's Devastating Bushfires." *Literary Hub*, January 22. https://lithub.com/battling-the-false-narratives-around-australias-devastating-bushfires/.

Newlands, Maxine. 2020. "Environmental Journalism in Australia and New Zealand." In *The Routledge Handbook of Environmental Journalism*, edited by David B. Sachsman and JoAnn Myer Valenti, 306–315. New York: Routledge.

North, Louise. 2012. "'Blokey' Newsrooms Still a Battleground for Female Journalists." *Australian Journalism Review* 34 (2): 57–70.

North, Louise. 2015. "Damaging and Daunting: Female Journalists' Experiences of Sexual Harassment in the Newsroom." *Feminist Media Studies* 16 (3): 495–510.

Our Watch. n.d. "Newsrooms and Gender Equality." Our Watch: Media Making Change. Accessed August 21 2021. https://media.ourwatch.org.au/reporting-violence-against-women/newsrooms-and-gender-equality/.

Pantti, Mervi, Karin Wahl-Jorgensen, and Simon Cottle. 2012. *Disasters and the Media*. New York: Peter Lang.

Rawson, Jane and James Whitmore. 2015. *The Handbook: Surviving and Living with Climate Change*. Melbourne: Transit Lounge.

Rawson, Jane. 2018. "The Invisible Extinctions." *Meanjin* (Spring). https://meanjin.com.au/essays/the-invisible-extinctions/.

Reilly, Tom and Aaron Cook. 2011. "North Queenslanders Batten Down for 'Monster Storm'." *The Sydney Morning Herald*, February 2. https://www.smh.com.au/national/north-queenslanders-batten-down-for-monster-storm-20110201-1acfs.html.

Robin, Libby. 2008. "The Eco-Humanities as Literature: A New Genre?" *Australian Literary Studies* 23 (3): 290–304.

Simons, Margaret. 2020. "Cry Me a River: The Tragedy of the Murray-Darling Basin." *Quarterly Essay* 77: 1–114.

Simpson, Inga. 2019. "Encounters with Amnesia: Confronting the Ghosts of the Australian Landscape." *Griffith Review* 63 (February): 272–81.

Steffen, Will and Lesley Hughes. 2019. *The Critical Decade: Climate Science, Risks and Responses.* Australian Climate Commission. Accessed November 14 2019. https://www.climatecouncil.org.au/resources/critical-decade-risks-responses/.

Sullivan, Helen. 2019. "Formal Night in Gunnedah." *The Monthly* (February). https://www.themonthly.com.au/issue/2019/february/1548939600/helen-sullivan/formal-night-gunnedah.

Tranter, Bruce. 2020. "Does Public Knowledge of Climate Change Really Matter in Australia?" *Environmental Communication* 14 (4): 537–554.

Tranter, Bruce. 2018. "Climate Scepticism in Australia and in International Comparative Perspective." In *Australian Social Attitudes IV: Australia in the Age of Insecurity*, edited by Shaun Wilson and Markus Hadler, 81–98. Sydney: Sydney UP.

Treddinick, Mark. 2004. *A Place on Earth: Nature Writing in Australia*. Sydney: UNSW Press.

Treloar, Lucy. 2020. "Writing the Apocalypse." *Meanjin* (Winter). https://meanjin.com.au/essays/writing-the-apocalypse/

Vidal John. 2006. "Australia Suffers Worst Drought in 1,000 Years." *Guardian*, November 8. www.guardian.co.uk/world/2006/nov/08/australia.drought.

Weldon, Annamaria. 2011. "Threshold Country." *Indigo: Journal of West Australian Writing* (Summer).

West, Brad and Philip Smith. 1996. "Drought, Discourse and Durkheim: A Research Note." *Australian and New Zealand Journal of Sociology* 32 (1): 93–102.

Wood Charlotte. 2019. "From Disbelief to Dread: The Dismal New Routine of Life in Sydney's Smoke Haze." *Guardian*, December 7. https://www.theguardian.com/australia-news/2019/dec/07/from-disbelief-to-dread-the-dismal-new-routine-of-life-in-sydneys-smoke-haze.

Wright, Judith. 1975. *Because I was Invited*. Melbourne: Oxford UP, 1975.

Wyld, Karen. 2019. "Thirteen Jetties and a Man on a Hill." *Meanjin* (Spring). https://meanjin.com.au/essays/thirteen-jetties-and-a-man-on-a-hill/

The Big Picture and the Small Scene: Anna Tsing's Assemblages vs. Paul Engle's Workshop, and the Nature In Between

Pablo Calvi

One of the most striking literary operations that Anna Tsing performs in *The Mushroom at the End of the World*,[1] and the one that struck me as the most foreign to the Anglo-American literary journalism I have gotten used to over the years, is the resort to what she calls an *assemblage*, an entity that ultimately becomes the central figure around which her narrative revolves. An *assemblage* is not a character in the Aristotelian sense: it is not the representation of an individual (human or non-human). Neither is it a collective. It is not a town, a community, or a nation. Yet, Tsing's ethnography builds a story around this unique entity; and that, I think, is what makes her book interesting, as well as illuminating—a trailblazing read for

[1] Ana Tsing, *The Mushroom at the End of the World. On the Possibility of Life in Capitalist Ruin* (Princeton: Princeton University Press, 2015).

P. Calvi (✉)
School of Communication and Journalism, Stony Brook University, Stony Brook, NY, USA
e-mail: Pablo.calvi@stonybrook.edu

P. Calvi (ed.), *The Literary Journalist as a Naturalist*, Palgrave Studies in Literary Journalism,
https://doi.org/10.1007/978-3-031-56634-9_17

literary journalists in their attempt to report on the potential effects of climate change.

It has to be said, however, that Anna Tsing is neither a literary journalist, nor does she claim to be part of any literary non-fiction cohort. She is, in fact, an anthropologist, and a professor at the University of California Santa Cruz. She wrote *The Mushroom*... in 2015 as part of her longer inquiry into how humans and non-humans adapt and change to new global climate patterns, an environmental shift, she proposes, that originated in the implementation of industrial agriculture, with the subsequent expansion of a capitalist productive matrix. To show all these connections, associations, and changes, her research and fieldwork required new analytical tools, which she finds in philosophical and literary concepts.

The *Mushroom*... is a beautifully written book that follows the foraging, commerce, and consumption of matsutake mushrooms, considered a culinary delicacy in Japan. Tsing trails pickers, foragers, exporters, importers, and forests from along the Pacific North West of the United States to Japan and Europe. She explains that matsutake only grows in disturbed pine woodlands—second growth—and that the mushroom doesn't follow the sowing and harvesting patterns of collection required by industrial farming. Matsutake as a crop is poorly suited for large scale agricultural operations. Still, due to its value on the international market, she argues, capitalist strategies overlapping with what she calls the supply chain have been put in place for matsutake's realization as a global commodity, allowing for its commerce to adapt and adjust to random patterns of growth, circulation, and consumption. Wherever people, urban areas, commercial routes, and legal infrastructure blossom and start intersecting around the collection and circulation of matsutake, Tsing proposes, the connecting links of this expanded network of mycelia, these *salvage economies*, have begun spreading underground. This is what she calls an assemblage.

> The concept of *assemblage* is helpful. Ecologists turned to assemblages to get around the sometimes fixed and bounded connotations of ecological "community." The question of how the varied species in a species assemblage influence each other—if at all—is never settled: some thwart (or eat) each other; others work together to make life possible; still others just happen to find themselves in the same place. *Assemblages are open-ended gatherings.* They allow us to ask about communal effects without assuming them [...] I need to see lifeways—and non-living ways of being as well—coming together. Nonhuman ways of being, like human ones, shift historically. For

living things, species identities are a place to begin, but they are not enough: ways of being are emergent effects of encounters [...] *Patterns of unintentional coordination develop in assemblages.* To notice such patterns means watching the interplay of temporal rhythms and scales in the divergent lifeways that gather. *Surprisingly, this turns out to be a method that might revitalize political economy as well as environmental studies.* Assemblages drag political economy inside them, and not just for humans. Plantation crops have lives different from those of their free-living siblings; cart horses and hunter steeds share species but not lifeways. *Assemblages cannot hide from capital and the state*; they are sites for watching how political economy works. If capitalism has no teleology, we need to see what comes together—not just by prefabrication, but also by juxtaposition.[2]

The idea of assemblage shows some similarities to what biologists Humberto Maturana and Francisco Varela had proposed, regarding the *autopoiesis* of living systems, in their 1970s work on cybernetics.[3] But *assemblage* as a notion goes further than the idea of open or closed systems, because the assemblage not only includes living and non-living entities, but also the overarching mechanisms, ideas, and the enunciations that bind and coordinate them together, whether these be rational, irrational, or haphazard.

The notion of assemblage is not new. It has been used both in literature and philosophy. French writer Georges Perec, member of the Oulipo (Ouvroir de Littérature Potentielle) built his novels, including his literary masterpiece, *La Vie Mode d'Emploi* (*Life, a User's Manual*)[4] as large assemblages. Based on Oulipian poetic theory, which proposed arbitrary restrictions to coerce and propel the creative act, Perec produced his work resorting to "formal constraint, systematic artifice, the ludic element and the integration of mathematical structure in the literary text."[5] The result in *Life, a User's Manual* is the bisection of a Parisian apartment house, narratively "exhausted" piece by piece in a series of vignettes that develop an implosive tale (from description to story) linked with one another in an intricate thread that reaches outside the text to the reader. A true

[2] Ana Tsing, *The Mushroom at the End of the World*, 22–23. Emphasis is mine.

[3] Humberto Maturana, and Francisco Varela, *Autopoiesis and Cognition: The Realization of the Living*. (Springer, 1991 [Editorial Universitaria:, 1972]).

[4] Georges Perec. *Life, A User's Manual* (Boston: David R: Godine, 1987 [Hachette, 1978]).

[5] Warren F. Motte Jr., "Perec on the Grid" In *The French Review*. American Association of Teachers of French (May 1984, Vol. 57, No. 6), 820–832.

assemblage, a puzzle locked together by random variables, woven into ideas, places, characters, and objects. In *Life, a User's Manual*, each piece of the puzzle is as much part of the assemblage as the rules that bind everything together, and the reader who interprets it all.

It was, however, French philosopher Gilles Deleuze who first championed the notion of assemblage in *Kafka, Toward a Minor Literature*.[6] His monograph proposes assemblages as techno-physical aggregations driven by desire, as well as an enunciation, which has the role to connect all the parts at a symbolic level. In Kafka's *The Trial*, Deleuze explains:

> [The] machine of justice is a machine metaphorically: this machine fixes the initial sense of things, not only with its rooms, its offices, its books, its symbols, its topography, but also with its personnel (judges, lawyers, bailiffs), its women who are adjacent to the porno books of the law, its accused figures who make up an interdeterminate material [...] That the technical machine is only a piece in a social assemblage that it presupposes and that alone deserves to be called machinic introduces another point: the machinic assemblage of desire is also the collective assemblage of enunciation.[7]

At a basic level, assemblages diffuse the Aristotelian demand of a character, and break the unities of dramatic tragedy: unity of action, time, and space. And when used in literature—Tsing now proves their value in nature writing too—assemblages propose a dialectic between detail and big picture, a seesaw between narrative, minute description and the general plane of understanding; a comprehensive theoretical panorama in which details and the map that connects them ebb and flow in perpetuity. The following quote, discussing the connection between the community Tsing calls Open Ticket and the machinic dynamics of supply chain capitalism, is a good example of this swing.

> Thinking through salvage rhythms changes our vision. Industrial work no longer charts the future. Livelihoods are various, cobbled together, and often temporary. People come to them for diverse reasons, and only rarely because they offer the stable wages-and-benefits packages of twentieth-century dreams. I have suggested we watch patches of livelihood come into being as assemblages. Participants come with varied agendas, which do their

[6] Gilles Deleuze, *Kafka:Toward a Minor Literature* (Minneapolis and London: University of Minnesota Press, 1986), 81–88.

[7] Deleuze, 81–82.

small part in guiding world-making projects. For Open Ticket mushroom hunters, these include surviving war trauma and negotiating a working relationship with U.S. citizenship. Such projects mobilize commercial foraging, drawing pickers into the forest to follow "mushroom fever." Despite differences across these projects, boundary objects have formed—and particularly a commitment to what the pickers call freedom. Through such imagined common ground, commercial picking gains coherence as a scene—and a gathering becomes a happening. Multidirectional histories become possible through its emergent qualities. Without top-down discipline or synchronization, and without expectations of progress, livelihood patches help constitute the global political economy.

In collecting goods and people from around the world, capitalism itself has the characteristics of an assemblage. However, it seems to me that capitalism *also* has characteristics of a machine, a contraption limited to the sum of its parts. This machine is not a total institution, which we spend our lives inside; instead, it translates across living arrangements, turning worlds into assets. But not just any translation can be accepted into capitalism. The gathering it sponsors is not open-ended. An army of technicians and managers stand by to remove offending parts—and they have the power of courts and guns. This does not mean that the machine has a static form. As I argued in tracing the history of Japanese-U.S. trade relations, new forms of capitalist translation come into being all the time. Indeterminate encounters matter in shaping capitalism. Yet it is not a wild profusion. Some commitments are sustained, through force.[8]

Following the root structure of a fungus, the assemblage in Tsing's narrative expands and projects under the surface of the forest, creating a network of contact, a series related by proximity, in which the emergent fruits are sometimes in the vicinity of one another, but other times they are receding and apparently disconnected. Just like in a rhizome, not all parts of the root are tangled with the others. The assemblage presents an interlocking pattern of humans, non-humans, and ideas, a system of connectivity and relations that delivers this narrative as a heterogeneous, but still cohesive, unit.

Featuring an assemblage, *The Mushroom* proposes a political economy of the matsutake, which aligns with the history of capitalism and its impact on people, the environment, international relations, migration, culture, and commerce. It also allows for Tsing to come up with new answers and concepts to understand the dynamics she is studying: salvage capitalism, a

[8] Tsing, 132–133.

peripheral mode of production brewing in the shadows of corporate-industrial capitalism, is a phenomenon that cannot be observed without the lens provided by the assemblage.

One of the most interesting ideas in Tsing's book is that working with an assemblage mitigates the narrative to place full agency in human individuals, displacing the active function to larger aggregates, corporate systems, international commercial flows, and sociohistorical trends. The decoupling between individuals and global phenomena posits the need for a holistic understanding of larger systems of meaning as complex entities of action. In a sense, the assemblage begs for the return of "grand narratives"[9] now grounded on different agents, less Western, less European, and less humanist. This is probably why Tsing's narrative yearns for a return to a general Marxist explanation, which should also be overcome and superseded. A post-humanism, something *other*.

Assemblages and Literary Journalism

Since reading Tsing's book, the appeal of the notion of assemblage to me has become double. On the one hand, after working on non-Western, non-White narratives for some time, I have come to appreciate the idea that many, if not most, stories outside of the West rarely feature a main character. It is in the West that narratives usually gravitate around the notion of a *hero*, a central rhetorical device that, in Hegelian fashion, carries on its proverbial shoulders the entire weight of the story/History.[10] In that sense, assemblages are less hubristic constructions, less burdened with the Western—rugged—individualism required for heroic narratives to move forward.

On the other hand, assemblages stand at the opposite end of Paul Engle and Wallace Stegner's new humanism and its emphasis on the symbolic. The Iowa Writer's Workshop and the Stanford writers workshops, which became a beacon for Anglo-American writers, fiction and non-fiction, and remained so after the Cold War due to US government funding and support, were, as historians have found, a front in the multi-pronged cultural

[9] Jean-François Lyotard, *The Postmodern Condition: A Report on Knowledge* (Manchester: Manchester University Press, 1984).

[10] John Yorke, *Into the Woods. A Five-Act Journey into Story* (New York: The Overlook Press, 2013), 123–141.

offensive of the West and the United States against the surgent socialism and communism of the Global South in the 1950s to 1980s.[11] Symbolism was the workshop's attempt to preserve a literature that appealed to the great themes of Western *humanity*, while also retreating from grand explicative narratives fostered by Marxism, structuralism, Third-World liberation movements, and dependence theory.[12]

"A writer, it was believed, should aim to create, on the one hand, an image of a life that could be no other, and, on the other, a means by which that singularity spoke to humanity," Eric Bennet summarizes in *Workshops of Empire*.[13] "No insignificant part of the obsession with craft, in the early writing workshops, reflected this pressure to make the text successfully autonomous in the economy of symbolic meaning. A writer had to come up with something new and convince the reader that it mattered."[14] That symbolic autonomy meant that the text should become devoid of theory, macro explanations, and comprehensive ideas that ranged past a vague European humanism. I quote Stegner via Bennet at length, because the paragraph illustrates this anti-theoretical sentiment rather vividly.

> In the *Yale Review* in 1950 Wallace Stegner, a novelist and founder of the Stanford program, offered a genealogy of Nazism and Stalinism that deemed the writer a crucial asset to democracy in the postwar world. It would be a mistake "to start considering men not as complex individuals, as little worlds each compact of love and honor and ambition and cheating and foolishness and cowardice and courage, but as units in a gigantic logistic problem, as ciphers from which large and meaningful statistics can be made."[15]

But unlike in Engle and Stegner's approach, assemblages require—even demand—big theories and the displacement of the individual to a secondary plane of agency. Assemblages are the anti-Lyotard, the flip side of the postmodern condition, which attempted to blow up larger

[11] Eric Bennett, *Workshops of Empire. Stegner, Engle, and American Creative Writing during the Cold War* (Iowa City: University of Iowa Press, 2015); Frances Stonor Saunders, *The Cultural Cold War. The CIA and the World of Arts and Letters* (New York: The New York Press, 1999); David Dowling, *A Delicate Aggression. Savagery and Survival in the Iowa Writer's Workshop* (New Haven and London: Yale University Press, 2019), 224.

[12] Bennett, 51–52.

[13] Bennett, 52–53.

[14] Yorke, 53.

[15] Bennett, 5.

explanatory theories into small micro narratives operating within the context of an atomized world. Assemblages ask for a larger understanding of how multiple systems, human and non-human, conform, expand, interlock, and end.

Assemblages and Literary Journalism

One of the most notable consequences of the neo humanist approach popularized by the workshops has been the retreat of journalism and narrative journalism from the realm of the thematic to the realm of the episodic. Joel Best and Michael Schudson have already pointed at these two "social-organizational dimensions of the news."[16]

Episodically framed news stories relate to specific events, presented as unique occurrences, and usually enacted by a main character. They are so individuated that any context, be it historical or political, becomes unnecessary and narratively irrelevant. Thematically framed stories are less individuated, have a larger temporal scope. They usually present multiple characters, take place over a longer time period, across different places, and offer a social, political, historical, and cultural background.

Although Best's and Schudson's categorizations are rooted in the analysis of televised news, it could be easy to argue that a similar line separates the daily coverage of climate news events (episodic) from the long-form approach found in books and magazines (thematic).

Yet, while still promoting a contextual view, Tsing's approach seems to expand the scope of the analysis by requiring the use of different tools. And using them, her nature writing reveals the inadequacy of the episodic frame, and the limits of the thematic approach.

Like Rachel Carson and many of the writers introduced in this anthology, Tsing opens the conversation about the climate crisis, and reframes it as a post-human event. An event that includes us; an event that we very well may be causing, but of which we are not the center. The agency of individuals or even large groups of individuals doesn't even come close to the agency of assemblages or systems, which flow in the undercurrents of individual narratives dispossessed of greater explanatory power.

[16] Michael Schudson, "News as Stories," in Media Anthropology, Ed. Eric W. Rothenbuhler and Mihail Coman (Sage, 2005), 121–128. Joel Best, *Random Violence: How We Talk About New Crimes and New Victims* (Berkeley: University of California Press, 1999).

> When humble commodities are allowed to illuminate big histories, the world economy is revealed as emerging within historical conjunctures: the indeterminacies of encounter. If conjunctures make history, everything rests on moments of coordination—the translations that allow Japanese investors to profit from American foraging, just as pickers take advantage of Japanese wealth.[17]

It is by re-focusing on those assemblages, zooming out of the small stories through the use of larger explanatory systems and theoretical tools, that what can bring back some sense of understanding to the literary environmental journalism and, eventually, open up the possibilities for a new collaborative approach with nature and the environment.

References

Bennett, Eric, *Workshops of Empire. Stegner, Engle, and American Creative Writing during the Cold War.* Iowa City: University of Iowa Press, 2015.

Best, Joel, *Random Violence: How We Talk About New Crimes and New Victims.* Berkeley: University of California Press, 1999.

Deleuze, G., *Kafka: Toward a Minor Literature.* Minneapolis and London: University of Minnesota Press, 1986.

Lyotard, Jean-François, *The Postmodern Condition: A Report on Knowledge.* Manchester: Manchester University Press, 1984.

Maturana, H. and Varela, F., *Autopoiesis and Cognition: The Realization of the Living.* Springer, 1991 [Editorial Universitaria:,1972].

Motte Jr., Warren F., "Perec on the Grid." In *The French Review*, Vol. 57, No. 6 (May 1984).

Perec, Georges, *Life, A User's Manual.* Boston: David R. Godine, 1987 [Hachette, 1978].

Schudson, Michael, "News as Stories," *Media Anthropology* (Eric W. Rothenbuhler and Mihail Coman eds.) Sage, 2005.

Stonor Saunders, Frances, *The Cultural Cold War. The CIA and the World of Arts and Letters.* New York: The New York Press, 1999.

Tsing, Ana, *The Mushroom at the End of the World. On the Possibility of Life in Capitalist Ruin.* Princeton: Princeton University Press, 2015

Yorke, John, *Into the Woods. A Five-Act Journey into Story.* New York: The Overlook Press, 2013.

[17] Tsing, 119.

Index[1]

A

B

C

[1] Note: Page numbers followed by 'n' refer to notes.

P. Calvi (ed.), *The Literary Journalist as a Naturalist*, Palgrave Studies in Literary Journalism,
https://doi.org/10.1007/978-3-031-56634-9

The manufacturer's authorised representative in the EU is Springer Nature Customer Service Centre GmbH, Europaplatz 3, 69115 Heidelberg, Germany. If you have any concerns regarding our products, please contact ProductSafety@springernature.com

Printed and bound by CPI Group (UK) Ltd, Croydon, CR0 4YY
01/10/2026
02225898-0007